About: Designing

Analysing Design Meetings

Editors

Janet McDonnell

Peter Lloyd

Associate Editors

Fraser Reid

Rachael Luck

Nigel Cross

CRC Press is an imprint of the
Taylor & Francis Group, an **informa** business

A BALKEMA BOOK

CRC Press/Balkema is an imprint of the Taylor & Francis Group, an informa business

reprinted form:
Intersections of Brainstorming Rules and Social Order
Ben Matthews
CoDesign Vol. 5 No 1 (March 2009), pp. 65–76, DOI: 10.1080/15710880802522403
'Does this compromise your design?' Socially Producing a Design Concept in Talk-in-Interaction
Rachael Luck
CoDesign Vol. 5 No 1 (March 2009), pp. 21–34, DOI: 10.1080/15710880802492896
Collaborative Negotiation in Design: A Study of Design Conversations between Architect and Building Users
Janet McDonnell
CoDesign Vol. 5 No 1 (March 2009), pp. 35–50, DOI: 10.1080/15710880802492862
Aspects of Language Use in Design Conversation
Friedrich Glock
CoDesign Vol. 5 No 1 (March 2009), pp. 5–19, DOI: 10.1080/15710880802492870
Performing Architecture: Talking 'Architect' and 'Client' into Being
Arlene Oak
CoDesign Vol. 5 No 1 (March 2009), pp. 51–63, DOI: 10.1080/15710880802518054

reprinted from:
Ethical Imagination and Design
Peter Lloyd
Design Studies Vol. 30 No 2 (March 2009), pp. 154–168, DOI: 10.1016/j.destud.2008.12.004
The Mechanisms of Value Transfer in Design Meetings
Christopher Le Dantec & Ellen Yi-Luen Do
Design Studies Vol. 30 No 2 (March 2009), pp. 119–137, DOI: 10.1016/j.destud.2008.12.002
Affect-in-Cognition through the Language of Appraisals
Andy Dong, Maaike Kleinsmann & Rianne Valkenburg
Design Studies Vol. 30 No 2 (March 2009), pp. 138–153, DOI: 10.1016/j.destud.2008.12.003
Analogical Reasoning and Mental Simulation in Design: Two Strategies Linked to Uncertainty Resolution
Linden J. Ball & Bo T. Christensen
Design Studies Vol. 30 No 2 (March 2009), pp. 169–186, DOI: 10.1016/j.destud.2008.12.005

Typeset by Vikatan Publishing Solutions (P) Ltd, Chennai, India.
Printed and bound in Great Britain by Antony Rowe (A CPI-group Company), Chippenham, Wiltshire.

Published by: CRC Press/Balkema
P.O. Box 447, 2300 AK Leiden, The Netherlands
e-mail: Pub.NL@taylorandfrancis.com
www.crcpress.com – www.taylorandfrancis.co.uk – www.balkema.nl

ISBN: 978-0-415-44058-5 (Pbk)
ISBN: 978-0-203-87551-3 (eBook)

Contents

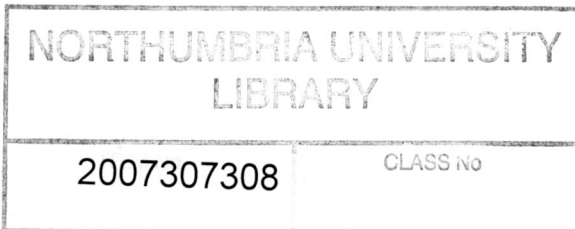

Part 4: Design Process Models

Part 5: Language, Discourse and Gesture

Part 6: Constructing Roles

Part 7: Objects, References, Context

Acknowledgements

There have been many people involved in producing this book, we'd like to thank some of them here. The book would not have been possible without the enthusiastic research efforts of all those who analysed the dataset, contributed at the workshop stage of the project, and produced the research findings which now comprise the chapters included here. Our associate editors Professor Fraser Reid, Dr Rachael Luck, and Professor Nigel Cross were invaluable during the reviewing phases of producing the material for the book. Many of the contributors also spent much time constructively reviewing the work of others, in addition to developing and revising their own work.

We would also like to thank Andrew Armes, Adrian Morrow, Rob Day and Tony Dunn who made access to the design meetings possible, and all the participants in the design meetings that formed the data for the project: Adam, Anna, Charles, Tony, Sally, Alan, Tommy, Jack, Chad, Todd, Sandra, Rodney, Stuart, Patrick and Roman.

The Open University and Central Saint Martins College of Art and Design provided many resources, not least our time as editors, to allow us to complete the book. The UK Arts and Humanities Research Council (Grant Reference: AH/E503284) provided support for data preparation and distribution and for the research network which underpinned the project as a whole.

Janet McDonnell and Peter Lloyd
November 2008

1
Introduction

Several years ago, and in conversations with fellow academics from the international community of design thinking researchers, we began to formulate the idea of providing a way of facilitating research collaboration and communication through sharing a common research dataset. We would invite internationally leading researchers to make use of the same dataset, share their findings with each other and at a workshop, subject their work to critical scrutiny by peers who were familiar with the same dataset, and produce a book from the results. The idea of distributing a dataset for analysis by researchers interested in design thinking had proved very successful in 1994 when Nigel Cross, Kees Dorst, and Henri Christiaans filmed designers at Xerox PARC to gather recordings of design problem solving behaviour in a laboratory setting. This data was distributed for wider analysis to test the potential (and limitations) of protocol analysis[1]. Indeed, the findings from that project, presented in the Delft protocols workshop, remain something of a landmark in design thinking research and we felt the time was right, 10 years on, to try and match this achievement by revisiting the shared data approach. Our conversations struck a chord, and this book is one of the outcomes of the project which has unfolded.

1. Cross, N., Dorst, K., and Christiaans, H. (1996) *Analysing Design Activity*, Wiley.

1 HOW THIS BOOK CAME ABOUT

In the 10 years since the Delft workshop and the start of our project the nature of design thinking research had also begun to change. Increasingly, studies were concentrating on designing in more naturalistic settings, if not wholly in design practice, and the scope of what was regarded as design activity broadened. The social aspects of design thinking were being emphasised; one trend was towards paying attention to the way that designing occurs between people trying to reach a common goal, rather than on individual design problem solving that can be studied in a lab-experiment. A richer, more contextual, understanding of design thinking was beginning to emerge; an understanding that shifted the imperative towards studying naturally occurring design activity in authentic settings. The common dataset on which the chapters in this book are based comprises material from real design meetings which form part of the routine practices of professional designers.

In 1994 the workshop had centred around a common method – protocol analysis – as well as a common dataset, but in recent years a wider variety of research methods have been used in response to our developing understanding of design thinking. Some methods are drawn from the social sciences and others finesse methods used previously in design thinking research. Some examples of the wide variety of approaches used to look at designing in context are: interaction

2. Tang, J.C. (1990) Findings from Observational Studies of Collaborative Work, *International Journal of Man-Machine Studies*, 34, pp. 143–160.

3. Dong, A. (2005) The Latent Semantic Approach to Studying Design Team Communication, *Design Studies*, 26, pp. 445–461.

4. Detienne, F., Martin, G., and Lavigne, E. (2005) Viewpoints in Co-design: A Field Study in Concurrent Engineering, *Design Studies*, 26, pp. 215–241.

5. Medway, P., and Clark, B. (2002) Imagining the Building: Architectural Design as Semiotic Construction, *Design Studies*, 24, 255–273.

6. Luck, R. (2003) Dialogue in Participatory Design, *Design Studies*, 24, pp. 523–535.

analysis[2], computational linguistics[3], viewpoint methodologies[4], semiotics[5,6], ethnography[7,8], functional linguistics[9], cognitive ethnography[10], and discourse analysis[11,12]. Additionally, many research disciplines aside from those primarily interested in design thinking have addressed the recording and analysis of professional practice. This has led to the development and application of methodologies which are equally appropriate for studying design practitioners and design practice. Relevant examples include work drawn from sociology[13], ethnomethodology[14], communication studies[15], naturalistic decision making[16], and social theories of learning[17]; and there are many others.

This plurality of methodologies presented a problem: how can common data support analyses which use different methodologies? Surely the focus of research enquiry comes first, then one selects an appropriate methodology and collects data in a certain way to meet the demands of the research, rather than starting with data and applying a methodology? Weren't we putting the cart before the horse? What we proposed is summarised in Figure 1 where it is contrasted with the procedure which is common in established subject disciplines where certain methods are unambiguously specified and are privileged, for example randomised controlled trials in some branches of medicine. Conventionally, scientific and social-scientific studies are carried out independently of one another but with a common methodology, leading to data being collected, analysed, and then results published. In some fields like experimental psychology, for example, there is broad agreement about what constitute legitimate research methods and what assumptions underlie them, providing at least some common basis for supporting discussion about findings and their validity.

Design thinking research and its research community has grown in a different way through contributions from researchers from different disciplines and fields. The contributors to this book, for example, include researchers with backgrounds in psychology, sociology, linguistics, philosophy, education, architecture, industrial and product design and a whole range of engineering disciplines. What became evident was that the research project that resulted in this book would keep the data constant, while accepting that methods would be various. This would mean that discussion about research findings would inevitably also invite and support discussion about method. The chapters in this book are intended to provoke methodological questions: if the authors make claims from the data, how does the research method they have used ensure their findings are valid and what means do they use to be convincing? Our hope is that as you read the chapters you will find it is instructive to compare methodologies, especially where different contributions focus on the same segments of data. We have tried to illustrate the potential for comparison that a common dataset can offer in Section 4 later in this introductory chapter.

7.
Cuff, D. (1992) *Architecture: The Story of Practice*, MIT Press

8.
Bucciarelli, L.L. (1995) *Designing Engineers*, MIT Press

9.
McDonnell, J. (1997) Descriptive Models for Interpreting Design, *Design Studies*, 18 pp. 457–473.

10.
Ball, L.J., and Ormerod, T.C. (2000) Putting Ethnography to Work: The Case for a Cognitive Ethnography of Design, *International Journal of Human Computer Studies*, 53, pp. 147–168.

11.
Reid, F.J.M., and Reed, S.R., (2005) Speaker-centredness and participatory listening in pre-expert engineering design teams, *Co-Design*, 1, pp. 39–60.

12.
Lloyd, P. (2002) Softening up the Facts: Engineers in Design Meetings, *Design Issues*, 17, pp. 67–82.

13.
Heath, C., Knoblauch, H., and Luff, P. (2000) Technology and Social Interaction: The Emergence of 'Workplace Studies', *British Journal of Sociology*, 51, pp. 299–320.

14.
Button, G. (2000) The Ethnographic Tradition and Design, *Design Studies*, 21, pp. 319–332.

15.
Holmes, J. (2005) Storytelling at Work: A Complex Discursive Resource for Integrating Personal Professional and Social Identities, *Discourse Studies*, 7, pp. 671–700.

16.
Klein, G. (1998) *Sources of Power: How People Make Decisions*, MIT Press.

17.
Wenger, E. (1998) *Communities of Practice: Learning Meaning and Identity*, Cambridge University Press.

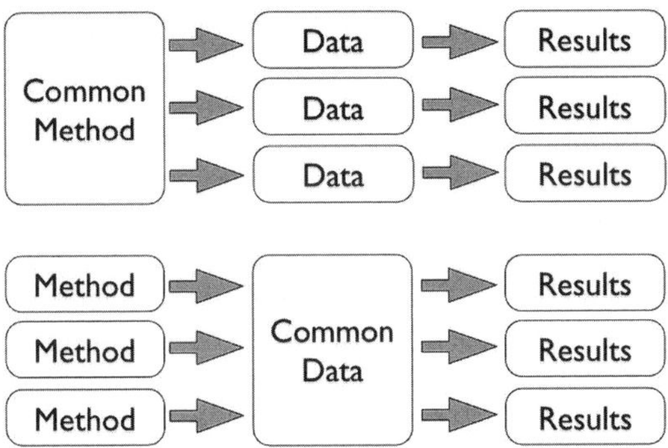

Figure 1. Conventional research approach (above) compared with a common data research approach (below).

Our interest was in providing a common dataset of material drawn from design activity taking place in natural settings. The collection of data from authentic professional design activity does present a number of problems however. There are confidentiality issues associated with products that are close to market. There is the timing and fragmented nature of real-world design processes which have hold-ups and delays, and periods of intensive, urgent activity, all for very good reasons. There is the limited ability of a researcher to 'be there' when significant activity takes place. There are also ethical concerns about obtaining permission to use recorded data for researchers at different institutions internationally.

The main problem, however, is in deciding what data to capture. The nature of design in practice means that even small projects take place in many different environments among a shifting set of participants. Designing occurs in many, often simultaneous, interactions, and is spread out over time and space. Designers work at computers, talk to other designers and clients, often solve problems away from formal workplaces, and are subject to many external influences which have critical implications for what they can do. Information is often forgotten or thrown away, and designers often work on a number of different design projects at any one time. How might it be possible to capture some real design activity as a single dataset to be shared? Aside from the fundamental issue of what a potentially huge and diverse dataset might be generated from even a modest design project there were two other matters to address, one very practical and one methodological. The practical issue we had to consider was the amount of research effort that would be required to analyse any dataset that we collected. Our intention was to focus analysis and discussion keeping the data analysed in common, so a manageable dataset was essential. From a methodological point of view, the dataset also had to be rich enough to support a number of different research agendas, and objective enough for all researchers to agree on exactly what constituted the dataset. For example, note-taking in organisations, a key method in ethnography, would, of necessity require presuppositions as the note-taker must decide what to attend to and what to record.

We decided to record a small number of design meetings as a way of sampling a much larger process. We reasoned that design meetings are not only a common form of naturally occurring design activity, but also, by virtue of their form, they expose what is happening at a particular point in a design project, helping to externalise decision-making. Meetings are also well-bounded, having a defined beginning and end, and typically last an hour or two, which meant that the size of the dataset could be kept manageable. Meetings also generate a variety of materials: drawings, sketches, notes and so on. Most importantly a meeting can be recorded objectively, and can be analysed in a number of different ways, depending on the interests of the researchers.

After some pilot studies, the material we selected to share comprised video-recordings of four design meetings from two different disciplines; two meetings in an architectural design process eight months apart, and two meetings in an engineering design process, several days apart. In total this amounted to around eight hours of video material. We also supplied transcripts of each meeting and images of all the materials that had been used or generated in the four meetings. This collection formed the complete dataset for distribution to research groups. Further details about the data collection follow in the next section. When the research groups received the dataset they were free to choose which meetings, or which segments of meetings to analyse according to their research interests and goals. Some chose to focus on only a small part of one meeting, while others analysed all four meetings.

This book presents twenty-one different analyses of data from the dataset, each in a separate chapter. In the remainder of this introductory chapter we first describe in Section 2 how the data was collected and prepared for distribution. We present an overview of the dataset in Section 3, briefly outlining each of the four design meetings, summarising the purposes of each meeting and introducing those who were present in each case. In Section 4 we attempt to show, through two examples, what sorts of comparisons working with a common dataset offers and to hint at the potential for constructive dialogue between researchers whose methods differ but whose interests overlap. We take two extracts from the data and examine how they have been interpreted and presented in different chapters of the book. In Section 5 we provide an overview of the rest of the book and provide some support for reading selectively across the contributions in a number of ways. We conclude this introductory chapter with an invitation to readers to approach each of the contributions with an enquiring mind and a critical eye. Our goal is to encourage reading of this book in ways which exploit two fundamental facts: that all of the work presented here, despite the diversity of methods, varied disciplinary allegiances, and epistemological differences of the researchers, is based on analysis of a common dataset; and that each of the individual contributions has something to add to our understanding of design thinking.

2 DATA COLLECTION AND PREPARATION

The architectural design meetings related to the design of a new municipal crematorium. The first meeting took place shortly after the conceptual work for the design had taken place, the second at the stage of preparing the project for the planning application. The engineering design meetings were the first meetings in the design of a new product: a 'digital pen' utilising novel print-head technology. The first meeting concentrated on the mechanical and formal aspects of the product, while the second meeting focused on the electronic and software aspects of the product. The two pairs of meetings differed in several other respects. The architectural meetings involved the architect(s) discussing an existing design with their clients, while the engineering design meetings concentrated on generating a wide range of ideas (and concomitant problems) concerning the formal, mechanical, and electronic workings of the digital pen. A typical plan for recording the meetings using multiple cameras is shown in Figure 2.

Small camcorders were attached to appropriate surfaces (walls, windows, ceiling fixtures) with clamps or suckers. This enabled the very quick setup time which is necessary when recording professional work *in situ*. Audio was recorded through the camera

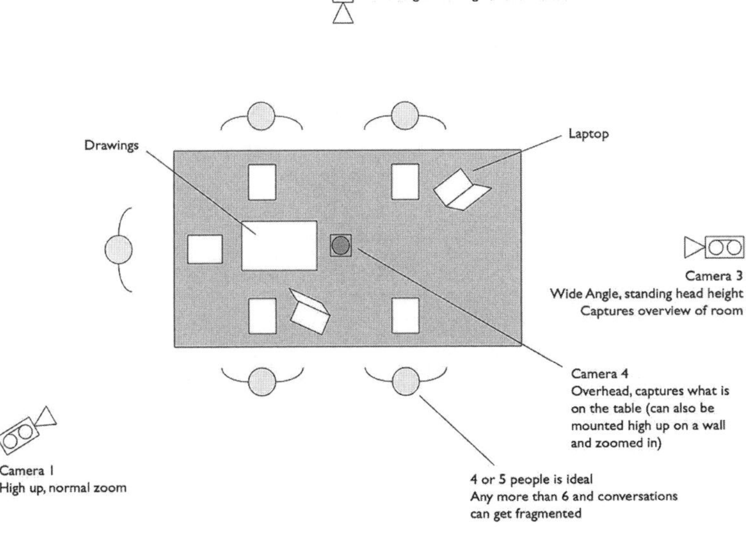

Figure 2. Plan of a typical, *in situ*, filming set-up.

microphones and although far from perfect, this was acceptable for the purposes of viewing and transcription. The independent camera streams were later synchronised and brought together in a four screen view with the clearest audio track added. Figure 3 provides an example of what the four camera view looked like.

The conversations in the meetings were transcribed. The simple transcript[18] notation used for this data has been retained for the extracts presented in this book which appear throughout in grey tint. The notation is summarised in Figure 4. Transcripts were provided to participating research groups without punctuation as is the convention for transcribed natural speech. Apart from rough indications of pause lengths, loudness contrasts and overlapping speech, the transcriptions did not attempt to characterise the vocal activities in detail to capture further aspects of the original speech (such as would be possible using a more elaborate notation such as Jefferson's[19]). This would have required far more transcription effort and in any case would have duplicated the data available to researchers in the audio stream of the video recordings.

There are no neutral conventions for identifying speakers in transcripts. We elected to anonymise the participants in the meetings via substitute given names which retain gender indications. This choice reflects the way the participants address and refer to each other in the data[20] and what can be seen plainly in the video recordings.

Some research demanded that the transcripts we supplied be augmented, for example work on gestures such as that in Chapters 15 and 16. Some necessitated segmentation of the data into units of analysis at sub-transcript line levels, for example part of the analysis

18.
This notation is a subset of that published in Vine, B., Johnson, G., O'Brien, J. and Robertson, S. (2002) *Wellington Archive of New Zealand English Transcribers' Manual*, http://qurl.com/6hjsr (accessed November 2008).

19.
Jefferson's notation is summarised in Atkinson, J.M. and Heritage, J. (1984) *Structures of Social Action: Studies in Conversation Analysis*, Cambridge University Press, pp. ix–xvi.

20.
This is the argument used by Schegloff (p. 566) in Billig, M. and Schegloff, E.A. (1999) Critical Discourse Analysis and Conversation Analysis: An Exchange Between Michael Billig and Emanuel A. Schegloff, *Discourse and Society,* 10, pp. 543–582.

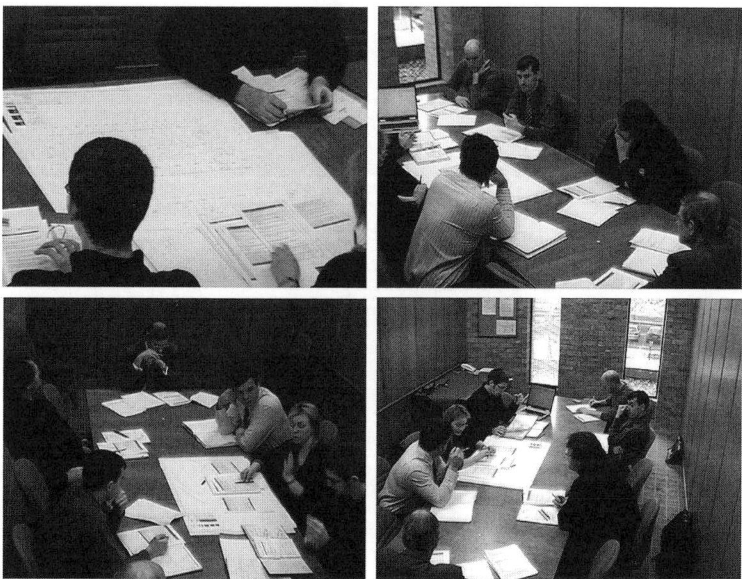

Figure 3. Typical four camera view of a design meeting in progress.

+	pause of about a second (++ two seconds, and so on)
.../....\... .../......\...	indicators of simultaneous speech
NO	emphatic stress indicated by capitals
ALBERT HALL	proper nouns: people, organisations, etc. indicated by small capitals
–	incomplete or cut-off utterance
......	material omitted
( )	unclear utterance
(over there)	unclear utterance, transcriber's best guess within brackets
[*points at map*]	transcriber's comments about additional activity
[*laughs*]	paralinguistic features

Figure 4. Simplified transcription notation.

presented in Chapter 12, while some required finer attention to the details of interaction than the transcripts we provided, for example the inferences drawn in Chapters 2, 13, 15, and 17 where nuances of overlapping talk and hesitations are significant. In all these cases researchers were able to supplement the transcript data provided by referring to the audio-video records of each meeting to support their particular needs.

All chapters in the book that quote extracts from the data (one or more lines from the transcripts) use the original transcript line numbers from the dataset to provide common anchor points. Quotations from the dataset appearing elsewhere in the text are similarly anchored by indicating which meeting and which line(s) of the transcript segment are being quoted. For example (A1, 345) refers to architectural design meeting 1, transcript line 345.

Transcript lines, unlike turns-at-talk, are to some extent arbitrary units. This can be seen in Extract 1, where a change in speaker prompts a new line of transcript, but a single turn may extend across several lines of transcript. Readers need to be aware of this when transcript lines are used as a unit of measure.

Extract 1, A1, Turns-at-talk vs. lines of transcript.

12	Anna	we've put these (images) up
13	Adam	I'm impressed
14	Anna	are you oh well I got those from the website for the-
15	Adam	KIMBELL
16	Anna	from the KIMBELL museum so that people could see what we were
17		trying to do so I copied them to show them what the idea was and
18		where it came from I put them up so that people could see and
19		everyone could sort of comment and have some sort of feedback from
20		that just to start with even just at this stage really
21	Adam	good well I'll look forward to hearing the feedback because that's the
22		purpose of the meeting after all yeah that's great

Some indication of how many turns at talk took place in each meeting is given in Tables 1 and 2, but these are approximations as closer examination of the talk at some points in the meetings show (for example, Extract 3 in Chapter 2).

The transcripts, the video-recordings of the meetings in a variety of formats, documentation relating to the meetings, and all drawings, sketches, presentations, briefs, flipcharts, and photographs of any objects shown at the meetings were compiled and sent on DVD to all research groups participating in the project. This material formed what is referred to throughout the book as: 'the DTRS7 dataset'[21].

21. Lloyd, P., McDonnell, J., Reid, F. and Luck, R. *DTRS7 dataset*. The workshop which formed part of the project from which this book originates was the 7th Design Thinking Research Symposium, http://design.open.ac.uk/dtrs7 (accessed November 2008).

3 OVERVIEW OF THE DESIGN MEETING DATA

We provide a brief introduction to the common dataset here. Tables 1 and 2 summarise the 'vital statistics' of each meeting.

3.1 *Architectural design meetings*

The architectural project brief was to design a crematorium comprising a chapel equipped with state of the art audio-visual equipment, seating up to 100 people; offices; cremation facilities; vestry; waiting rooms; covered entrance; and parking and landscaping on a site at which a similar facility was already in operation. The two meetings were at a stage in the project when the design concept and the main features of the program were already fixed and the major functions allocated. The clients were already broadly familiar with the plans, sections and elevations brought to the two meetings by the architect(s). Figures 5 and 6 show two drawings that were discussed during the first meeting. These are an early concept sketch of the crematorium (Figure 5), and a plan

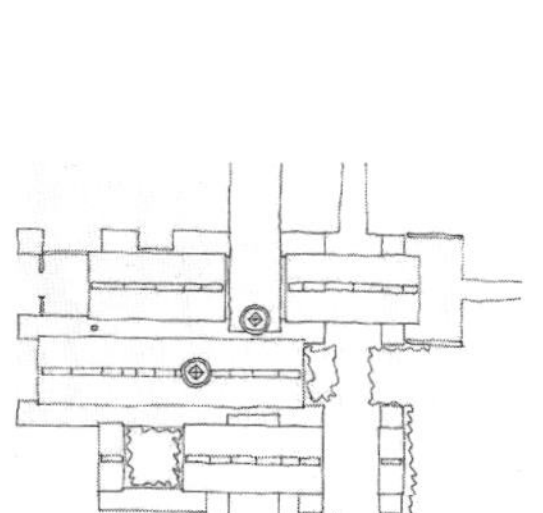

Figure 5. Early conceptual sketch of crematorium plan.

	Meeting Length		
	Time	Transcript Lines	Speaker Turns
Architectural Design Meeting 1 (A1)	2 hours 17 minutes	2342	987
Architectural Design Meeting 2 (A2)	1 hour 36 minutes	2124	1314
Design Brief	Design a crematorium with chapel set in a landscaped site		
Time Between Meetings	8 months (project at pre-planning application stage)		
Participants	Architect(s) and clients		
Additional Data Supplied	Plans at different scales, elevations, sketches, orthographic projections referred to during each meeting, 30-minute video of an informal interview with the principal architect describing the background to the project		

Table 1. Summary of architectural design meetings.

	Meeting Length		
	Time	Transcript Lines	Speaker Turns
Engineering Design Meeting 1 (E1)	1 hour 38 minutes	2019	1483
Engineering Design Meeting 2 (E2)	1 hour 41 minutes	1867	1193
Design Brief	Design a digital pen to exploit novel print-head technology		
Time Between Meetings	3 days (project at conceptual stage)		
Participants	Multi-disciplinary groups of engineers and product designers		
Additional Data Supplied	Task briefs supplied to participants, sketches and flip charts produced during the meetings, photographs of products and prototypes used during the meetings, engineering drawing of prototype		

Table 2. Summary of engineering design meetings.

of the building layout and surroundings (Figure 6). The two architectural design meetings, referred to throughout the book as A1 and A2, were intended to support the refinement of the design to sufficient detail for a planning application.

Meeting A1 takes place at the existing crematorium adjacent to the site of the proposed new building, a layout for the meeting is shown in Figure 7. The municipal architect, Adam, meets Anna, manager of the existing facility, and Charles, an officer from local government representing the municipality's interests. Also present at the meeting is a researcher (Peter Lloyd) who does not participate materially in the design discussions. Adam uses plans of the proposed building at different scales to talk Anna and Charles through the design. The topics which need discussion flow from one to another in smooth transitions. Figures 5 and 6 are examples of the eight documents referred to and drawn on to support the meeting's needs. Adam states the purpose of the meeting at its outset as being to obtain the clients' and building users' feedback to the design. (See Extract 1 above which is drawn from the opening remarks in the meeting.) During the course of the meeting, a range of topics are raised by Adam, Anna and Charles and a course of action is agreed. Chapter 14, Table 2 indicates the topics discussed in this first architectural design meeting.

Meeting A2 is similar to A1 in terms of setting and its layout is shown in Figure 8. In addition to those present at A1 a second architect, Tony, is also present as is Sally whose role is to observe and liaise between the architectural practice and its clients to ensure client satisfaction with its municipal design processes. Sally does not materially participate in the meeting. A2 takes place eight months after A1 and is the follow up to it. In the intervening

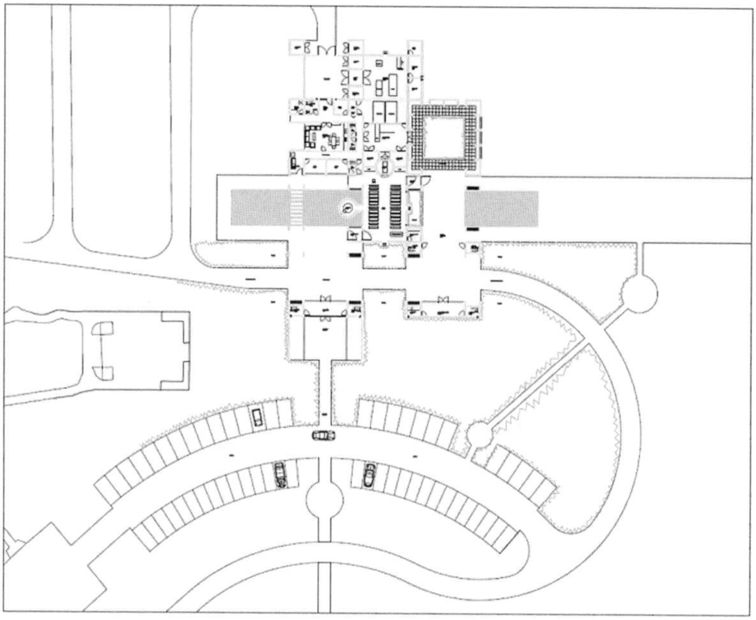

Figure 6. A scale plan of the proposed scheme showing buildings, pond, car parking, vehicular access routes and landscaping.

period some communications between the parties has taken place but no formal meeting. The aim of the second meeting is for the architects to show how they have responded with design changes to matters raised in A1 and subsequently. It is also an opportunity for both parties to raise further matters, either new issues that have come to light in the intervening months, or issues arising from the proposed changes to the design.

3.2 *Engineering design meetings*

The two engineering design meetings, E1 and E2, are held to discuss possible ways of addressing a range of issues associated with the early stages of designing a digital pen. This will allow

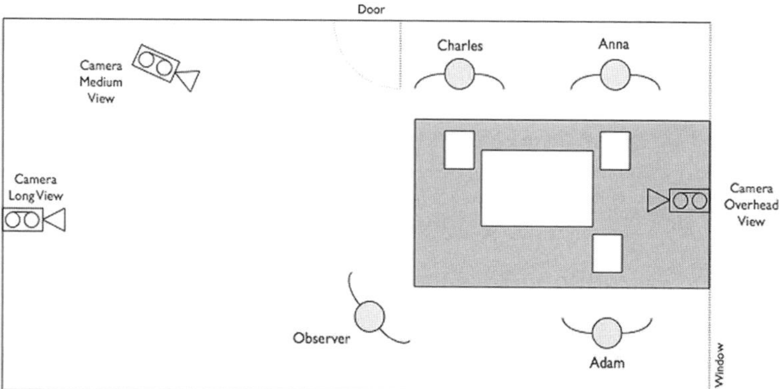

Figure 7. Plan of first architectural design meeting (A1).

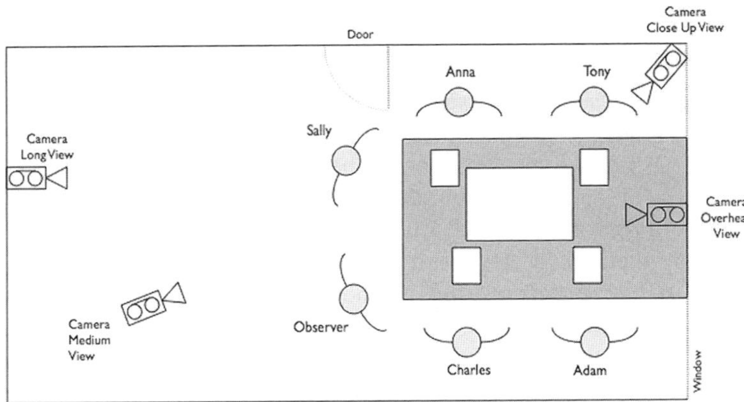

Figure 8. Plan of second architectural design meeting (A2).

the company, which has developed a novel print-head technology, to exploit commercially its intellectual property. The pen might be a toy or a sort of artist's tool, this and many other matters are open for discussion at this initial, idea generation stage of the project. The two meetings take place three days apart. There are seven members of staff present at each meeting, four participants attend both meetings to provide continuity and represent overlapping interests. The three other participants in each meeting attend to provide relevant specialist technical input in support of the different concerns being addressed in each meeting. Figure 9 shows a technical drawing for an early prototype for the digital pen made

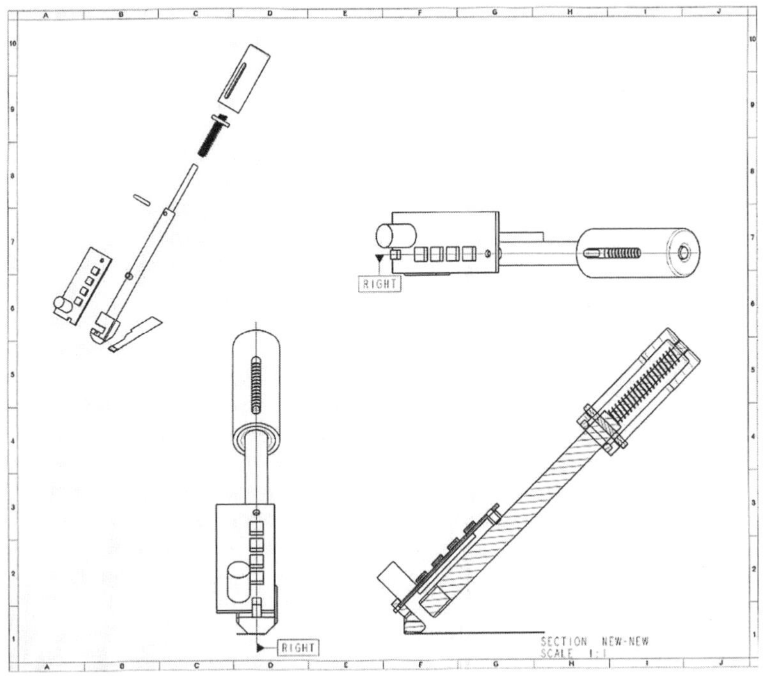

Figure 9. Technical drawing of digital pen prototype.

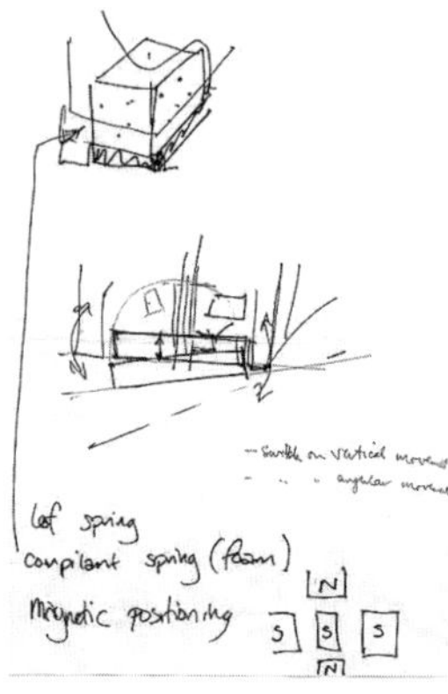

Figure 10. Sketch of design concept from meeting E1.

available to participants in E1, while Figure 10 shows a sketch of a design concept that is drawn during the same meeting.

The design issues addressed in E1 are more focused on mechanical challenges, whereas in E2 the focus shifts to electronics and software and the sort of features the pen, which is conceived as a child's toy, might incorporate. At the start of each meeting one member takes the lead in summarising the issues to be addressed. In both meetings attempts are made to tackle the task as a whole via a sort of brainstorming protocol. The extent to which this matches 'official' brainstorming protocol is analysed in Chapter 2, and some of the differences between the way the briefs for each event are set in place is the subject of the analysis in Chapter 11. The meeting layout for E1 is shown in Figure 11.

Those present at meeting E1 are Alan, business consultant and meeting facilitator; Tommy with expertise in electronics and also business development; Sandra with expertise in ergonomics and usability; Rodney, an intern with the company who is managing the project with a background in industrial design; and three mechanical engineers, Chad, Jack and Todd. The meeting ranges over a series of topics oriented towards ideas for 'solving' a series of issues with the mechanical aspects of the technology. These include keeping the print head level, ergonomic and other issues related to constraining the way the print head meets potential writing media, protecting the print head mechanically, and preventing overheating of the print head.

At meeting E2 the focus shifts from mechanical to functional aspects of the project. Those present include Tommy, Jack, Rodney and Sandra from the first meeting; these four are joined by Stuart, Patrick and Roman whose expertise lies in electronics and software systems. Figure 12 shows the meeting layout. This second meeting is organised in a similar way to E1 in that the (written) agenda is presented as a 'series of problems we have to solve with the print

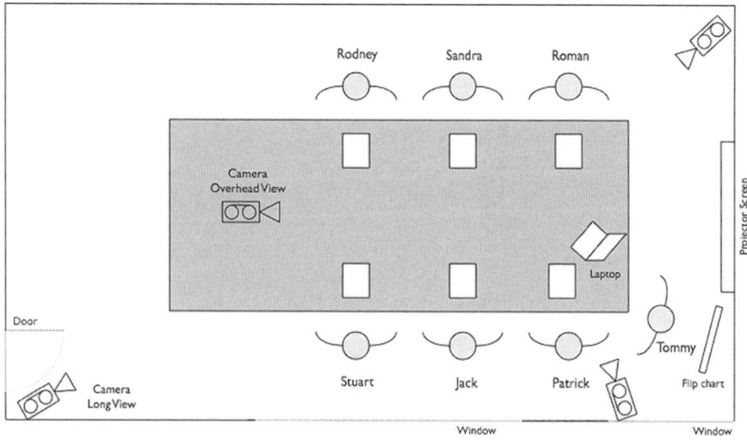

Figure 11. Plan of first engineering design meeting (E1).

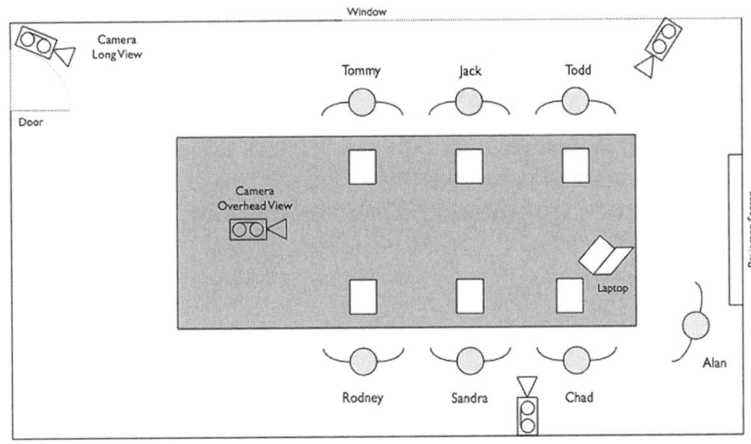

Figure 12. Plan of second engineering design meeting (E2).

head system, features control, and the form of the pen's user interface'. The meeting focuses on three main areas: potential product features, how user features might be selected and controlled, and what feasible architecture the device might have to be saleable at low cost and high volume.

4 ANALYSING COMMON DATA

To illustrate the potential value to research of using different methods to analyse a common dataset (see Figure 1) we have selected two extended extracts from the DTRS7 dataset to compare the approaches and findings of a number of different authors. We draw from a series of chapters to show how the data was used to support researchers' claims and arguments in a number of different ways. We also try to indicate how dialogue among researchers – both consensus and debate – is supported through using common data.

Reading through the two extracts in this section (Extracts 2 and 3) gives some feel for the meetings themselves. The architectural meetings were reflecting on a design proposal that had already been produced whereas the engineering meetings were focused on producing new ideas. The rhythm of the architectural meetings, then, was more measured than that of the engineering meetings, with participants generally talking for longer about a particular topic as they pursued the consequences of design decisions. The rhythm of the engineering meetings was much more quick fire, with participants building on one another's contributions and suggesting new ideas with short interjections. Both extracts are quoted in full below.

Quoting segments of what the authors have written about these data shows how descriptions and interpretations of the same event can both support one another and differ. Some treatments of the data tend be more descriptive, and focus on drawing our attention to various phenomena, while others interpret the data as a kind of text which presents opportunities for other types of discourse. Some researchers take the opportunity for both, such as Luck in Chapter 13. A further alternative approach, exemplified in several of the studies is to code these data, often explicitly linking them to particular theories of designing, thinking, or other types of action.

4.1 *Analysing design discourse*

Extract 2 is taken from shortly after the beginning of A1, the first architectural meeting. The architect (Adam) talks over one of his plan layouts for the crematorium, describing how people would enter the building, and where they would wait before going into the chapel area. However, he is not sure that he's got the size of the waiting area quite right and asks the client (Anna) to comment on what he has done. Five chapters (5, 14, 16, 17, 20) make reference to what is happening in this particular extract, and here we bring together some of what they have to say about it. Several other chapters make reference to the same episode, and we leave the reader to explore these further.

Glock, in Chapter 16, takes the basic transcript of Extract 2 and adds further descriptors to it. Extract 2a shows how lines 113–117 of Extract 2 have been amplified to incorporate the actions of the participants (for example gazes and head shaking), brief pauses in the conversation, and simultaneous speaking. Again these are relatively objective qualities of these particular data.

Through adding more detail to the transcript Glock is able consider a wider range of data to build his argument. In the following quotation (from Chapter 16) he comments on the vague nature of the language used when assessing the crematorium waiting area.

> "Although the precise size of the waiting room can be read from the plan, participants in this episode do not discuss or negotiate the size of the waiting room in terms of precise numbers but in rather vague, qualitative terms: "small", "relatively small" (A1, 112), "big enough" (A1, 117). Moreover the vagueness of the expression is even enhanced: "kind of + small" (A1, 93). Whereas the statement: "at the moment it's exactly on brief" (A1, 93) conveys some state of affairs which constitute the truth conditions of the proposition, the vague expression used in the assessment relies on a resemblance relation: 'small … in relation to the size of the project'."

Chapter 7, by Dong, Kleinsmann and Valkenburg, also examines assessment and appraisal during the architectural design meetings using a formal coding scheme based on a functional-semantic

Extract 2, A1, The crematorium meeting area.

Adam: to answer your question it's a relatively small room the area is similar to the existing waiting room and as such you are unlikely to get more than about eight to ten maximum twelve seats in there I would have thought probably because the building well the room is doing so much it's allowing people through into the porch area it's also allowing access to the loos so my first question to you is is waiting the room big enough and would you like us to increase it

Anna: I would say although the time spent sitting and waiting might not be very long and eight seats seems enough at some stages even our waiting room is too big so it's slightly I'm also thinking of the fact that if we've got a flow of people walking through that then restricts us we can't put seats through that because in a sense we need to keep an access open and so the seating will be against the wall

Adam: what I'd recommend is that we look at doing something like that extending it which will give you seating areas here seating areas here seating areas here as well as here and here which effectively will double the seating capacity from what I was just saying

Anna: yes I mean people waiting for cabs or for people waiting to be picked up as well for services you know it might not sort of eight might be more than enough for funerals for the majority of the time but I would think it's nice to give them a bit more space as well because we might get people waiting for the eleven o'clock funeral erm and people at the ten o'clock perhaps arrive and so they keep in their little groups they don't want to mix with other people so the feeling of keeping them segregated just because they don't know the other people might also be there and we do have problems with families like that during funerals they don't want to be [*laughs*]

Chris: police attendants [*laughs*]

Anna: police attendants quite often you know you'd think it would bring them together but it actually makes it worse

Adam: really gosh

Anna: yeah and they sit separately in the chapel as well it's all to do with money and you know they've left someone something wonderful that's most of the time what it is or the other family are cross because one family has arranged it and they used they never visited her while she was alive and how dare they get involved with this and it all escalates

Peter: it's like east EASTENDERS [*all laugh*]

Anna: indeed I mean yes it can escalate to sort of violence at times not here so far but threats of it at times so the idea that number one we have people for other services arriving perhaps at the same time they want to keep separate families like to keep separate and also the fact that perhaps people are waiting for lifts or other things as well perhaps they've had the service and they come back in the waiting room to wait for someone to pick them up or just because they're waiting for the taxi or something

Adam: ok

Anna: so I'd like it a little bit bigger I think + not hugely because there is it is a wasted space most of the day really

Adam: yeah well I would have thought another couple of metres on there [*writes on drawing*] would do the trick so shall we agree a two metre extension yes or thereabouts hmm

Extract 2a, A1, Glock's fine-grained transcription.

<table>
<tr><td></td><td></td><td>| presents drawing | points</td></tr>
<tr><td>113a</td><td>Adam</td><td>to the existing waiting room / | + and as such + erm you are really | unlikely</td></tr>
<tr><td></td><td>Anna</td><td>/ (right)</td></tr>
<tr><td>113b</td><td>Adam</td><td>to get more than</td></tr>
<tr><td>114</td><td></td><td>about eight to ten maximum twelve seats in there I would have thought +</td></tr>
<tr><td>115</td><td></td><td>erm probably because the building well the room is doing so much it's</td></tr>
<tr><td>116</td><td></td><td>allowing people through into the porch area / it's also allowing access to the loos</td></tr>
<tr><td></td><td>Anna</td><td>/ yeah [low voice]</td></tr>
<tr><td></td><td></td><td>| gazes to Anna | short gaze to Charles</td></tr>
<tr><td></td><td></td><td>| | looks at drawing | | gazes to Anna | Anna headshake</td></tr>
<tr><td>117</td><td>Adam</td><td>|so + | my first question to | you | is is the waiting room | big enough</td></tr>
<tr><td></td><td>Anna</td><td>/ no</td></tr>
</table>

analysis of the language used. In Chapter 17, Oak, working from a compatible methodological approach to that of Glock, also draws on data from Extract 2. She attributes the openness to possibilities and the ambivalence of the discourse more to the client, Anna, than the architect and suggests that through this Anna 'performs' her client role. She writes:

> "In this extract the architect makes a clear request for information about room size (A1, 117–118), but the client's answer leaves open for interpretation exactly how big the new waiting room should be. Later, when the architect suggests possible dimensions for the room (A1, 126–128) the client's reply is also ambiguous: the current room may or may not be big enough (A1, 129–141). Eventually the client says that she would like the waiting room: "a little bit bigger I think + not hugely because … it is a wasted space most of the day" (A1, 158–159). The architect responds to the client's ambivalent comments by choosing dimensions ("a couple of metres" (A1, 160)) and writing on the drawing, minuting their talk as a design decision made (see the second meeting, A2: "one of the items on the minutes last time was to increase the size of the waiting room" (A2, 49–50)). However, despite the architect's decision to extend the space, the client has not actually made a definitive response about her preferred room size. Instead, she answers his last question about waiting-room size (A1, 161–162) by changing the topic to suggest seating outside (A1, 163–164)."

Oak draws attention to the disparity between the terms used by the architect and those used by the client, interpreting their significance as means through which they perform their roles during the discussion:

> "In contrast, Client 1 [Anna] performs her role as client through replying in ways that tend not to directly answer the architect's questions; that is, she does not reply using the kind of terms in which the

questions are presented (terms associated with spatial dimensions or room measurement)."

The language of the client and the language of the architect is something that McDonnell also addresses in Chapter 14. Before she goes on to analyse the data from Extract 2, she first summarises the 'waiting area' episode, sketching in more detail and constructing a kind of narrative:

> "The first episode, early in meeting A1, concerns the size of the waiting area. [...] The topic is initiated by Adam with a question about whether the size of the waiting room should be increased: "the first query I have [...] about whether you wanted the size of the waiting room increased" (A1, 90–92). Anna's response shows that, looking at the plan, she cannot tell how many seats there will be in the area. The architect compensates for the representation (the plan) by outlining a skeletal scenario of use, focusing on the functional aspects of the space thus: "the room is doing so much it's allowing people through into the porch area so it's also allowing access to the loos" (A1, 115–117). Anna gives a functional affirmatory response to Adam's question, i.e. in terms of number of seats. [...] But then she moves on to give a rich account of what people might be waiting for, different kinds of 'waiting', and she demonstrates a requirement to be able to separate waiting people for a whole series of reasons: "we might get people waiting for the eleven o'clock funeral erm and people at the ten o'clock perhaps arrive [...] they don't want to mix [...] and we do have problems with families during funerals" (A1, 132–137). After further rich elaboration, she moves on to say: "people like to smoke at funerals [...] and the seat that we've got out by the car park [...] even if it's cold and not very nice [...] people feel more happier out there than they do sometimes in the waiting room" (A1, 164–168)."

McDonnell concludes that both architect and client have expertise in their fields which they bring to bear on the design task, and that one way in which they are able to express expertise and effectively to collaborate with each other is through the frame of 'building use':

> "This episode shows how the architect's non-expert notions about waiting at funerals, once expressed, offer a conversational entry point through which the building user's extensive understanding of what waiting is about can be explained, and thus emerges what adequate provision for waiting might entail. The information she volunteers gives the architect the opportunity to understand 'waiting' in the crematorium context and thereby the requirements to cater for the practical, social and psychological needs associated with 'waiting' and separation."

Reymen, Dorst and Smulders build on this observation by McDonnell in Chapter 4. It is interesting to contrast the approach of embellishing the extract through narrative construction with, for example, Glock's embellishment of the transcript with additional descriptors drawn from the (video) data. Both approaches enhance the transcript data, Glock taking arguably a more objectively grounded approach than McDonnell. McDonnell's conclusion about this episode focuses on an understanding of what kind of activity 'waiting' is in the context of the crematorium.

Lloyd's approach, in Chapter 5, is initially similar to that of McDonnell, attempting to frame episodes in terms of a narrative, and it is interesting to compare the two descriptions:

> "Early on in the first architectural meeting the waiting area in the new building is discussed and we see directly how architectural form can affect human behaviour. As services at the crematorium take place throughout the day, often one after another, there needs to be a waiting area where people can gather before their service begins, and while the previous service ends. The size of the waiting area that Anna has specified is of concern to Adam, who thinks it looks: "kind of small" (A1, 93). Initially Anna is unconcerned, drawing on her experience of the existing crematorium she concludes that the: "eight to ten, maximum twelve seats" (A1, 114) that Adam suggests are maybe too many. However, after Adam suggests increasing the space Anna nuances her position, acknowledging that while the space would be adequate for the majority of the time she could envisage situations that would require more room particularly when: "people … arrive and … keep in their little groups, they don't want to mix with other people" (A1, 134–135). A larger space would allow people to keep themselves separate from one another. Anna suggests that the space should have a: "feeling of keeping [people] segregated" (A1, 135)."

However Lloyd uses the meeting transcripts as a jumping off point to ask more theoretical (and hypothetical) questions about the ethical nature of the design process. From this description of the episode Lloyd goes on to speculate about the ethical implications of what might happen following the meeting:

> "Although an uncontentious design issue here – Adam is clearly willing to respond to Anna's request – this is nevertheless a decision point where, safely back in the office, an architect might intervene for a number of reasons. Having a space where people could hide from one another might go against the aesthetic vision of the architect or the architect might feel it their job to 'improve' behaviour through design, that people *should* see and talk to each other. The architect, then, has to decide whether to support existing behaviours with space – behaviours

> that they might personally think strange or reactionary – or whether to try and use design to change or modify the behaviour of building users."

Lloyd uses the transcript more as a text of underlying assumptions about the design process rather than focusing on the particular words themselves.

The final use of the data provided by Extract 2 that we will feature briefly here comes from Stacey, Eckert and Earl in Chapter 20. They take a rather different approach to Extract 2, locating in it an example of an 'object reference'. Their chapter is concerned with classifying the external references used in the development of design ideas during design meetings to better understand the pragmatics of object references. In their work they use object reference categories to illuminate the types of references used and the functions object references play. They find that the majority of the design-relevant object references in the architectural design meetings are to individual buildings (other than the one being designed) and they discuss the functions these and other categories of references play in the design process. According to their analysis, Extract 2 contains the sole example in the architectural design data of a reference type they classify as 'human behaviour as part of the system being designed':

> "The other three design-relevant object references in the crematorium meetings mapped to other parts of the product-user-environment-activity socio-technical system; two of these were exemplars of the range of music mourners might ask for. The remaining case is shown in A1 (143–152). Here the mapping is to users of the crematorium, with difficult interpersonal relationships, interacting with each other in the physical environment provided by the crematorium."

The above quotations illustrate, however, the wide range of possibilities for analysing the transcript and how, in considering these, the methods and coherence of different analyses can be compared.

4.2 *Debating concept definitions*

Extract 3 comes from about halfway through the first engineering meeting, E1, and is concerned with the form and use of the pen yet to be designed. The people involved in the discussion suggest a number of ways to hold and use the pen, drawing on a range of familiar forms and mechanisms with which people make marks on paper – in sketching, writing, painting, drawing, etc. In the following discussion we draw attention to the ways in which two chapters make use of data drawn from this extract as they test and extend schemes

Extract 3, E1, The form and use of the digital pen.

Todd	if you've got if you've got a switch there there there and there and it only works if all four switches are down and those switch positions guarantee that the head is down then-
Tommy	yeah that's the control ( ) approach isn't it
Jack	yeah that's the yeah and you design it in such a way that that's a reasonably comfortable position and if you don't put it in that position it doesn't work
Jack	it'd be nice though if you naturally pick it up and it works
Todd	yeah but that's you can do the two together can you
Jack	yeah we could have a look
Alan	the other thing to to think about is in almost all cases when I look at pens the apart from re-wired sort of micro-pens the th- tip is actually the narrowest part of the product whereas in what we're looking at it could actually be as wide or wider-
Tommy	mmmm
Alan	than-
Chad	so it's more like a paintbrush isn't it like a DIY paintbrush
Alan	yeah
Tommy	it's more like a roller like a roller yeah
Alan	++ mmm
Todd	it's there's a prototype-
Alan	if you had like if you if you had it connected to one side like that or a- it would be here the print head you could have then going to there then going to there so-
Jack	that you don't smudge it
Tommy	yeah
Alan	it would have flexibility because of the metal on which it's hinged if you like
Tommy	yeah I think ( ) actually I wonder if you intuitively know how to use a roller because you're older than eleven and ( ) intuitive if you've never seen it before
Jack	my two year old rolls paints
All	[*laugh*]
Jack	OK I'll shut up

for coding concepts concerning the use of analogy in reasoning and the pragmatics of object references during design conversations.

Stacey, Eckert, and Earl whose chapter (20) we have already referred to and Ball and Christensen whose work is presented in Chapter 8 address the use of analogies in the design process and make explicit reference to the material which appears in Extract 3. It is interesting to consider the dynamics of how these chapters both analyse the data and cross-refer to each other, since it highlights another aspect of using a common dataset: to support dialogue between researchers who have similar interests but differences of opinion. These differences are revealed clearly and can be discussed constructively using the data itself as a conduit for communication. Showing the data also allows readers of these researchers' accounts of their analyses to make their own judgement about the validity of particular interpretations.

Stacey, Eckert and Earl suggest that many of the references to external objects in the engineering meetings – in this instance paint brushes and paint rollers – are intended to be contrasts pointing up distinctive features of the design. They direct the reader to a segment of the data as evidence:

> "A large fraction of the object references in the meetings are *contrasts* (for instance sledge and skis): mappings between very similar elements force mappings between alignable differences and direct attention to them – often to communicate the point that the way that the design is different from the contrasting situation is important for the success of the design, as in E1 (819–827)."

They go on to note that most object references aren't usually one-to-one relationships with some feature of the design, but stem from a number of related analogical associations which they refer to as 'complex structures':

> "Many of the object references were simply mapped to some aspect of the design or product-user-environment-activity system. However, some mappings were not binary, but created a more complex structure comprising mappings between three or more objects, as in E1 (819–827)."

They cite the same segment of data as exemplifying such a 'complex structure' of 'three or more objects' (in this case a micro-pen (820), a paint brush (825) and a paint roller (827)).

Ball and Christensen focus on exactly the same segment in explaining how they have coded the data; firstly, in terms of the existence of an analogy, secondly, in terms of the relatedness or 'distance' between the two objects referred to:

> "The transcripts were coded for presence of analogies by applying Christensen and Schunn's[22] approach. Any time a designer referred to another source of knowledge and attempted to transfer concepts from that source to the target domain then this reference was coded as an analogy. All analogies were also coded for ANALOGICAL DISTANCE using a binary categorisation scheme where *within-domain* analogies involved mappings from sources that related to tools, mechanisms and processes associated with graphical production and printing (E1, 819–822), whilst *between-domain* analogies involved mappings from more distant sources."

22. Christensen, B.T., and Schunn, C.D. (2007) The Relationship of Analogical Distance to Analogical Function and Preinventive Structure: The Case of Engineering Design, *Memory & Cognition*, 35, pp. 29–38.

The objects cited by the designers in Extract 3 are, according to Ball and Christensen 'within-domain', all being examples of graphic

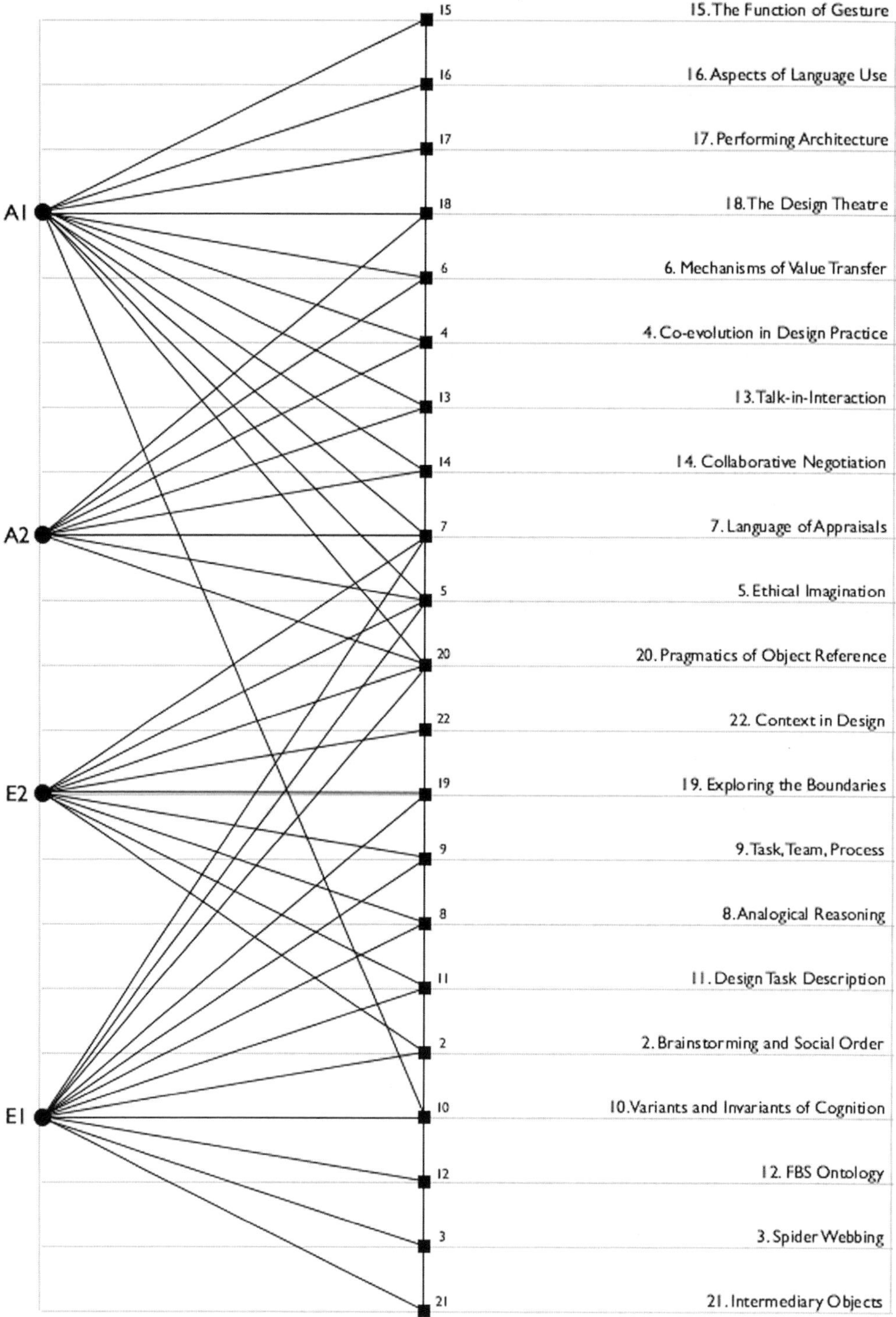
A1
A2
E2
E1
15
16
17
18
6
4
13
14
7
5
20
22
19
9
8
11
2
10
12
3
21
15. The Function of Gesture
16. Aspects of Language Use
17. Performing Architecture
18. The Design Theatre
6. Mechanisms of Value Transfer
4. Co-evolution in Design Practice
13. Talk-in-Interaction
14. Collaborative Negotiation
7. Language of Appraisals
5. Ethical Imagination
20. Pragmatics of Object Reference
22. Context in Design
19. Exploring the Boundaries
9. Task, Team, Process
8. Analogical Reasoning
11. Design Task Description
2. Brainstorming and Social Order
10. Variants and Invariants of Cognition
12. FBS Ontology
3. Spider Webbing
21. Intermediary Objects

Figure 13. Analysis of design meetings by chapter.

understand what was happening. This is rather a crude distinction, and we are using the notion of a 'theory' rather broadly, incorporating any sorts of models, ontologies and frameworks brought to the dataset and imposed upon it. Despite shortcomings, we think it helps to differentiate some essentially different types of analysis. For example we consider that Chapter 2 (Matthews), which looks at the relationship between brainstorming and social order uses a data-driven approach, pointing out to us something that happens in this particular dataset, namely how the 'rules' of brainstorming are circumvented by the 'rules' of social order. Although Matthews theorises about what is happening in the engineering design meetings, it is a theory that has arisen from his interpretation of the dataset itself. In contrast, Chapter 9 which concerns the relationships between the task, the team, and the process takes an existing theory of design activity around the idea of 'team mental models', which has been built up over a number of years and through a number of studies, and tests the extent to which it applies to the DTRS7 dataset.

We have also used Figure 14 to show whether or not a coding scheme has been used in the analysis of the data. The creation of a coding scheme – either with a data-driven approach or a theory-driven approach – is often a matter of some debate in design research. Debate focuses on the multiplicity of schemes and the robustness of application. To some extent the variety of schemes simply reflects the variety of phenomena that are the foci of researchers' attention. However, many schemes are criticised for being too poorly documented to be reliably used by others and in many cases independent rating and reporting of coding reliability are omitted from the reports of studies. In the chapters where a coding scheme appears, explicit references to coding categories usually appear in SMALL CAPS to emphasise that what is being talked about is a coding category and not something else, although some schemes are hierarchies or networks of constructs in which case other typographical styles are incorporated to clarify terms.

Some authors make explicit attempts to compare their characterisation of phenomena with the characterisations of others. For example, Akın, in Chapter 10 makes some proposals for mapping from his coding scheme, motivated by many years of studying design cognition, to that developed for capturing semantic design information oriented to distinguishing between function, behaviour and structure, again developed over a long period by Gero and colleagues (applied here to part of one of the engineering design meetings, and presented by Kan and Gero in Chapter 12). Dong, Kleinsmann and Valkenburg in their work on the language of appraisals (Chapter 7) make use of some of the data categorisations (types of design values and human values) identified in the

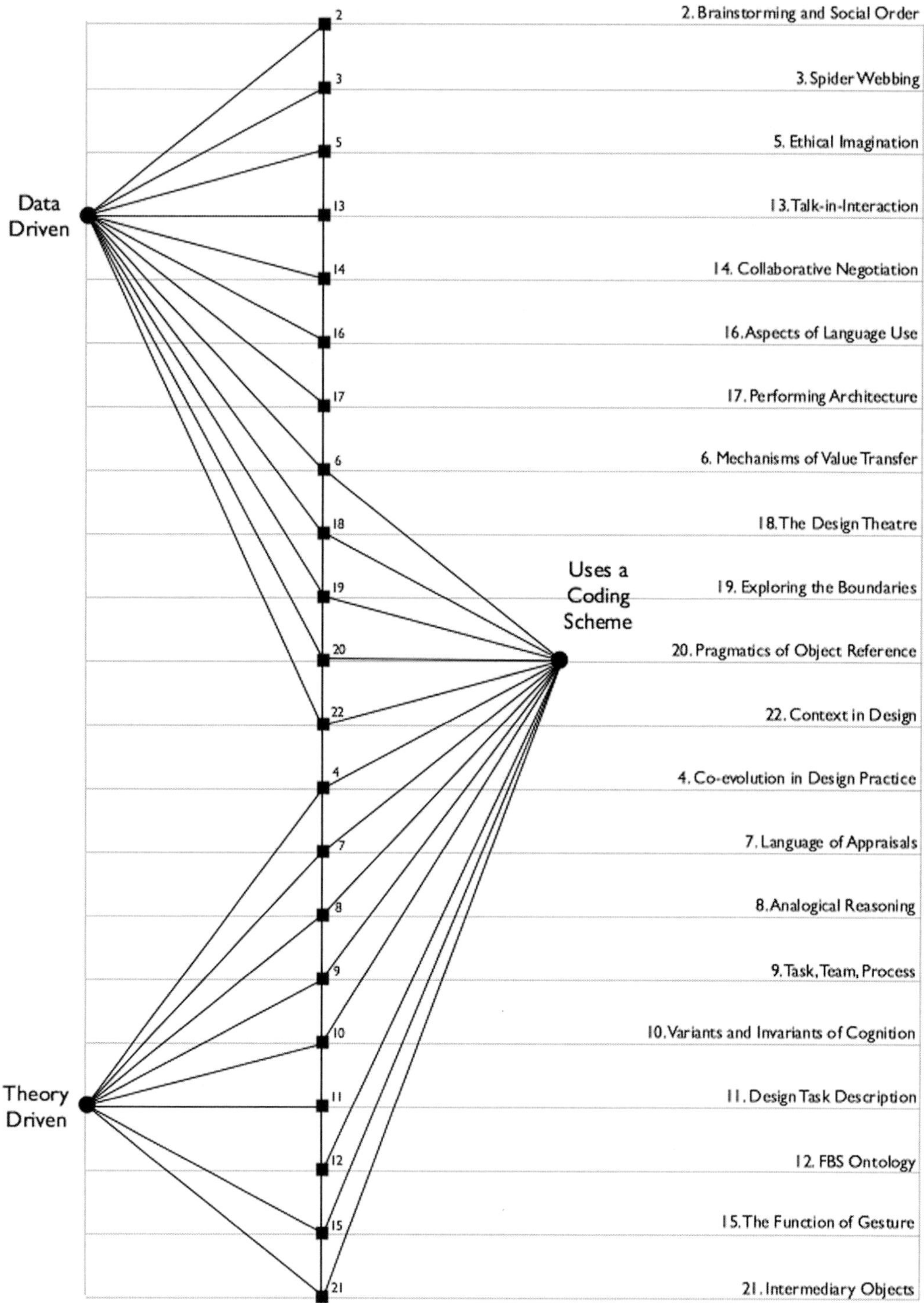
Data Driven
Theory Driven
Uses a Coding Scheme
2
3
5
13
14
16
17
6
18
19
20
22
4
7
8
9
10
11
12
15
21
2. Brainstorming and Social Order
3. Spider Webbing
5. Ethical Imagination
13. Talk-in-Interaction
14. Collaborative Negotiation
16. Aspects of Language Use
17. Performing Architecture
6. Mechanisms of Value Transfer
18. The Design Theatre
19. Exploring the Boundaries
20. Pragmatics of Object Reference
22. Context in Design
4. Co-evolution in Design Practice
7. Language of Appraisals
8. Analogical Reasoning
9. Task, Team, Process
10. Variants and Invariants of Cognition
11. Design Task Description
12. FBS Ontology
15. The Function of Gesture
21. Intermediary Objects

Figure 14. Research approach taken by chapter.

dataset by Le Dantec and Do (in Chapter 6) drawing comparisons with their own identification of appraisals in the dataset.

In addition to the threads through the book supported by Figures 13 and 14 we use two means of highlighting the many more specific connections between chapters. Firstly, Figure 15 provides an overview of which chapters make links to others. Secondly, within each chapter, internal references to other chapters in the book appear in **bold type**. Exploring these linkages can reveal surprising relationships. A quick glance through the book will show that links between chapters are often quite dense. There are many more specific themes and relationships that deserve exploration; we leave readers to identify and pursue these for themselves.

6 DISCUSSION

We believe that the sharing of a common dataset promotes research in a number of ways. The DTRS7 dataset has provided the basis for an eclectic collection of studies, each of which makes its own claims for being a distinctive contribution to our understanding of design thinking. In the most straightforward way, therefore, the common dataset has supported 'business as usual' for research teams and individual researchers by providing access to a scarce research resource, namely data from authentic design practice. From this point of view, considered together, the chapters present a good cross section of the different approaches to the study of design thinking currently taking place; they are representative if not comprehensive.

Many of the chapters not only present new understandings of a particular aspect of work in their field but also provide critical summaries of previous relevant work, and thus provide the reader with a valuable account of what is the current state of understanding of the particular topic. For example, Chapter 8 summarises the state of current knowledge about the uses of analogy in reasoning in design contexts; Chapter 20 makes reference to a broad range of prior work on the use of object references in talk about design, including some of the authors' own work on this topic. Chapters 10 and 12 present and extend the authors' own prior findings in design cognition and the development of design ontologies respectively, which serve to introduce new audiences to their work.

Many other chapters situate the findings they present based on the DTRS7 dataset in the context of (often their own) extensive earlier studies and in this way offer focused primers on a whole range of relevant work which underpins their research agendas. Examples of these include Chapter 22 which extends prior work on designers' consideration of context during designing; Chapter 4 which

Figure 15. Cross referencing between chapters, chapters on the left reference chapters on the right.

takes a critical look at the notion of co-evolution in design practice; Chapter 21 which continues work to identify objects critical for recovering the rationale for design decisions; and Chapter 9 which tests the implications of models of what participants collaborating on a design task share in terms of team cognition.

In all cases, the researchers who accepted our invitation to make use of the dataset have pursued their own established research agendas, using the DTRS7 data to test their own or others' current beliefs about aspects of designing, to push their work in new directions or probe the boundaries of its applicability. Each contribution can therefore be evaluated separately and it is left to the reader to make a judgement as to whether the data has been appropriate for the type of investigation undertaken and whether and the arguments made in each chapter are plausible for the claims made. Similarly, if a coding scheme has been used, the reader must consider critically whether its use is convincing, whether it clarifies something, or whether it obfuscates. When any interpretation is presented, whether through coding, or by some other device such as a narrative description, we invite you to judge whether this is compelling because sufficient evidence from the data is provided, or because you are persuaded from a clear explanation of the process through which the interpretation has been developed.

A common dataset can provide a focal point for a diverse community of researchers and, although not a consensus building mechanism, the shared dataset can operate as a boundary object between differently principled communities of design researchers. At a very practical level, people listen to one another because they have looked at the same things. Researchers with very different points of departure have paid attention to each others' findings and have been able more easily to grasp the methods that others use to reach their goals. Some have set out with research questions or hypotheses to pursue like the approaches taken for Chapters 4 and 11 for example, some have immersed themselves in the data, arriving by a process of induction at accounts of the patterns they have observed, a style of working exemplified in Chapters 6 and 19. Where there are fundamental differences in approach, sharing data opens up a new channel of communication for the broader research community, perhaps not one sharing a common language or disciplinary values, but one which conveys common interests and goal orientations and better understandings of what constitutes sound research in each other's frames of reference. This is important for the quality of design research generally and, as a consequence, for what research contributes ultimately to our collective understanding of design thinking.

Part 1

Understanding Design Processes

2
Intersections of Brainstorming Rules and Social Order*

Ben Matthews

This chapter examines how designers manage their accountabilities to the rules of brainstorming and the normative social order. It argues that designers' adherence to the rules of brainstorming is noticeably tempered by their orientation to social order, a finding that suggests a reassessment of the nature and use of methods in design.

For many years, design research has exhibited a keen interest in design methods. Yet this interest in methods has been diverse. For some researchers, methods for doing design offered a way of formalising design practice. This was seen as a key to understanding how designers work (a way of describing designers' activity), to automating design (e.g. by means of algorithms), and to educating design practitioners. In many ways, Simon[1] remains an archetype of such research programs and their ambitions. Over time, however, the interest in methods shifted; ultimately they were challenged as a means to each of these ends. Problems were encountered when attempts were made to account for the activities of designers in terms of the steps prescribed by design methods[2]; the inherent limitations (philosophical and practical) of formal logic and its derivatives to produce solutions to genuine design problems has been the basis of a wide ranging discussion in design research[3,4,5,6,7]; and the very idea that good design work is, or can be, the straightforward outcome of the application of a method was not something ever vindicated by the results of methods-based design programs at universities and design institutes. Other bases were sought for design education[8]. Whatever it does take to be a successful designer, it is surely more than having facility with design methods.

Nevertheless, methods are still taught, still found useful, and remain a fundamental component of many educational programs and professional design studios. But designers' situated and practical *use* of methods has not been a topic of study for design research, with few exceptions. Perhaps most notably is Bucciarelli's memorable account of one project manager's 'disaster meeting' with the Pugh method[9]. The Pugh method is a common engineering tool used to evaluate and compare different design alternatives according to project-specific criteria[10]. Bucciarelli's analysis shows how the method is founded on a presumption that the establishment of project-specific criteria is itself unproblematic. In examining the method's use, however, he found that negotiating what the criteria

1.
Simon, H.A. (1981) *The Sciences of the Artificial*, MIT Press.

2.
Hales, C. and Wallace, K. (1988) Detailed Analysis of an Engineering Design Project, *International Journal of Applied Engineering Education*, 4, pp. 289–294.

3.
Coyne, R. and Snodgrass, A. (1993) Rescuing CAD from Rationalism, *Design Studies*, 14, pp. 100–123.

4.
Rittel, H.W.J. and Webber, M.M. (1973) Dilemmas in a General Theory of Planning, *Policy Sciences*, 4, pp. 155–169.

5.
Schön, D.A. (1990) The Design Process, in Howard, V.A. (ed), *Varieties of Thinking: Essays from Harvard's Philosophy of Education Research Center*, Routledge pp. 110–141.

6.
Simon, H.A. (1973) The Structure of Ill-structured Problems, *Artificial Intelligence*, 4, pp. 181–201.

7.
Winograd, T. (1991) Thinking Machines: Can there be? Are we?, in Sheehan, J.J. and Sosna, M. (eds), *The Boundaries of Humanity: Humans, Animals, Machines*, University of California Press.

8.
Schön, D.A. (1987) *Educating the Reflective Practitioner: Toward a New Design for Teaching and Learning in the Professions*, Jossey-Bass, San Francisco.

*Reprinted from: CoDesign Vol. 5 No 1 (March 2009), pp. 65–76,
DOI: 10.1080/15710880802522403

were, and delineating exactly what should be circumscribed by each criterion was a tough job – establishing consensus on these matters proved laborious. And *until* criteria can be established, the Pugh method is unable to be the help it is intended to be to designers, which is to enable them to decide among competing design alternatives. In this sense, the method provided little more than a (albeit still useful!) framework for negotiating what the criteria for the project should be.

It is in this spirit that I have focused on engineering designers' use (and non-use) of the rules of brainstorming in the course of the two engineering meetings provided as part of the DTRS7 data, meetings E1 and E2. Brainstorming is a ubiquitous method for generating ideas that hardly needs introduction. Although there are local variations in how brainstorming is conducted, it is founded on a number of basic ideals, including suspending judgment during the session, allowing one speaker the floor at a time, encouraging 'wild' ideas, and building on the ideas of others. It is in common use in countless design consultancies, including some of the most prestigious (such as IDEO). But although there are many extant studies of brainstorming, these have typically focused on its effectiveness, evaluating it in terms of things like the number and quality of ideas generated, and other instrumental payoffs of its use[11,12,13]. In contrast, I simply want to examine how the method itself comes to be of use to designers in the course of their work. In particular, I closely examine the transcript of these two brainstorming meetings to see what considerations designers actually employ when taking into account (or not) brainstorming rules. To do this, I will draw on the field of conversation analysis (CA), which is an investigation of social order – of the concerns which members of society visibly and concretely orient to in their interactions with each other. My analysis of how a method like brainstorming is actually used enables a reassessment of the potential for design methods and other formats for interaction to be useful to design[14]. In the course of this chapter, I also hope to have gone a small way towards demonstrating some of the potential of the tools of conversation analysis to illuminate what designers do and how they do it, and to have shown how social order is foundational to design activity.

1 ON SOCIAL ORDER

Studies of ordinary conversation have shown how speakers and listeners co-produce a normative social order in interaction. An obvious example of social order in conversation is the expectation created by the asking of a question. That is, in a normal conversation the next turn at talk following a question will be scrutinised

9.
This case is discussed in Bucciarelli, L.L. (1994) *Designing Engineers*, MIT Press, Cambridge, pp. 151–164

10.
Pugh, S. (1981) Concept Selection – A Method that Works, in *Proceedings of the International Conference on Engineering Design (ICED'81)*, pp. 447–506.

11.
Dennis, A.R. and Valacich, J.S. (1994) Group, Sub-group and Nominal Group Idea Generation: New Rules for a New Media?, *Journal of Management*, 20, pp. 723–736.

12.
Jablin, F.M. (1981) Cultivating Imagination: Factors that Enhance and Inhibit Creativity in Brainstorming Groups, *Human Communication Research*, 7, pp. 245–258.

13.
Sutton, R.I. and Hargadon, A. (1996) Brainstorming Groups in Context: Effectiveness in a Product Design Firm, *Administrative Science Quarterly*, 41, pp. 685–718.

14.
At the DTRS7 symposium (September 2007) where the work described here was first presented, Nigel Cross pointed out that these brainstorming meetings are probably not good examples of how brainstorming ought to be done. He is probably right.

(by the conversationalists themselves e.g. the question-asker) for its answer-ness to the question just asked. The relevance of 'next turns' to immediately prior ones is a normative matter that conversationalists themselves exhibit and orient to[15]. That this is so can be seen by consideration of many facets of conversation, including just how delicate topic shifts within conversations actually are for conversationalists[16]. This is just one example of the emergence of social order in interaction, but it is one that has application to the analysis that follows. One important aspect of the program of conversation analysis has been its demonstration of how finely detailed and richly textured is the production of social order. For example, very small silences in conversation (tenths of seconds) can be shown to have interactional relevance – they may, for example, prefigure misunderstandings or other 'trouble', or they may mark a place for speaker transition. The point is that social order is a very finely organised phenomenon, and is observable in the details of interaction.

There are a number of ways in which such observations might prove relevant for design research. In the context of the present data of the engineering brainstorming meetings (E1 and E2), there is the potential to identify possible intersections between different 'orders'. The purpose of this chapter is to explore the relationship between the rules of brainstorming and social order. I will do this by juxtaposing of some of the 'rules' for conversational interaction with the rules for brainstorming to see how they play off each other, as in how they intersect for the participants in the meeting.

At first glance there are several possible intersections of social order and brainstorming rules. Firstly, the normativity of *relevance* in turns at talk appears that it may conflict to some extent with the brainstorming injunction for novelty. While there is an expectation in conversation that talk should be relevant to the immediately preceding talk, there is also the brainstorming injunction that ideas should 'break existing mindsets' (as Alan instructs the group in E1 (E1, 31)), that divergence not convergence is desirable (E1, 40–41), and that all possible avenues of ideas should be explored (E1, 41–42). Secondly, we might expect some overlap between how interruptions arise and are handled in interaction and the brainstorming rule not to interrupt the current speaker (E1, 15–31). Finally, it may be instructive to compare the organisation of criticism in interaction with the brainstorming rule not to criticise ideas in the session including one's own ideas (E1, 7–8). These are not hypotheses so much as circumstances in which an analyst might be able to see how engineering designers manage their accountability to both social order and the rules of brainstorming. It is these three 'intersections' that are the focus of the analysis presented below.

Certainly, one possibility available to analysts would be to hold designers' actions up against an external standard (such as how brainstorming is supposed to be conducted). Of course, it is also possible, and it is the approach adopted here following Garfinkel, that participants' actions *not* be held up against "a rule or a standard obtained outside actual settings" (p. 33) in order to clearly see the local standards that are for participants in play. This rather simple difference actually conceals a host of methodological consequences for inquiry, discussed at length in Garfinkel. Cross's point, however, was that it is problematic to draw implications for the utility of methods to design from such a case. However, it remains that this is how brainstorming was actually done (at least at this consultancy) and how the method was actually used. Implications about 'the use of methods' drawn from such an analysis may be problematic, but only to the extent that the case is atypical of other actual engineering brainstorming meetings. That is something to be judged by practicing engineers and against further studies of practice.
Garfinkel, H. (1967) *Studies in Ethnomethodology*, Prentice-Hall, New Jersey.

15.
Normativity here is not deterministic. It is not a guarantee that every question asked will receive an answer. That questions invite answers, creating a place for and expectation of them, does not entail that answers are somehow automatically forthcoming.

16.
See for example Lecture 5 (Spring 1970) or Lecture 9 (Spring 1971) in Sacks, H. (1995) *Lectures on Conversation*, Blackwell, Oxford.

2 METHOD

In most analyses of design, the categories used by the analyst are drawn from common sense, from relevant theoretical work or from other authoritative sources (e.g. dictionary definitions), or are inductively generated from the data itself. In conversation analytic studies, however, the point is not to build theory but to unpack how events are organised and ordered – to see how social actions are structured and accomplished. This is not done by employing e.g. theoretically-derived concepts or definitions, but rather it is attempted through an analysis of the data on the basis that each turn at talk displays *the speaker's understanding* of what is going on. It is in this way that an effort can be made to recover the understandings the participants exhibited during the interaction, and how their actions were recognisable to co-participants as they were produced. So in this way, each utterance in the meeting can be viewed as an analysis of the preceding talk – as the speaker's own in situ analysis – and *its* study can direct us towards a recovery of the understanding that the conversationalists possessed (and exhibited) in the course of the interaction. A short example (Extract 1) from the transcript of E1 is illustrative:

Extract 1, E1, Speakers' turn-by-turn understandings

1413	Alan	erm potentially feedback how quick the how quick the head cools down if
1414		it's on the actual paper it's supposed to be on versus if it was on some
1415		different media sort of-
1416	Tommy	I think that's I think that's really for Monday
1417	Alan	ok

Tommy's utterance at 1416 displays an analysis of Alan's prior turn. Tommy frames this as an opinion ('I think that…') that questions the relevance of Alan's comment to *this* meeting by suggesting that the discussion belongs in Monday's upcoming meeting. In a similar way, Alan's turn at 1417 acknowledges and appears to accept Tommy's suggestion, also displaying Alan's understanding of Tommy's turn *as a suggestion to drop the topic*. My analyses in this study draw on the general approach of and findings from conversation analysis in order to explicate the practices of designers in the meeting. However, it is not intended as an example of or contribution to conversation analysis; it is simply a selective application of some conversation analytic insights to designers' interactions[17]. Although I transposed the excerpts into Jefferson's transcription notation for my analysis, here I present extracts using the simplified notation from the DTRS7 transcripts for consistency with the other analyses in this volume.

17.
One mainstream conversation analytic study of design interactions is Button and Sharrock (2000). Other studies, such as this one, have selectively applied conversation analytic insights to study design (Bowers and Pycock 1994, Luff and Heath 1993, Matthews 2007, Oak (Chapter 17)).
Bowers, J. and Pycock, J. (1994) Talking Through Design: Requirements and Resistance in Cooperative Prototyping, in *Proceedings of Human Factors in Computing Systems CHI '94*, ACM Press, pp. 299–305.
Button, G. and Sharrock, W.W. (2000) Design by Problem-solving, in Luff, P., Hindmarsh, J. and Heath, C. (eds), *Workplace Studies: Recovering Work Practice and Informing System Design*, Cambridge University Press, pp. 46–67.
Luff, P. and Heath, C.C. (1993) System Use and Social Organisation: Observations on Human-Computer Interaction in an Architectural Practice, in Graham Button (ed), *Technology in Working Order: Studies of Work, Interaction and Technology*, Routledge, pp. 184–210.
Matthews, B. (2007) Locating Design Phenomena: A Methodological Excursion, *Design Studies*, 28, pp. 369–385.
Oak, A. (Chapter 17) Performing Architecture: Talking 'Architect' and 'Client' into Being.

3 ANALYSIS

3.1 *On not being critical*

In introducing the first brainstorming meeting (E1), Alan declares the most important rule of brainstorming is not to criticise others' ideas (E1, 7–8). However, apart from its appearance here along with other rules of the session, it is very seldom that the participants explicitly invoke any of the rules of brainstorming during the meetings. The following sequence (Extract 2) from E1 is a rare case, occurring a little over an hour into the meeting.

Extract 2, E1, Invoking the rule not to criticise ideas

1355	Todd	[*laughs*] like a quill
1356	Rodney	like a quill that means it's not something that you could put in your pencil case
1357	Alan	would it be-
1358	Sandra	yes sure /yes\
1359	Alan	/would\ it be- you're not supposed to be negative (Rodney)
1360	Tommy	('d you wanna put it in your pencil case)
1361	Alan	well I think there's an issue actually there as well with children sometimes

At E1 1359 Alan actually produces a version of the 'no criticism' rule, following Rodney's turn in 1356. Again, reference (oblique or otherwise) to one of the rules of brainstorming during the session after Alan's introduction where the rules were established was very rare. Looking at this sequence, we may be able to see what some of the consequences of invoking a rule in interaction actually are. Alan's use of the rule at this point paints Rodney's earlier turn as a violation of it; not only that, however, it also censures Rodney for having transgressed the rule[18]. As an analyst, it is difficult to see exactly what triggers this particular censure from Alan, as there are numerous other comments made at other times that appear equally 'negative', but which patently did not occasion a reiteration of the rule not to be critical. It may be noteworthy that following this censure from Alan, the transcript does not record Rodney attempting another turn at talk for more than 500 lines of transcript, or about 25 minutes of the meeting. This may also lend significance to the scarcity of occasions that brainstorming rules such as 'don't be critical' are invoked – that participants themselves are aware of the possibility that censuring criticism closes down dialogue much more effectively than permitting the occasional critical comment. This raises the possibility that *not* invoking the rule in the meeting is actually in keeping with (the spirit of) the rule. That is, if designers appreciate that the point of the rule 'don't criticise' is for dialogue to remain open so that more ideas can be aired and more participants can contribute, and if the censure of a participant risks

18. Wieder, D.L. (1974) *Language and Social Reality: The Case of Telling the Convict Code*, Mouton, The Hague.

ostracising him or her, then *disattending* to the rule is actually a delicate means of keeping it. And such an analysis suggests that if these designers are adhering to the rules of brainstorming, they are not doing so in obvious ways, but in ways mediated by other subtle concerns.

As intimated earlier, much more can be seen to be going on in the details of the interaction than designers simply following the rules of brainstorming or transgressing them. One interesting aspect is exactly how and in what circumstances a rule is seen to be locally relevant by participants; another is what the interactional consequences of producing the rule are. Scrutiny of the two meetings show that designers only selectively orient to the tenets laid down as rules at the start[19]. What then are designers orienting to, if (by and large) not the rules of brainstorming? To what standards can they be seen to hold each other accountable to? Inspection of other intersections between social order and brainstorming rules shed some light on this.

19.
There were possible references to other rules of brainstorming that did not appear to be such delicate matters for the participants, such as the rule to write down all ideas. But this was not one of the rules Alan listed at the start of E1, and I have not looked in detail at its use.

3.2 *On interruptions and unfinished turns*

In their seminal paper on the organisation of turn-taking in conversation, Sacks, Schegloff and Jefferson[20] lay out a number of the various features of real-time interaction that speakers produce, recognise and orient to in conversation. One of these components that they identify from recordings of conversations is the 'turn-constructional unit' (TCU). One of the 'problems' conversationalists face is the management of turns at talk in a conversation. That is, in spite of the fact that conversation is a distributed affair involving multiple parties each of whom speak at one time or another, there is surprisingly little overlapping talk. Speakers shift from one to another, but in a highly regulated way that exhibits (and allows for) remarkably little simultaneous talk. It is in the course of scrutinising how it is that speakers pull off such impressive coordination of their own actions with others' that Sacks, Schegloff and Jefferson[21] identify different types of turns at talk. So one type of turn component is a sentence, another is a clause, another is a phrase, another is a word, and others may simply be a non-lexical item such as 'mmm', or 'uh huh'. This aspect of the shape of turn components is relevant to the organisation of turn-taking by virtue of the fact that conversationalists can quite clearly be seen to wait for the possible completion of a TCU before beginning their own turn at talk. This place at the end of a TCU is termed a transition-relevant place (TRP), because it is where in conversational interaction next speakers try to begin next turns. It is a striking feature of the organisation of conversation that speakers who are waiting for the floor do not just begin talking at any point in the current speaker's turn, but that they wait to begin their turn until the next possible completion point of a TCU. One example Sacks et al. provide is given in Figure 1.

20.
Sacks, H., Schegloff, E.A. and Jefferson, G. (1974) A Simplest Systematics for the Organization of Turn-taking for Conversation, *Language* 50, pp. 696–735.

21.
ibid.

Tourist:	Has the park changed much?
Parky:	Oh:: ye:s,
	(1.0)
Old man:	Th' *Fun*fair changed it'n awful lot didn' it.
	[[
Parky:	Th- That-
Parky:	*That* changed it,

Key to notation:

::	extension of the sound of a syllable	(1.0)	time between utterances (one second)
[	start of overlapping utterances	*italic*	emphasis or stress in pronunciation

Figure 1. Turn taking in talk[22].

The old man's turn here has two possible completion points. 'The funfair changed it' could stand alone as a complete turn, as could 'The funfair changed it an awful lot'. It is only at *these particular points* in the turn that overlap occurs; these are the transition relevant places, where the next speaker has anticipated (incorrectly) that the old man's turn was complete and has started his own turn.

22. This example is taken from Sacks, Schegloff, and Jefferson, *op. cit.*, p. 721.

This aspect of conversation is relevant to the examination of designers' uses of the rules of brainstorming since the first engineering meeting began with an explicit request that participants do not interrupt each other, initially listed among the other rules at the start of the first meeting (E1, 15). This rule was reiterated a second time as a courtesy to make the transcription of the meeting easier (E1, 22–23). It should be said, however, that while we (as analysts) are free to define 'interruptions' for analytical purposes any way we find appropriate, it is also possible to inspect the transcript to see how and when the participants themselves interpret an interjection as an interruption. This is the particular tack I want to take here, to see how speakers treat others beginning a new turn before they have finished a TCU.

One of the ways we might detect that participants are interpreting an interjection as an interruption is to see if they go back to finish their turn after the interjection. Consider Patrick's turns in the sequence in Extract 3.

Extract 3, E2, Patrick is interrupted.

1062	Sandra	although you've got to uh you've got still to have some way of
1063		selecting it on the pen (which wouldn't be the eas/iest)
1064	Patrick	/(tryin to) print yeah that's true ++ could you-
1065	Sandra	do you have to select colour as well ++
1066	Patrick	/could you have\
1066a	?	/yeah that's\ on the paper
1067	Roman	it's already in colour I think
1068	Tommy	yeah the colour comes from the paper
1069	Sandra	/(I see)\
1070	Jack	/could you have a\ knob to control writing speed ( )

Patrick initially doesn't get to complete a TCU (a possible sign that there has been an interrupting turn) because of Sandra's question. At the end of her question (which is the first next transition relevant place) he tries again, starting his turn by repeating the same phrase 'could you'. This attempt is also interrupted with an answer to Sandra's question from Roman. Sandra acknowledges the two answers she receives (Roman's and Tommy's) with 'I see', and then Jack picks up with a question that appears to be his own continuation of Patrick's (now) twice aborted turn, as it starts with the same three words as Patrick's last attempt: 'could you have…'. It is on account of Patrick's repeated attempt to ask the same question that we can see he treats Sandra's intervening turn as an interruption – that he didn't get to finish his initial turn. It is of interest (and will be followed up in a moment) that his unfinished turn is taken up shortly afterwards by someone else (Jack), who may or may not be asking the same question Patrick was going to. A participant's repeated attempt to say the same thing on either side of an interjection constitutes one possible way of identifying an interruption in the transcript. Another, considering Sacks, Schegloff and Jefferson's[23] identification of turn constructional units, might simply be to find turns that are cut off by another speaker before a transition relevant place, and to see how these are handled. This is what happened twice to Patrick in the previous extract[24].

It was in the course of looking for interruptions that I came across an interesting feature of the organisation of turn taking in these meetings. That is, there were a surprising number of unfinished turns – turns that did not complete a sentence, clause or phrase such as Sacks, Schegloff and Jefferson[25] found; and therefore these were turns that did not end at a transition relevant place. A number of ideas just seemed to stop short, to be left unfinished. Sometimes they were completed by another participant, sometimes they were interrupted by another speaker and not returned to, and sometimes they were simply abandoned by both the speaker and the group. Initially, I thought that this signalled that there were many interruptions in the meetings (and consequently, multiple breaches of the brainstorming rule not to interrupt). My inclination to this interpretation was suggested by a conversation analytic finding that has shown that speakers in the middle of a TCU can often pause without another speaker taking the floor (something that is not the case at a transition relevant place, which needs no pause at all for another speaker to begin his/her turn)[26]. Thus I expected that a mid-TCU pause that was used by another speaker to begin a new turn would be treated as an interruption by the first speaker (e.g. in the way that Patrick treated Sandra's turn as interruptive by trying for a second time to ask his question and trying to regain a turn at talk at the next transition relevant place).

23.
ibid.

24.
My discussion of 'interruptions' is, at best, cursory from a CA perspective. There are various conversational phenomena that may be relevant to the types of speaker transition I am considering here. For instance, Jefferson's (1983) work suggests that 'overlap' and 'trailing off' are analytically distinguishable from 'interruptions', and may account for some of the forms of speaker transition I describe here.
Jefferson, G. (1983) On a Failed Hypothesis: 'Conjunctionals' as Overlap-vulnerable, *Tilburg Papers in Language and Literature*, 28, pp. 1–33.

25.
op. cit.

26.
The transitions here are occurring at points of 'maximum grammatical control' (Schegloff 1996), such as "after a preposition but before its object [or] after the infinitive marker but before the verb" (p. 94), which have elsewhere been discussed as places where a pause in a turn does not invite transition (Schegloff 1982). An exception to this has been described in an analysis of teachers' practices of stopping mid-sentence as a means of prompting students to provide the correct completion of it (Lerner 1995).
Lerner, G.H. (1995) Turn Design and the Organization of Participation in Instructional Activities,

But these expectations were not realised in the analysis of the transcript. The interesting thing was that even when there was speaker transition at unfinished TCUs, these instances were often *not* treated as a problem by the participants. Two examples of these unproblematic speaker transitions follow in Extracts 4 and 5.

Extract 4, E1, An unproblematic transition that is *not* at a transition relevant place.

Alan	which means each blade can independently of housing can move a little bit
	that way a little bit as well can't it so that's uhm so that's uhm +
Tommy	how's that achieved then is that +++ is that achieved by a plastic that's
	bendable or is it achieved by
Todd	no it's basically a spring isn't it
Alan	a spring that goes to surface
Todd	there's a piv- uh it pivots

Alan's turn at 296-7 trails off in the middle of a TCU. Tommy comes in with a question and candidate answer that also trails off, only beginning another unfinished turn with 'or is it achieved by…'. Todd adds a counter suggestion with a questioning tag 'isn't it?' and the discussion continues. Neither of these two interjections (speaker transitions before the completion of a turn) is treated as an interruption by the participants. For example, we don't have the participants giving any indications that an interruption has occurred, such as repeats of turn beginnings, or the 'interrupted' speaker attempting to regain the floor at the next possible transition relevant place. Instead, we have something like, but not identical with, a collaborative completion of a turn[27].

Extract 5, E2, Another unproblematic transition not at a transition relevant place.

Tommy	erm + err the oth- ah the other thing that we've seen in the past is erm ++ er
	people program certain toy to be programmed from barcodes so you
	end up swiping it over barcodes
Patrick	mmm
Sandra	/oh yeah\
Tommy	/to build up se\quences and bits and pieces which
Sandra	(you mean) it's (kind) of library of patterns so you could scan the ones-
Patrick	yeah you could publish a book with patterns in with barcodes
Tommy	yeah
Patrick	you can scan the right barcode

In Extract 5 of interest is Tommy's turn at 1023, which is cut off at the beginning of a phrase starting with a preposition 'which'. Sandra creates one continuation of the thought by talking about a library of patterns that could be scanned in, but this thought too is

Discourse Processes, 19, pp. 111–131.
Schegloff, E.A. (1982) Discourse as an Interactional Achievement: Some Uses of "uh-huh" and Other Things that Come Between Sentences, in Tannen, D. (ed), *Georgetown University Roundtable on Languages and Linguistics*, Georgetown University Press, Washington, pp. 71–93.
Schegloff, E.A. (1996) Turn Organization: One Intersection of Grammar and Interaction, in Ochs, E., Schegloff, E.A., and Thompson, S.A. (eds), *Interaction and Grammar*, Cambridge University Press, pp. 52–133.

27.
Lerner, G.H. (1991) On the Syntax of Sentences-in-progress, *Language in Society*, 20, pp. 441–458.

unfinished in the turn. Patrick takes up an extension of this notion, interrupting Sandra with a token of agreement 'yeah', and developing the idea into a book. It is noteworthy that although we have speaker transition taking place at a juncture in the turn other than a paradigmatic transition relevant place, the 'interruptive' turns at 1024 and 1025 are again not treated as interruptions by the participants[28].

Taken together, these examples suggest that turn transition in brainstorming meetings may not exactly replicate turn transition in ordinary conversation[29]. More interesting for design research is the indication that leaving turns unfinished is a methodical practice that co-opts others into idea generation. These examples show that this is one recurring consequence of leaving ideas hanging in the middle of a turn. Stopping or trailing off mid-turn like this can induce others to add to or alter the idea in possibly significant ways without the speaker needing to make any explicit request. No question is asked, no overt invitation to comment is made; rather, this is simply achieved with the grammar of an incomplete thought.

3.3 *On speaking relevantly and generating novel ideas*

One enduring feature of ordinary conversation is that there is an overwhelming normative expectation to speak relevantly. In conversation, each turn is scrutinised by conversationalists for its relevance to the last. This is exhibited through and through by the work that speakers do within conversation to close down current topics and introduce new ones. It is also plainly evident that conversationalists request clarification when the relevance of a speaker's turn is not apparent. Furthermore, it is observable in the work (such as prefacing a story) that speakers do to introduce a coming topic *as relevant* when they anticipate that its relevance may (otherwise) be questionable.

McHoul and Rapley[30] use an unpublished draft manuscript of Harvey Sacks' to illustrate a pragmatic use of the phrase 'I just had a thought'. In their analysis of this segment of Sacks', the phrase was used as a way of linking the story to come to the immediately previous stretch of talk. It served to mark the coming account as being 'on topic'; it is a way of suggesting that what the speaker is about to say, which may first appear to be irrelevant to the current topic, actually has relevance[31]. The work that speakers do to demonstrate relevance is significant – speakers are accountable to fellow conversationalists for the relevance of their contributions[32]. In conversation there is an expectation that each 'next turn' should be relevant to the last with its relevance displayed in the turn itself or if not that its irrelevance will be accounted for (also) within the turn itself. The use of 'on

28.
One of Sacks' observations about sentence or turn completion by others is that in certain settings it can be used to show joint authorship or ownership of the matter being discussed (Levinson 1988, pp. 201–203); in this case the matter is design ideas that are being handled by participants. But this is not to say that these engineers are *doing* 'showing joint authorship' each time they take up an unfinished utterance. See also note 29.
Levinson, S. (1988) Putting Linguistics on a Proper Footing: Explorations in Goffman's Concepts of Participation, in Drew, P. and Wootton, A. (eds), *Erving Goffman: Exploring the Interaction Order*, Polity Press, pp. 161–227.

29.
This is a suggestion and not a claim as its demonstration would require (at least) a considerably larger corpus of data that examines how these unfinished turns are treated by those who take them up. Furthermore the notion that 'brainstorming meetings' exhibit a distinct modification of the structure of ordinary conversation requires empirical study. Not everything that is called a 'brainstorming meeting' may share these structural features, and other interactions that have little or nothing to do with brainstorming may also exhibit them. Hester and Francis' (2001) discussion is important here – conversational structures do not neatly map to our ordinary stock of categories of events.
Hester, S. and Francis, D. (2001) Is Institutional Talk a Phenomenon? Reflections

topic' markers is one way that speakers accomplish the relevance of their talk.

on Ethnomethodology and Applied Conversation Analysis, in McHoul, A. and Rapley, M. (eds), *How to Analyse Talk in Institutional Settings: A Casebook of Methods*, Continuum Books.

In the transcript we can see a use of a similar construction employed as a topic marker, though a little more elaborately. About twenty minutes into the first meeting Todd talks about a ghost toy his son has; Extract 6 shows how he gets to start talking about it.

Extract 6, E1, Todd's use of an 'on topic' topic marker.

391	Alan	OK fine OK erm right anybody bring any other-
392	Todd	I only brought a magic marker cos ( )
393	Alan	/OK\
394	Todd	/I did\n't think there was a problem
395	All	[*laugh*]
396	Todd	cos if you ever go with (that) it's not too bad but it uh it made me think
397		of uhm a toy uh I don't know if any- (pri- prob- wel-) everyone's se- well I don't
398		know if everyone's seen it you can get ah my son has a uhm + like a ghost uh it's
399		a little ghost it's about you know + an inch high

Todd responds to Alan's request (at 391) with a joke about only having brought a magic marker. He then uses the marker as a means of discussing an idea related to an object he *hasn't* brought. The connection is made by use of 'it made me think uh- of a toy' (lines 396–397). Todd begins his idea with an 'on topic' marker – suggesting that this idea relates, at least in 'toy-ness', to an earlier turn (the magic marker). The phrase 'it made me think' is itself interesting. It is a passive-voice construction which suggests (at least metaphorically) a causal link between the preceding object and the idea Todd is introducing – '*it made me* think…'. This phrase defers the responsibility for what Todd thought on account of what the magic marker has made him think about[33]. And it is in this way that he creates a link for the rest of the group between his idea and his preceding turn. To claim that one thing has made one think about something else is to claim an association between the thing and what was thought. This is one way in which we can see that participants orient to the 'relevance rule' of conversation – that they actually do this kind of work to introduce new ideas with reference to the local context of talk. This appears to be the case even to the extent that Todd here creates a local context himself in order to get to talk about what he wants to. By producing an 'uninteresting' object (magic marker), Todd creates a way of having an answer to Alan's request (concerning what objects people brought to the meeting); this conversational move gains him the floor. He then uses the marker to provide a link to the cereal box toy that he proceeds to talk about.

30.
McHoul, A. and Rapley, M. (2003) What can Psychological Terms Actually Do?, *Journal of Pragmatics*, 35, pp. 507–522.

31.
ibid, pp. 510–512.

32.
Something demonstrated throughout Sacks' (1995) lectures, *op. cit.*

33.
I owe this observation to Max Eckardt.

Shortly following this is another example, shown in Extract 7. Todd has been sketching and describing this particular toy. Note Sandra's remark at E1 419.

Extract 7, E1, Sandra's use of an 'on topic' topic marker.

Todd	yeah an- it's a self an- a yeah its got its got a pad an inky pad here an- its ghost
Jack	blimey does that go all over the walls and everything
Todd	well yeah y-
All	[*laugh*]
Todd	you walk along on the um on the page
Sandra	yeah
Todd	and it makes footprints
Sandra	that's interesting cos I was lik- I was thinking + pastry brush you know (a) pastry cutter-
Todd	oh ye- yeah uh huh
Sandra	and white line machines but I couldn't think of anything that sort of quite tied into this but the idea of a wheel-
Alan	yeah
Sandra	pushing a wheel along

Consider 'that's interesting cos I was thinking pastry brush'. We might suppose that a different beginning (such as 'what about a pastry brush' or 'did you think about a pastry brush') wouldn't do the same work of showing the relevance of this idea to the preceding talk. But 'that's interesting' offers a positive assessment of the previous idea, and the 'because I was thinking' projects a link to what is about to appear. Continuity with the preceding topic is being handled here.

Of course, one may ask how a pastry brush is actually relevant to Todd's son's ghost toy. Interestingly, this particular relevance is not demonstrated by Sandra – instead she switches from a pastry brush (something which is relevant to one aspect of the task, namely a product that follows an uneven contour, but not clearly relevant to the toy Todd has been talking about), to a pastry cutter, then to a line-marking machine. These latter two share an obvious principle of operation with Todd's son's toy, and the ultimate relevance of Sandra's turn is not questioned by her co-workers. What is also of note is that the idea of the pastry brush, which is actually a novel idea that these engineers haven't considered as a way of addressing this problem in the meeting yet, is not pursued by Sandra or the others. Rather, the conversation continues along a more obvious extension of the current topic, concerning the use of wheels to guide the print head.

One further example, Extract 8, shows (a) that designers also explicitly orient to the topic-relevancy of their contributions, and (b) the amount of work designers actually invest in order to successfully introduce a discontinuous topic or idea.

Extract 8, E1, The work required to introduce something 'off topic'.

Todd: [*laugh*] alright yeah yeah I'm just wondering about the jus- sort of the ++
erm + I know ah it is probably off off topic but it is sort of the usefulness
I do- well I'm jus- I'm jus tryin to think about what erm +++ what the pen does
really erm + if it's a if it's a f- if it's fun ++ it's gotta do- +
Alan: sort of wider range of features
Todd: yeah

Todd, after a number of false starts, draws attention to the 'off topic' (line 949) character of the coming idea. The multiple pauses, uhms, ahs, and restarts here are features of talk that often accompany delicate situations – situations where e.g. word choice is important, and/or where the action the speaker is trying to pull off has the potential to cause discomfort, embarrassment, offence, etc. These kinds of features can be indicators of interactional 'trouble' of some kind. Todd bookends this little conversational detour with an apologetic 'and again off topic' at line 971. The point here is to highlight the remarkable work it takes to introduce an idea that doesn't have clear relevance to the *local* talk. I mean 'local' in terms of immediately prior turns and the current topic, not in terms of other possible scales of 'local' such as 'today', 'this project', 'this meeting', 'this task', etc. We can quite clearly see that this immediate sense of 'local' is the scale at which the participants themselves are trying to display the relevance of their talk[34]. In these three examples neither Todd nor Sandra are working to demonstrate the relevance of their idea to the company, to the project, or to the task, but are working to show its relevance to the current topic of conversation.

34.
c.f. Schegloff, E.A. (1997) Whose text? Whose context?, *Discourse & Society*, 8, pp. 165–187.

At this point we might speculate about the relationship between the two 'requirements' of *relevance for conversationalists* and *novelty for brainstormers*. Having seen the work required to shift topics, and having seen the attention designers pay to creating demonstrable links to the current topic, we might wonder at the extent to which the normative expectation that conversationalists speak with local relevance is detrimental to the introduction of ideas that are truly novel, or that genuinely break existing mindsets. After all, Sandra's pastry brush seemed to get short shrift. Clearly, the introduction of groundbreaking ideas is unlikely to be a matter of 'you know, I think…', or 'what about…'; rather it is likely to need to find hooks on the current topic from which to justify its appearance in conversation just here. However, one cannot draw the conclusion that design is impoverished by virtue of this conversational commitment to topical relevance. For instance, it is wholly unclear what kind of conversation would transpire in the absence of this conversational 'requirement', and there is a significant possibility that interaction without it would fail to be recognisable as a conversation about anything. Furthermore, it is difficult to see how other important aspects of brainstorming, such as building on the ideas of others, would be possible without a shared and accountable expectation that talk be locally relevant.

4 DISCUSSION

While there are many different formats for interaction (e.g. interviews, ceremonies, legal testimonies and presentations are each distinctive in important ways), there are no time outs from social order. Brainstorming sessions may have a character of their own, yet it is a character dependent in essential ways on the same order from which the other arenas of our interactions with each other in daily life are built. In this chapter, I have tried to show some of the ways in which designers handle the rules of brainstorming in light of the other orientations they clearly display in interaction, and I have argued these are orientations to the production and maintenance of a normative social order. But what are the lessons for design?

We can and do alter formats of interaction which create different frameworks for participation. Designers are not condemned to accept the format of a meeting or any other social encounter if it does not offer them the flexibility or structure that they require of it. Furthermore, while we have seen that orientations to social order are not suspended on account of the rules of brainstorming, we have also seen that the structure and rules of brainstorming may have had influences on the structures of action within the meeting – notably with the work done by unfinished sentences to invite other-speaker completion of an idea, but also by the fact that participants do occasionally invoke the rules of brainstorming, and of course that they are actually generating ideas throughout. That is, the rules of brainstorming *do* have an effect on the proceedings, though it may be a milder effect – and one severely modulated on account of social order – than is typically assumed. Methods such as brainstorming are not impotent to structure designers' actions, but nor are they simply and unproblematically 'obeyed' or universally oriented to. Much, much more is going on in interaction.

The 'rules' of a session or a method (or the stages in a model of the design process) do not easily account for designers' activities, yet they can be (as they were here) among the participants' own accounting devices within the activity itself. So, rather than become an analyst's resource to account for what designers do[35], they can be

35.
c.f. Dorst's deployment of Simon and Schön.
Dorst, K. (1997) *Describing Design: A Comparison of Paradigms*, PhD Thesis, TU Delft.

an analytic topic for study[36], since, as we have seen here, they are a *participant's* resource, and since many participants' actions cannot simply be seen as actions in accord with brainstorming rules. That is, the method's rules are an *occasioned* resource on hand for the participants to assign sense, meaning and order to the proceedings in their course[37]. It is largely in this way that methods come to be of use to designers, to the extent they are deemed by participants to have local relevance. Invoking a rule is a means of sustaining social order. Yet although social order is normative (as are rules), it is by no means deterministic. And it is in this space that we may find a practical application for this kind of understanding of design practice. The very identification of designers' normative orientations (e.g. to the local relevance of talk) is one important step towards the creation of formats of interaction that might be able to 'tamper' with social order, in similarly mild ways, so as to be more conducive to design objectives.

36. Zimmerman, D.H. (1971) The Practicalities of Rule Use, in Douglas, J.D. (ed), *Understanding Everyday Life: Toward the Reconstruction of Sociological Knowledge*, Routledge and Kegan Paul, pp. 221–238.

37. This paper is heavily indebted to an ethnomethodological and Wittgensteinian understanding of rules. Related work would include, at the very least the following list:
Garfinkel, H. (1967) *Studies in Ethnomethodology*, Prentice-Hall, New Jersey, Chapter 6.
Sharrock, W.W. and Button, G. (1999) Do the Right Thing! Rule Finitism, Rule Scepticism and Rule Following, *Human Studies*, 22, pp. 193–210.
Suchman, L.A. (1987) *Plans and Situated Actions: The Problem of Human-machine Communication*, Cambridge University Press.
Wieder, D.L. (1974) *Language and Social Reality: The Case of Telling the Convict Code*, Mouton, The Hague.
Zimmerman, D.H. and Pollner, M. (1971) The Everyday World as a Phenomenon, in Douglas, J.D. (ed), *Understanding Everyday Life: Toward the Reconstruction of Sociological Knowledge* Routledge and Kegan Paul, pp. 80–103.

ACKNOWLEDGEMENTS

The analysis presented here would not have been possible without the help of Max Eckardt, who also spent a great deal of time in the data and commented on earlier drafts. I also received helpful pointers to relevant work in CA regarding unfinished turns and sentence completion by others from Jakob Steensig, Johannes Wagner, Kristian Mortensen, Søren Beck Nielsen, Trine Heinemann, Monika Buscher and Tine Larsen. In particular, Trine Heinemann provided incisive and critical comments on an earlier draft, and did much to nurse this paper's (and its author's) sensitivities to CA. Naturally, I alone am responsible for the errors that remain.

3

Spider Webbing: A Paradigm for Engineering Design Conversations during Concept Generation

Ade Mabogunje, Ozgur Eris, Neeraj Sonalkar, Malte Jung & Larry Leifer

Measuring critical team interaction variables can lead to interventions that will improve design performance. The identification and definition of these variables begins with a search for underlying patterns of design conversations that adequately describe the observed behaviours. In this chapter we present an analysis of the DTRS7 data from the first engineer meeting, E1, which revealed two distinct patterns of conversation: one corresponded to a linear approach to problem solving, and the second was similar to a pattern of conversation previously described as spider webbing. We discuss a design behavior, termed *resumption*, which describes the second pattern and we suggest an analogous relationship between the outcome of *resumption* and the geometry of spider webs. Specifically, the arcs and sectors in a spider web-like representation serve as a proxy for the solutions and the requirements that are discussed in the design conversation. The relationship was considered as a potential intervention tool for sensitising designers to the structure of design conversations and increasing their design performance.

The engineering design process can be viewed as a sequence of conversations – encounters between people and artefacts – during which thoughts, ideas, and verbal descriptions are transformed into concrete products and experiences. This transformation can be studied from an information handling viewpoint by considering the flow of information during the design process[1,2]. Beginning with a design brief, designers pose information-seeking questions, clarify and evaluate the responses to these questions, and then synthesise these responses to produce new associations and ideas. Information in the form of ideas is represented and expressed between the designers through gestures, sketches, and text[3], words – noun phrases and verb phrases[4,5], and physical objects[6]. The new information is processed on the basis of pre-existing knowledge and converted into a concrete product through iterative cycles of ideation, representation and evaluation[7,8].

Despite the considerable amount of research done in this area there is still very little consensus amongst researchers about the governing parameters of design conversations. This is especially true during early stage activities such as concept generation.

1.
Kuffner, T., Ullman, D. (1991) The Information Requests of Mechanical Engineers, *Design Studies*, 12, pp. 42–50.

2.
Baya, V. (1996) *Information Handling Behavior of Engineers in Conceptual Design: Three Experiments*, Doctoral Dissertation, Stanford University.

3.
Tang, J. (1989) *Toward an Understanding of the Use of Shared Workspaces by Design Teams*, Doctoral Dissertation, Stanford University.

4.
Fowler, T.C. (1990) *Value Analysis in Design*, Van Nostrand Reinhold, New York.

5.
Mabogunje, A., Leifer, L.J. (1997) Noun Phrases as Surrogates for Measuring Early Phases of the Mechanical Design Process in *Proceedings of the 9th International Conference on Design Theory and Methodology, ASME*, Sacramento, California.

6.
Harrison, S., Minneman, S. (1996) A Bike in Hand, in Cross, N., Christians, H., and Dorst, K. (eds) *Analysing Design Activity*, Wiley, pp. 417–436.

7.
Schon, D. (1983) *The Reflective Practitioner: How Professionals Think in Action*, Basic Books, New York.

8.
Cross, N. (1989) *Engineering Design Methods*, John Wiley & Sons.

By governing parameters, we mean the variables that directly shape the conversation and influence its outcome. One way of thinking about those variables is to categorise them as: drivers, regulators, and constraints. For example: the design brief is a driver; the rules of brainstorming are a regulator; and the duration of the activity, the knowledge of the participants, and the availability of tools that enable them to express, manipulate, and internalise knowledge are constraints. Understanding the governing parameters and their inter-relationships is important to researchers concerned with improving the performance of engineering design teams.

At the Stanford Centre for Design Research, in our attempt to develop this type of understanding, we have been focusing on the development of an instrumentation framework for design activity[9]. Our canonical premise is that any attempt to systematically improve design learning, thinking and performance hinges on our ability to make accurate measurements of the governing parameters, which, naturally, requires the development of metrics and instruments[10]. The metrics often emerge from understanding patterns of conversation. The first step in this direction is to know the shape of the conversation.

9.
Leifer, L.J. (1991) Instrumenting the Design Process: For Real-time Text-graphic Design Process Records in *Proceedings of the International Conference on Engineering Design, (ICED'91)*, Zurich.

10.
Leifer, L.J. (1998) Design Team Performance: Metrics and the Impact of Technology, in Brown, S.M., Seidner, C. (eds) *Evaluating Organizational Training*, Kluwer Academic Publishers, Boston, Massachusetts.

In this study, audiovisual records of the engineering meetings provided insight into some of the team interaction patterns during conceptual design. These include interpersonal interactions, responses to the design brief, and other interactions mediated by the artefacts and tools that were present in the environment. This chapter seeks to contribute to an understanding of the nature of the conversations that took place during those interactions, and to describe them in a form that can lead to the development of formal indicators of performance.

1 APPROACH

We are still in the early stages of developing reliable measures of design activity and this dictates that our methods of inference will be inductive rather than deductive. Therefore, we will begin with some key observations of the data and end with a design representation, which will serve as a proxy for the discovered relationships. We used a grounded approach to conduct the research[11]. A team of five researchers watched the meeting videos jointly. Each researcher was allowed to stop the video at points where he or she noticed an item of interest, and this was then discussed by the group. While the primary aim of this exercise was to determine which among the four videotaped meetings was to be analysed, it also served to develop a shared context amongst the researchers and develop ideas about different points of view for analysing

11.
Strauss, A., Corbin, J. (1990) *Basics of Qualitative Research: Grounded Theory Procedures and Techniques*, Sage Publications.

the data. This period of joint viewing was followed by a period of individual viewing where different frameworks were developed and tested. In time, one framework was chosen and developed further, and then critiqued and improved by the entire team.

Our approach can be described as a four-step process. It begins with a set of observations, which is followed by the recognition of patterns, then the formulation of a hypothesis, and finally the description of a theory. In this chapter, we will cover the first two steps in detail, and offer some preliminary thoughts on the latter two.

2 DATA ANALYSIS AND OBSERVATIONS

In this section we set out four observations grounded in the data and in each case we present the associated data analysis.

2.1 *Observation 1: There is a correspondence between the noun phrases extracted from the transcript and the concepts listed on the flip chart*

We initiated the analysis by viewing the audiovisual data and developing a broad understanding of the flow of the conversation. Based on the ensuing discussion, we decided to focus on the engineering meetings. We later narrowed this down to the first meeting, E1, which was focused on mechanical engineering design. This was followed by reviewing the transcripts of this meeting in order to identify specifics of the conversation.

Given that the primary goal of the brainstorming meeting was ideation (we know this from documentation supplied as part of the E1 data which included a briefing document for the meeting participants), we wanted to focus the analysis on the concepts that were generated. To this end, we extracted noun phrases and verb phrases from the transcript. In layman's terms, a noun phrase is a group of words that can function as the subject or object of a verb. Technically, a noun phrase is a group of words functioning as a single grammatical unit with a noun as its head. The head is modified by the other words in the phrase. Examples from the transcript are, 'thermal pen' and 'soldering iron', where the words 'pen' and 'iron' are the heads of the two phrases. Similarly a verb phrase is a group of words functioning as a single grammatical unit with a verb as its head.

Our previous research has shown that noun phrases denote the subcomponents of the final product, and evolve from verb phrases which denote the different sub-functions intended by the designer[12]. In other words, the verb phrases, in general, represent the design

12.
Mabogunje, A. (1997) *Measuring Conceptual Design Performance in Mechanical Engineering: A Question Based Approach*, Doctoral Dissertation, Stanford University.

Reduced data set for E1			Image of part of flip-chart produced during E1
137	Jack	sledge	*SLeigh – Wide base – force in middle. – STABILISER. – UNIVERSAL BASE BUT HOLD ONTO DIFFERENT ANGLES – WINDSURF
141	Jack	sledge keeps level by having a wide base and a main force in the middle	
150	Jack	snowboard or skis	
151	Alan	guiders down the side	
154	Jack	to keep pen at right angle – use a set of stabilisers	
156	Jack	idea of sledge	
157	Alan	stabilisers like a bicycle	
160	Chad	flat base with a universal joint – like a windsurf mast	

Figure 1. An illustration of the correspondence between the reduced data set and the ideas listed on the flip chart by the team during the first engineering meeting, E1.

13.
Feland, J. (2003) Innovation Impact Map: An Opportunity Evaluation Tool, in *Proceedings of the 14th International Conference on Engineering Design (ICED'03)*, Stockholm.

requirements or needs, and the noun phrases represent the design propositions or concepts. This is similar to the concept of Need-Solution pairs proposed by Feland[13]. Therefore, we extracted noun phrases and any corresponding verb phrases from the transcript. This gave us a *reduced dataset* of requirements and propositions. In effect, we went from a 59-page meeting transcript to an 11-page semi-filtered data document. Information about the line numbers, begin times and end times of utterances, and speakers were preserved. To check our work, we compared these extracted phrases with the concepts written on the flip chart during the meeting by those participating. All the 37 concepts documented by the engineering team could be found in the reduced data set (see Figure 1 for a sample comparison). We interpreted this finding as validation of our data reduction approach. We then proceeded to identify the overall structure of the conversation during the session.

2.2 *Observation 2: There were two distinct phases in the conversation*

The conversation consisted of two major phases. Phase 1 lasted roughly 70 minutes during which each designer spoke about a concept for short periods of time. Phase 2 lasted about 40 minutes during which the designers spoke about specific design configurations for slightly longer periods of time than in Phase 1. When we plotted the number of words per idea over time using the reduced dataset, we observed a spike in the plot at about 1 hr 10 minutes into the meeting (Figure 2).

This prompted us to re-examine the data more closely, and in so doing, we found that the phase change was deliberate and initiated by one of the participants, Jack, over a period of roughly 4 minutes as can be seen by Jack's words in the following extract (Extract 1).

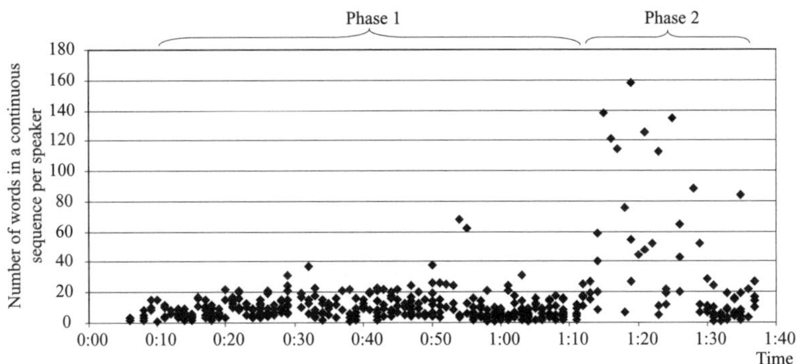

Figure 2. The two phases of engineering meeting E1 using data from the reduced dataset.

Extract 1, E1, An extract showing both the beginning and the intention of Phase 2.

1394	Todd	yeah
1395	Jack	OK want to do some more detail a bit more mechanism
1396	Alan	have we got enough on this the last one here exactly when ( )
1397	Jack	well we we've got various things haven't we we just have to contact it
1478	Alan	shall we move onto sorry Jack ( )
1479	Jack	I was just going to say yeah I think we should concentrate on what the
1480		mechanism might look like
1481	Alan	OK yeah
1487	Jack	yeah I just wondered if you wanted to organise it in some sort of more
1488		sketches rather than what's
1489	Tommy	Yeah
1490	Alan	yeah OK
1491	Tommy	well do you want to try
1492	Alan	yeah go on
1493	Tommy	does does anyone see anything very much wrong
1494	Alan	yeah that's a good idea

As noted earlier, the reduced data set was specifically focused on isolating the concepts generated during a brainstorming session. Clearly, the team decided to do some detailed design, and we were able to detect this change given this specific filter we used on the data. In order to give us a rough idea as to how such a shift might manifest itself graphically had we not used the reduced dataset, we extracted the portions of the conversations in which a concept was proposed, counted the number of words for each speaker turn, and plotted the word count against time. The resulting graph is shown in Figure 3. In this graph it is not possible to detect easily the phase change we observed in the reduced dataset (Figure 2), this finding supports the case for the utility of the noun phrase analysis. The focus on concepts which the reduced data set produces means that conversation is filtered on the basis of each distinct concept introduced.

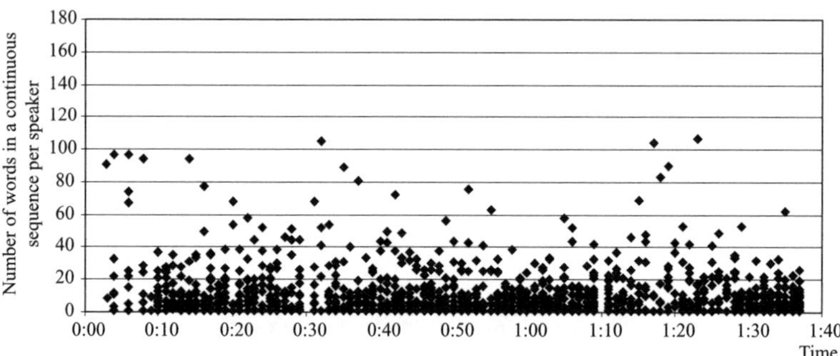

Figure 3. The two phases of E1 are more difficult to detect numerically with the full dataset.

In Phase 1, several questions were asked and answered, comments and interjections were added; there was a constant give and take. Most members of the team participated actively. The designers appeared to be operating in a mode of information gathering and requirements elaboration. In the following example (Extract 2), which is quite typical of this phase, the discussion of one solution, 'suspension', led to the discussion of a new *requirement*, 'children doing things jerkily quickly', which led to the discussion of a new solution, 'using gyroscopes'.

Extract 2, E1, An example of rapid turn taking behaviour during Phase 1.

480	Alan	well that's that's the other thing I suppose we haven't considered
481		anything like erm air pockets or anything like that that could be like a little
482		suspension
483	Sandra	suspension
484	Alan	almost
485	Chad	that's the sort of thing about children they do things jerkily quickly
486	Todd	yeah ( )
487	Tommy	so it didn't get to take a
488	All	[*laugh*]
489	Todd	/but\
490	Tommy	/sorry\
491	Alan	gyroscope
492	Tommy	stick a big gyroscope in it so they can't jerk it around

In the exchange, the designers also further elaborated and re-interpreted the given requirements. For example, while the original requirements identified the problem of 'wobbly arm movement of the user (5–11 year old)', a member of the team raised the issue of left handed users, and a number of approaches were discussed for this class of users. In this phase, team members appeared to act as triggers of new concepts (information attribute) and advocates (social attribute). We also observed several jokes and good camaraderie between the participants.

Phase 2, by contrast, consisted of monologues by a few team members on specific design configurations. During this phase the designers appeared to be doing the work of composing and simulating. By composing, we mean a tendency for the designer to describe a device as if the various parts and ideas from the preceding conversation had been built and were now being configured. In the following section (Extract 3), Jack describes a particular configuration while Alan supports him with 'mmm-mmm' and Todd with 'yeah'.

Extract 3, E1, A monologue illustrating composition during Phase 2.

Jack	but what you can't do is it can't the print head can't tilt back or forward
	because these slots are in
Alan	they're in a fixed position yeah
Jack	this this stays at a fixed angle it can move up and down
Alan	mmm-mmm
Jack	and side to side it's fixed that way
Alan	mmm-mmm
Todd	yeah
Alan	the print head relative to the pen
Todd	yeah

By simulating, we refer to the tendency of the designer to describe his or her composition as now functioning together with some other object in the environment (behaviour). In the following example (Extract 4), Todd describes an interaction between the pen and the case work. During these monologues, other team members seemed to serve as verifiers of the feasibility of the speaker's ideas (information attribute), and validators (social attribute).

Extract 4, E1, A monologue illustrating conceptual simulation in Phase 2.

Todd	that when you come too far over the pen starts moving back in and so
	( ) back in and therefore the switch comes off
Alan	mmm
Todd	before it looses contact you know it depends on the shape of the case work
	for it
Alan	mmm-mmm
Todd	the outside bit

The overall two-phase process can perhaps be seen as one of requirement (re)interpretation/concept generation followed by concept composition, simulation, and reasoning. Eris[14] has linked this type of thought progression, which designers often exhibit, to the

14.
Eris, O. (2004) *Effective Inquiry for Innovative Engineering Design*, Kluwer Academic Publishers, Boston.

process of divergence and convergence, and has emphasised that the modes of questioning are markedly different between the two phases. In the divergence phase, questions of type ‘generative design questions’ are instrumental. In the convergence phase, questions of type ‘deep reasoning questions’ are instrumental. The data analysed in this study provides further corroboration for this observation especially when we see that all 37 concepts written down on the flip chart were generated in roughly the first 69 minutes of the protocol, namely Phase 1.

2.3 *Observation 3: Design conversations in Phase 1 exhibited a unique pattern of behaviour we have termed resumption*

Perhaps the most interesting aspect of the conversation was the way the designers *resumed* an old discussion thread, and further developed a concept in that thread or explored it from a different point of view based on new insight. We term this behaviour resumption. The following example (Extract 5), illustrates the designers reinitiating the discussion on holding the pen 18 minutes after its initial discussion. The topic is first discussed (E1, 361–367) 21 minutes into the meeting and is resumed 18 minutes later (at E1, 715).

Extract 5, E1, An example of resumption – on the topic of holding the pen.

361	Alan	erm yeah something that’s much more of a quite a different shape that
362		doesn’t require them to have that difficult sort of hold things in place like
363		this erm because that’s one of the challenges children have isn’t it
364	Tommy	yeah
365	Alan	holding the pen at the right angle
366	Alan	yeah I agree
367	Todd	( ) that’s probably one of the things you want to teach a child though isn’t
715	Todd	there are pens that you do have to hold in a certain orientation
716	Sandra	yeah I think that’s what I tried its really hard because I don’t hold my pen
717		in the way that they expect you do
718	Alan	they expect you to some people write like this don’t they
719	Sandra	that’s what I join my fingers up at the top where as that one expects that
720		you do more of that which I don’t do

A phenomenon similar to resumption has been observed in other design conversations in the past. Bergner[15], while studying decision making processes of engineering teams, developed a coding scheme which included a ‘return’ code. The code was used when designers referred to an idea or concept that had been mentioned earlier. Goldschmidt[16] identified a type of link which she called a ‘back-link’ – a link of moves that connect to previous moves. According to her framework, a link exists when the contents of two

15.
Bergner, D. (2006) *Dialogue Processes For Generating Decision Alternatives*, Doctoral Dissertation, Stanford University.

16.
Goldschmidt, G. (1992) Criteria for Design Evaluation:

moves have enough in common. Guindon[17] observed that designers were opportunistic and did not seem to follow the linear waterfall model prescribed by most theorists. This observation is now common knowledge, and the observed non-linear behaviours are often explained in terms of iteration.

A Process–oriented Paradigm, in Kalay, E. (ed) *Evaluating and Predicting Design Performance*, Wiley, pp. 67–79.

17.
Guindon, R. (1990) Designing the Design Process: Exploiting Opportunistic Thoughts, *Human-Computer Interaction*, 5, pp. 305–344.

However, the resumption behaviour we observed in Phase 1 was noticeably different from the more linear behaviour that was characteristic of Phase 2. More importantly, it was also distinct from iteration. Resumption entailed initiating, suspending, and resuming multiple threads in parallel in the design conversation.

In other words, we take the position that the notion of iteration is associated with a goal-oriented process which is driven by *a* distinct and explicit goal. When new insights result in the recognition that the goal is not yet met or that it should be redefined, iteration occurs since it is necessary to repeat the parts of the process that were previously executed. On the other hand, resumption is intrinsic to a process that does not yet have a clearly defined goal. In the absence of a clearly defined goal, the process is driven by multiple potential goals that are pursued in parallel.

2.4 *Observation 4: Resumption manifests itself in other types of discourse*

The resumption behaviour we observed in the design meeting was similar to other behaviours we have come across in other areas of our research seemingly unrelated to the study of engineering design.

The first description comes from an ongoing project aimed at co-developing sustainable models of entrepreneurship with preliterate villagers in rural India. After several frustrating attempts to elicit information about the aspirations of a group of these villagers, our colleague on the project, a folklorist, explained to us that one reason for the difficulty we were experiencing could be related to the differences in how people organise their memories. He described our questioning style as very linear, which he attributed to our over reliance on textual information. He asserted that, for the villagers, memories appeared to be organised in a circular manner – a feature he attributed to their large dependence on oral information[18]. He challenged us to carry out a project without taking written notes. Under this condition, he claimed that our information storing and recall strategies would change. For the villagers, the absence of an external mechanism to manage memory meant that they were essentially relying on sound for information storage and primary associations; vocal repetition could be instrumental in forming associations and facilitating memory formation and retrieval.

18.
From a personal conversation with S. Reddy on Jambapuranam plot structure, Hyderabad, India (2007).

The second description comes from psychology research on marriage, which has been of interest to us because it yields insights on the emotional functioning of a 2-person team, and we see potential in extending some of those insights to the 3–4 person design teams we typically study. The specific insight that is relevant to our observations on resumption came from Eggerichs[19], he writes:

19. Eggerichs, E. (2007) *Cracking the Communication Code*, Thomas Nelson Inc., Nashville.

> "You won't find a formal definition of this term in the dictionary, but most married couples will recognise what spider webbing is. *Someone starts with this point and goes to that point but doesn't finish that point before going to another point, not finishing that point but doubling back to an earlier point.* Multitasking women are masters of this art. They can get together and start talking about things. They never finish one point because that reminds them of some other point. They can go on for half an hour, but somehow they always bring the conversation full circle and eventually finish all the points! For husbands, however, this kind of conversation is usually not that simple."[20]

20. *ibid.*, p. 220, italics added.

The analogy Eggerichs made between resumption in conversation and spider web construction inspired us to consider spider webs as potential dynamic representations of design conversations. In the next section, we will articulate the relevance of this dynamic representation by using data from the transcript of E1.

3 MAPPING THE TRANSCRIPT DATA ON RESUMPTION TO A SPIDER WEB

While there are several types of spider webs, the one implicit in our discussion up to this point is the orb web (shown in Figure 4). To build the orb web, a spider begins by constructing a rectangle-like frame. Then, from the web's centre, it deposits numerous rays of sticky silk called radii, radiating toward the frame. Finally, it lays down many spirals of silk across the radii. In nature, no two spider webs are alike – even if they are created by the same spider.

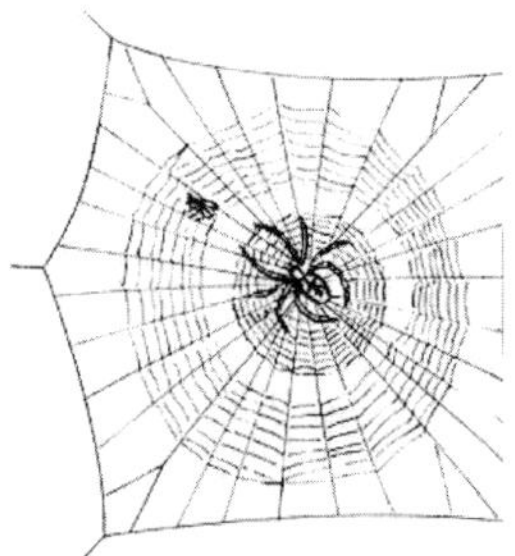

Figure 4. A spider web of the orb type.

The geometry of this simple web structure consists of spokes, arcs, and sectors, and there is a clear mapping between the actions that are taken by the spider and the resulting geometry. In other words, the construction of a spider web, similar to a design conversation, has its own set of drivers, regulators, and constraints.

In order to demonstrate the analogy we envisioned between the resumption behaviour in design conversations and spider web construction, we needed to represent the actions of the designers, as manifested in the discourse data, appropriately. In our analysis that led to the observation on resumption, we have relied on: the meeting transcripts, which contain information about time and the noun and verb phrases that were spoken during the design conversation in the meeting E1; a Powerpoint presentation provided as part of the data set, which documented for the meeting participants the design requirements that were identified; and the flip charts, which document the concepts that were generated during the meeting itself. We chose to use the design requirements (Table 1) as a basis for organising the data because the team used it to structure their brainstorming session. It should be noted that although the wording of the requirements was fixed, how members of the team interpreted them during the conversation varied.

In Table 1 the first column defines the requirement, the second column refers to the context of the requirement, specifying a key observation behind the requirement. The third column gives an abbreviation (label) for each requirement. For instance, the first requirement is about the wobbly arm of a user, and is labelled 'Arm', thus 'Arm RC' refers to the context of that requirement, whereas, 'Arm RD' refers to the definition of that requirement. The meeting moderator suggested that the team should focus on the first two requirements (i.e. the first two rows of Table 1) during the first engineering meeting on which we focus here, and on the latter two requirements during the second meeting, E2. The participants did adhere to this suggestion, but not strictly.

In reviewing the reduced data set, we observed periods of conversation during which the discussion focused on a particular interpretation of one or more of the requirements. This resulted in framing elements that served as a generative mechanism for new concepts to address the requirement(s) under discussion. Each of these conversation periods was identified as a solution-segment, and numbered chronologically. Table 2 outlines the solution-segments, the generative framing elements, the number of concepts that were generated, and the requirements that were addressed within that

FOCUS : GENERATING CONCEPTS TO OVERCOME SPECIFIC PROBLEMS

Requirement Definition (RD)	Requirement Context (RC)	Abbreviation
Print head needs to stay in contact with thermal paper to print	Wobbly arm movement of the user (5–11 year old)	Arm
The print head needs to activate the paper at the right angle to ensure good quality printing	Keeping the print head within an optimum angle range	Angle
The print head is susceptible to damage if pressed or bashed too hard onto a surface, breaking the product	Over pressuring the print head on the paper and/or abuse of the pen	Press
Consider ways to detect when the print head should (should not) fire	Overheating of the print head and possible print head damage if not in contact with media	Heat

Table 1. Requirements supplied as Powerpoint presentation to brief meeting participants.

Solution Segment	Start h:mm	Stop h:mm	Framing Element	No. of Concepts	No. of Reqs.	Requirement(s) under consideration
1	0:10	0:15	Sleigh	4	2	Guidance: keep pen at right angle; keep pen flat (level)
2	0:17	0:19	Shaver	2	1	Guidance: keep the pen flat
3	0:21	0:25	Shape of train	3	1	Holding the pen at right angle
4	0:26	0:28	Tram	3	2	Guidance: keep the pen in contact with the paper despite movement of child's hand
5	0:29	0:29	Tram	1	3	Built in steering to maintain angle and keep pen flat
6	0:32	0:33	Tram	3	1	Sensors and switches to activate printing and determine thickness
7	0:35	0:38	Laser levellers	5	2	Make sure things are level; make sure things are at the right angle
8	0:39	0:42	Battery on wrist	1	2	Holding the pen (1)
9	0:43	0:43	Right position	1	1	Comfortable and at right angle
10	0:46	0:50	Ring	1	1	Holding the pen (2)
11	0:51	0:53	Drag pen	1	0	Left handedness
12	0:56	0:58	Sheath	1	1	Protect print head: sheath solution
13	0:59	1:00	Syringe	2	1	Protect print head: syringe solution
14	1:00	1:04	Cap	5	1	Protect print head: cap solution
15	1:05	1:07	Cradle	2	1	Protect print head: cradle solution (with cleaning mechanism)
16	1.07	1.09	Dots	2	1	Overheating: Detecting the media

Table 2. Chronologically ordered solution-segments in the first 69 minutes of the meeting.

REQUIREMENT

Solution Segment	Arm RCon	Arm RDef	Angle RCon	Angle RDef	Press RCon	Press RDef	Heat RCon	Heat RDef	Misc RCon	Misc RDef
1		1		1						
2		1								
3				1						
4	1	1								
5		1	1	1						
6				1						
7		1		1						
8					1	1				
9				1						
10			1							
11										1
12						1				
13						1				
14						1				
15						1				
16								1		

Table 3. The requirements mapped against the solution-segments.

solution-segment. We also observed that new requirements were identified that were not in the original list such as a device for left handed people.

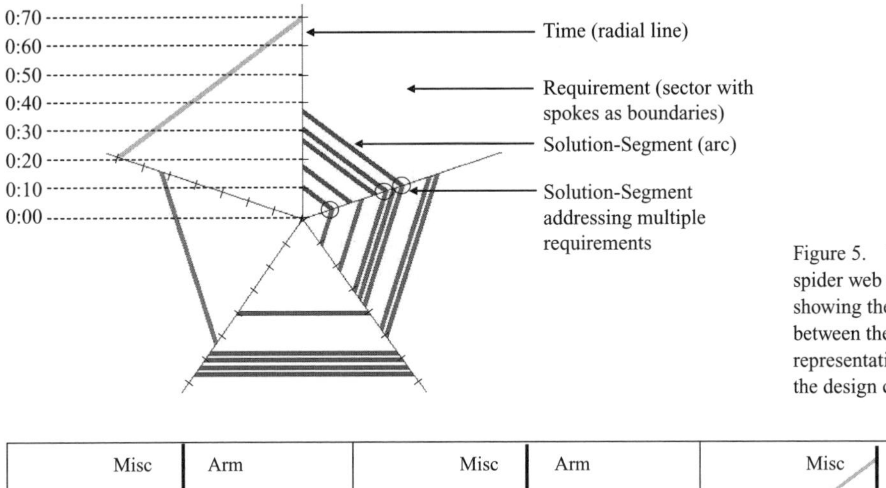

Figure 5. The annotated spider web at 69 minutes showing the correspondence between the parts of the representation and aspects of the design conversation.

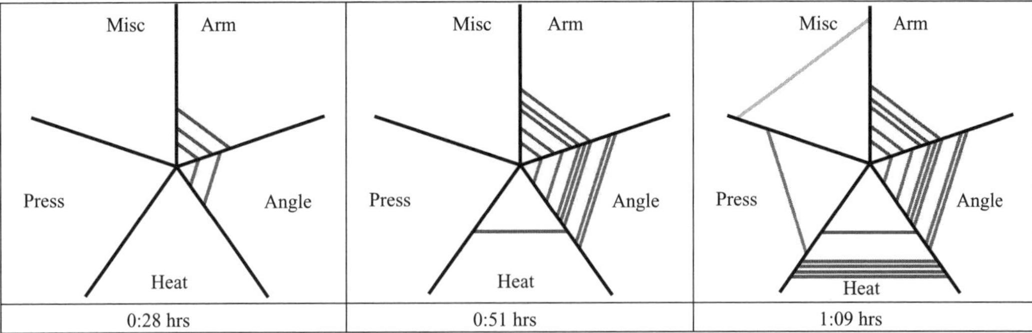

Figure 6. Spider web representation of requirements and solution-segments at three points in time during Phase 1 of the conversation.

Table 3 is another way of representing the relationship between the requirements that were considered and the solution-segments. In particular it further articulates the relationship at the resolution of the requirement context and requirement definition distinction. An additional pair of columns has been included to capture the miscellaneous requirements that were identified outside of the original scope.

The data summarised in Tables 2 and 3 are represented visually in the form of a spider web graph in Figure 5. In this, time increases in a radial direction outward from the centre. The sectors represent the requirements (as identified in Table 1). The radial lines (spokes) represent the boundaries of the sectors. The solution-segments are represented by arcs. The graph also showcases situations where one solution-segment attempts to satisfy multiple requirements, or where one requirement is satisfied by multiple solution-segments. Figure 6 shows the spider web representation of requirements and solution-segments at three points in time during Phase 1 of the conversation. The ordering is clockwise and reflects the order of appearance of the concepts listed on the flipcharts produced during the meeting. The labels correspond to the abbreviated requirement labels in Table 1.

4 DISCUSSION

The spider web plot presented in Figure 5 provides a temporal understanding of the sequence of the solution-segments in which requirements are addressed. It also allows us to see where a specific proposal addresses multiple requirements. The further away from the centre an arc is, the later in the meeting that solution-segment occurs. The representation captures the behaviour of designers moving from one requirement to another when considering specific proposals. Furthermore, it shows that during these movements, they often *resume* previous conversations that have been suspended.

In reflecting on our interpretations of the spider web representation, we realised that the term 'requirement' can have multiple meanings, which can be confusing. In particular, how does the term 'requirement' relate to the terms 'need', 'problem', and 'solution'? For the purposes our analysis, we have defined a need to designate a *gap* between current reality and a desired reality. A solution is an attempt to close this gap. A problem is said to exist when any one of the following four conditions is true:

1. A solution that attempts to bridge the gap does not exist;
2. A solution exist but fails to bridge the gap completely;
3. A solution exists and overshoots the boundary of the gap in such a way that it is considered sub-optimal;
4. The boundaries of the gap changes such that previously optimal solutions become sub-optimal.

The relationship between needs, problems and solutions is shown in Figure 7. Based on these definitions, each sector of the spider web represents a need space. The spider web representation depicts the optimality of each solution-segment visually as a spatial gap between the end-point of an arc and the boundary of

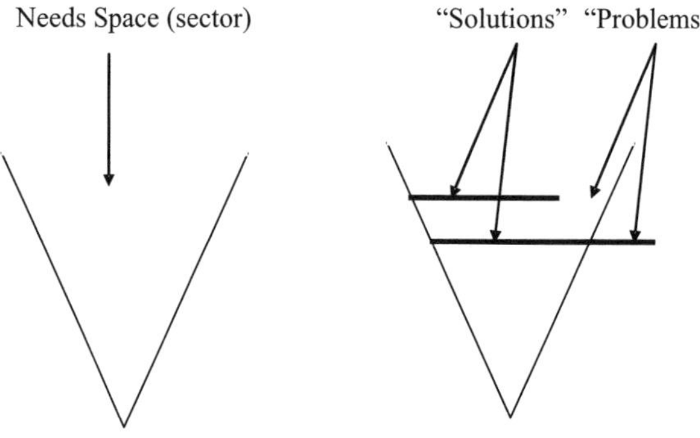

Figure 7. A sector of a spider web showing how needs, solutions and problems are represented.

the sector. A solution that does not fully address a need does not bridge the gap, and therefore, constitutes a problem. A solution that addresses a need in a way that requires more resources than necessary or results in unnecessary trade-offs bridges but overshoots the gap, and therefore, also constitutes a problem. In other words, the spider web representation has the potential of not only helping us to visualise the temporal aspects of the relationship between solutions and requirements, but also of enabling us to visualise the degree to which solutions are optimised. The representations depicted in Figures 5 and 6 assume an ideal state whereby the solutions that are part of this dataset are all considered to be optimum. Thus, this articulation assigns a specific meaning to the spokes of the spider web.

This dimension of the representation clearly requires further development if it is to be useful. First, the definition of an optimal solution for a given need has to be further developed so that it leads to an objective metric that can be applied to determine the relative length of a solution-segment. Second, the spider web representation can be extended to more accurately illustrate the needs that emerge during the conversation as opposed to the needs provided in the design brief. We aimed to achieve this by creating a miscellany requirements category (see Table 3). During conceptual design activity, if the spider web representation was automatically generated through natural language processing or manually generated by a dedicated human agent, it would enable the designers to visualise their concept generation process in a more comprehensive way, and eventually, to assess the ratio of needs (sectors) for which they have developed new solution-segments (arcs) to the total needs.

5 BROADER IMPLICATIONS FOR DESIGN PRACTICE AND EDUCATION

In a more abstract sense, the spider webbing analogy can help us think about the behaviour of designers in the engineering brainstorming meeting in two ways. First, we see the constraints introduced into the brainstorming session as serving to change the shape of the spider web. By this, we mean the number of arcs and sectors, and their order of appearance. These constraints include the written list of requirements, the listing of concepts on the whiteboard as they emerged, and the framing elements that are listed in Table 2. They have the effect of anchoring, extending, and/or limiting the shape of the spider web. Thus, these variables can be seen as being distinct and separate from the spider, and determine the properties of the web. If we imagine a universal

21.
Gericke, K., Schmidt-Kretschmer, M., Blessing, L., (Chapter 11) The Influence of the Design Task Description on the Course and Outcome of Idea Generation Meetings.

22.
Brereton, M.F., Cannon, M.C., Mabogunje, A., Leifer, L. (1996) Collaboration in

spider that possesses generic design expertise, then it can be seen that the context will determine the type of web it will spin. The implication here is that learning to do design is learning the methods to spin different types of webs. This is a task-based approach and is very similar to the method of analysis adopted by Gericke, Schmidt-Kretschmer and Blessing[21] to explore the influence of the design task description (DTD) on concept generation. Elsewhere, we have also described how the social interaction shapes the product[22].

Second, the constraints described above can be seen alternatively as extending the designers' memory and serving as 'glue' between the memory fragments of individual designers – they transform the designers from separate individuals into a single cognitive entity. In this case, we can imagine a local spider, whose specific nature determines its habitat and the type of web it comes equipped to spin. This idea builds on Shaw's notion[23] that the span of a representation accounts for its ability to draw a group of participants into proximal interaction. Another way to think of this idea is to imagine a group of designers who do not gesture, sketch, or speak a common language, and who do not have a shared white board or have any physical objects to show each other – there are no shared constraints or 'glue' between them. That would limit their ability for coordinated action that is required to design. Therefore, the design team and its constraints can be said to become a specific type of spider, spinning a particular type of web. The implication is that learning to do design is like learning to become different types of spiders. This is a individual-based approach and is explored in a very concrete way by Oak[24]. Using Membership Categorisation Analysis, Oak was able to demonstrate how the interaction between two people allowed the roles of architect and client to be talked into being, and how such roles may unconsciously constrain participant behaviour.

Design Teams: How Social Interaction Shapes the Product, in Cross, N., Christians, H., Dorst, K. (eds) *Analysing Design Activity*, Wiley, pp. 319–341.

23. Shaw, B. (2007) *More than the Sum of the Parts: Shared Representations in Collaborative Design Interaction*, Doctoral Dissertation, Royal College of Art, London.

24. Oak, A. (Chapter 17) Performing Architecture: Talking 'Architect' and 'Client' into Being.

6 CONCLUSION

In pursuing our long term research goal of instrumenting the design process, we approached the DTRS7 data in search of one or more descriptive phenomena that can serve as the basis for new team interaction variables. The data showed how design conversations might exhibit two different patterns: a linear pattern which corresponds to a process driven by a distinct and explicit goal, and a non-linear pattern which corresponds to a process driven by multiple potential goals that are pursued in parallel in the absence of a clearly defined goal. We have argued that the second pattern relies heavily on a behaviour we termed resumption. To further explore resumption, we proposed and constructed an analogous relationship between resumption behaviour in design conversations and spider web construction. We explored several variables which may be used to describe a spider web such as the number of sectors, number of arcs, the appearance of new sectors, and the temporal order of appearance of new arcs. In addition, we discussed how the metaphor of spiders and spider web construction can be used to explore questions in design education and practice. In conclusion, spider web representations have the potential of effectively illustrating the timing and manner in which solutions and requirements develop during conceptual design activity, and of portraying the phenomenon of resumption.

ACKNOWLEDGMENTS

The authors wish to thank Ben Shaw for reviewing the paper and providing very insightful feedback. We would also like to thank Kate Deibel for her help with data visualisation. This research is funded in part by the National Science Foundation under Grant No. 0230450 and the Kozmetsky Global Collaboratory's Sub-Project on Socially-Supportive Workspaces.

4
Co-evolution in Design Practice

Isabelle Reymen, Kees Dorst & Frido Smulders

The concept of co-evolution is considered a key characteristic of designing. Several authors have described design thinking processes as the co-evolution of design problem and design solution. The theoretical grounding of co-evolution is, however, still in an early stage. In this chapter, we develop the concept by analysing real world design meetings of an architect and a client. Thirteen co-evolution episodes are identified in the two architectural meetings and we focus in detail on the utterances of two co-evolution episodes. We find that modelling co-evolution in terms of problem and solution is difficult for this data, and we develop a revised model of how co-evolution in a multi-actor setting might work. Conversation in an area in between problem and solution, for example about 'use', would seem to more accurately describe how designer and client reach agreement.

Over the last ten years, several authors have described design thinking processes as the co-evolution of design problem and design solution[1,2]. In fact, this process of co-evolution has quickly become part of the 'conventional wisdom' about design – it is considered by some design researchers to be a vital and unique part of design thinking, and even to be one of the key characteristics that discerns design from other forms of human endeavour. For example Mabogunje et al.[3] refer implicitly to co-evolution of problem and solution when they refer to the characteristic of designers being multi-tasking, 'developing the problem and solution simultaneously' focusing on 'requirements' and 'solutions' instead of the co-evolution terms of 'problems' and 'solutions'.

In particular, the work of one of us (Dorst), in collaboration with Cross[4], has been widely referenced in recent years as providing evidence for the co-evolution model. In that work the introduction of the term 'co-evolution' is based on the observation that:

> "...creative design is not a matter of first fixing the problem and then searching for a satisfactory solution concept. Creative design seems more to be a matter of developing and refining together both the formulation of a problem and ideas for a solution, with constant iteration of analysis, synthesis and evaluation processes between the two notional design 'spaces' – problem space and solution space. In creative design, the designer is seeking to generate a matching problem-solution pair. The model of creative design proposed by Maher et al. is based on such a 'co-evolution' of the problem space and the solution space in the design process: the problem space and the solution space

1.
Maher, M.L., Poon, J., and Boulanger, S. (1996) Formalising Design Exploration as Co-evolution: A Combined Gene Approach in J.S. Gero and F. Sudweeks (eds) *Advances in Formal Design Methods for CAD*, Chapman and Hall, London.

2.
Dorst, K., and Cross, N. (2001) Creativity in the Design Process: Co-evolution of Problem-Solution, *Design Studies*, 22, pp. 425–437.

3.
Mabogunje, A., Eris, O., Sonalkar, N., Jung, M. and Leifer, L. (Chapter 3) Spider Webbing: A Paradigm for Engineering Design Conversations during Concept Generation.

4.
Dorst and Cross (2001) *op. cit.*

co-evolve together, with interchange of information between the two spaces."[5]

5. Dorst and Cross (2001) *op. cit.*, p. 434.

This notion is then illustrated with an analysis of an existing set of empirical data obtained from a set of protocol studies of nine experienced industrial designers. From the protocol data aspects of creativity in design are identified, related to both the formulation of the design problem, and to the concept of originality. These observations are also related to a proposed model of creative design as the co-evolution of problem/solution spaces. This led to an initial confirmation of the validity of the model.

This, however, is only the beginning of building a framework for describing design as co-evolution, and for establishing a solid foundation of design as an activity involving co-evolution. There are many aspects of the co-evolution process that need critical scrutiny before it can be claimed that co-evolution is the kernel of design thinking.

In this chapter we aim to contribute to the discussion about co-evolution in design by studying it in a real world setting, with multiple actors, and looking from a broader theoretical perspective. This is in contrast to previous studies which were lab-based, single-actor, and using a purely cognitive approach to modelling design. We will be investigating the co-evolution perspective for a number of actors with different roles in the design process that make up the DTRS7 data gathered in actual design practice. This will allow us to bring the concept of co-evolution closer to real-world design activity, and reflect critically upon its merits.

The research questions that drive our current study include:

1. Can co-evolution be discerned in conversations between designer and client?
2. How much of a typical design conversation in the conceptual stage of a design project is it possible to describe as co-evolution?
3. How does a co-evolution episode start?
4. How are the issues finally resolved within a co-evolution episode?
5. What are the detailed patterns of interaction between 'problem space' and 'solution space' during the co-evolution episodes?
6. Is it possible to identify during the co-evolution episodes individual strategies that aim to clarify the problem and solution, for example by making explicit the implicit knowledge the different actors might have?

To answer these questions, we first describe our research method. We then introduce a co-evolution model for a multi-actor design setting in order to look at the data. We describe our detailed analysis of the data, addressing the research questions. Finally, we discuss our findings and end with some conclusions about co-evolution in practice.

1 RESEARCH METHOD

In this empirical study we concentrate on the two architectural meetings (A1 and A2) that are part of the DTRS7 data. In these meetings a preliminary design concept is discussed between the architect, the client, and a representative of a regulatory body. We carried out a qualitative analysis of these protocols, focusing on the discussions that take place in the meeting. The unit of analysis is the utterances of the actors in the project meetings, grouped in episodes that are distinguishable by subject. The data were independently coded by two researchers from differing backgrounds (industrial design and architecture) to ensure reliability.

There are five stages to our analysis. First, we identified the episodes in the meetings where co-evolution was seen to take place. To do this the transcripts were first divided into episodes, where each episode is a discussion about a specific issue. A more formal definition of episodes and the phases in which they occur is given by McDonnell[6].

6. McDonnell, J. (Chapter 14) Collaborative Negotiation in Design: A Study of Design Conversations between Architect and Building Users.

Second, the episodes involving co-evolution were coded based on the indications of learning in parts of the episode – learning is defined here as the architect and/or the client getting a new insight into the subject matter (research questions 1 and 2). To determine the relative proportion of co-evolution in the meetings, for each co-evolution episode, the beginning transcription line and end transcription line were noted. The total numbers of lines of all co-evolution episodes were then compared to the total number of lines for each meeting.

Third, we analysed the opening and closing utterances of co-evolution episodes (research questions 3 and 4). In this chapter the utterances are paraphrased for reasons of clarity. We identified how co-evolution episodes start – with a problem, a solution, or otherwise – and the means by which the issues in the episodes were resolved – through an emergent solution, or through other means such as social interaction strategies.

Fourth, we mapped the interaction patterns within co-evolution episodes (research question 5). To do this two particular co-evolution episodes were selected for more detailed analysis; identifying the 'jumps' from problem space to solution space and back. All utterances in the selected co-evolution episode were coded as problem, solution, or 'other'. We counted the number of transitions between problem and solution and looked at what kind of transitions there were.

Fifth, this allowed us then to study the way in which individual strategies clarify the problem and solution (research question 6).

2 OBSERVING CO-EVOLUTION IN A MULTI-ACTOR DESIGN MEETING

Co-evolution has hitherto been described as a single-person activity, so dealing, in this chapter, with a designer-client conversation is already an enrichment of the concept of 'co-evolution'. It also provides us with an opportunity to both extend the theory of co-evolution, and also come up with a model of how co-evolution might work in a multi-actor design situation. Figure 1 shows our initial model. In comparison with this, Luck[7] takes a slightly different approach: concentrating on 'moves' within a combined design problem-solution space defining a design space as one "in which participants move and modify the properties of a design".

7.
Luck, R. (Chapter 13) "Does this Compromise your Design?" Socially Producing a Design Concept in Talk-in-Interaction.

The starting point for our model is that there are not only several actors involved within these design conversations, but also that these actors play different roles within the design situation. It is these roles that we see as directly relevant to the process of co-evolution.

The building blocks of the model in Figure 1 are: the problem as viewed by the architect (Pa), the solution as viewed by the architect (Sa), the problem as viewed by the client (Pc), and the solution as viewed by the client (Sc). We assume that in a design project meeting, both the architect and the client have their own 'image' of the problems and solutions within the design situation. The architect 'owns' the 'factual' solution (i.e. knows more about the solution than the client) and the client 'owns' the 'factual' problem (i.e. knows more about the problems that are associated with the current situation).

Both the architect and the client have their own initial problem-solution (P-S) combination when the discussion starts. This gives us

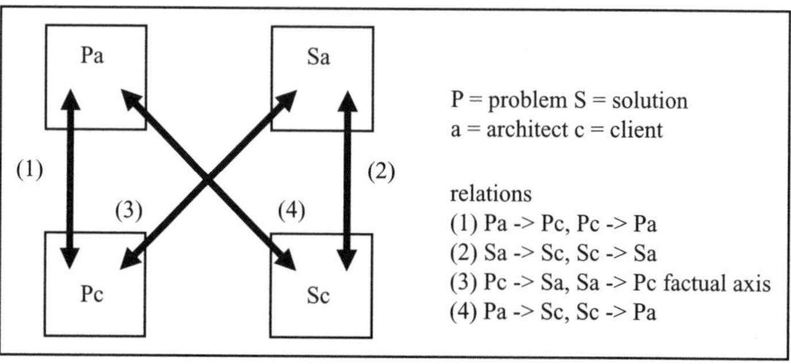

Figure 1. A model of co-evolution in a multi-actor design situation.

four possible lines of communication when co-evolution occurs and 8 different vectors (alternating between actors, problems, solutions and problem-solution pairs). These lines of communication are the relations, represented by arrows, in Figure 1. Co-evolution can also, however, take place in the cognitions of just one person – architect or client.

So the possible moves within a co-evolutionary conversation are:

1. Relation 1: The comparison of the images that the architect and client have of the problem area at hand. For instance, the architect has to understand the problems that the client has encountered with the current building.
2. Relation 2: An information exchange around the images that architect and client have about the solution. The clearest example of this is when the architect presents a possible solution to the client by showing a new drawing and talking about the new design solution. The client then has to construct an image in their head of what the solution entails. This is a vital process because failure to build up an image of the design in the client's mind will let the design process run out of control.
3. Relation 3: The client uses their knowledge of the problem area to question the solution proposed by the architect. This is a possible locus for co-evolution: here both problem and solution can start to evolve in the conversation. The proposed solution might put a different slant on the initial problem statement, causing it to be developed or even reframed.
4. Relation 4: The opposite of Relation 3 can also happen. The client comes up with possible solutions that actually upset the initial view the architect had of the problem, causing it to be redeveloped or reframed. This might easily happen as a result of a misunderstanding in the briefing process or when the architect introduces a problem that the client tries to solve.

A design conversation is aimed at the construction of an agreed problem-solution pairing, through an exchange of information and value judgments about possible problem-solution pairings. The goal is to work towards a correspondence of the matching Pc-Sa and Pa-Sc pairs, through a shared understanding or even the complete resolution of the issues. If this goal is achieved then the P-S pair disappears, and becomes Pca-Sac (i.e. Pa = Pc and Sa = Sc) as there is consensus over the factual problem and the quality of the factual solution. The test as to whether this has been achieved comes when the architect is able to defend Pca within their own office, and when the client is able to defend Sac to other stakeholders in the process (for example the funeral services companies).

The initial model that we propose for helping to analyse the dataset is *prima facie* relatively straightforward, although there may be assumptions that could limit our view and understanding of what is happening in the data when we apply this model to the data. In our data analysis we will be open to the possible occurrence of events that fall outside this framework, and these will be dealt with separately. For the conversation patterns that do fit within this model, our data analysis will focus on trying to identify the patterns and mechanisms of co-evolution in multi-actor design.

3 CO-EVOLUTION EPISODES

A more operational definition of co-evolution is a discussion concerning problem or solution in which actors provide insights into producing problem-solution pairs. In the following sections we will first answer research questions 1–4 from the design meetings in general, and second, answer research questions 5 and 6 by looking at two specific co-evolution episodes. Detailed comments on the proposed model described in Section 2 are given when answering research question 5.

3.1 *Co-evolution in the design meetings*

Research question 1: can co-evolution be discerned in conversations between designer and client? When looking at the analysis of the data in Table 1, we can easily see that we could discern co-evolution in the data.

Research question 2: how much of a typical design conversation in the conceptual stage of a design project can be described as co-evolution? In A1 64% (1488 out of 2342) of transcription lines can be identified as 'co-evolution lines'. For A2, this figure is 24% (511 out of 2124). The amount of co-evolution in design meetings may thus depend on the stage of the project (besides many other things of course). As we know, the second meeting comes some months after the first one and concentrates more on elaborating the design. A possible hypothesis might be that in early phases of designing conversations among different actors tend to have more co-evolution episodes than later phases. There is no evidence to suggest that the co-evolution episodes are longer than the non-co-evolution episodes in the data, and the length of co-evolution episodes seems not to depend on the phase of the design process.

Research question 3: how does a co-evolution episode start? Table 1 shows that, at the beginning of A1, the architect raised issues that had still to be resolved as well as presenting (partial) solutions to issues that had come up earlier in the project. Later in A1 the client introduced the new subjects of discussion. In A2 both architect and client introduced new subjects. It seems that the architect had prepared a list of questions about issues that still need to be discussed

Meeting	Start	End	Total	Start utterance (paraphrased) A = architect, C = client	End utterance (paraphrased) A = architect, C = client
A1	85	198	113	(A) Introducing present state of the design in relation to previous meeting.	(A) Summing up conclusions regarding the waiting room.
	198	326	128	(A) Introducing subject of previous meeting: porte cochere.	(A) Summing up conclusions regarding the porte cochere.
	327	548	221	(A) Introducing entrance to chapel by describing use-process.	(C) Ending with an 'OK' regarding an analogous situation that is representing the solution that was just discussed.
	549	974	425	(A) Summing up the subjects that have been discussed in the meeting so far, moving on to the audio visual room.	(A) Concluding that a (new) brief is necessary for that room.
	1130	1207	77	(C) Introducing a concern regarding a design feature: the pond.	(A) Discussing how to finalise the design on that point.
	1305	1435	130	(A) Introducing the issue of materials in the final design.	(C) Ending with an analogous situation in order to make sure the solution is shared enough.
	1758	1882	124	(C) Introducing a new feature by client: signage and getting lost.	(A) Ends when the problem seems resolved.
	1883	1999	116	(C) Referring back to a subject from previous meeting: the ancient hedge.	(C) Ends by summing up conclusions.
	2000	2154	154	(C) Introducing a subject that still needs to be discussed: book of remembrance.	(A) Ends by briefly mentioning that it is OK as discussed.
			1488	**Total lines of co-evolution identified in Meeting A1**	
A2	119	279	160	(A) Starts with 'OK' regarding the last subject and fluently goes over to the next subject, guided by the drawings.	(A) Ends when the situation is clear and the issue resolved.
	354	408	54	(C) Mentioning a concern regarding the staff room.	(C) An 'OK', indicating that the concern is resolved.
	409	551	142	(C) Starts after a pause with discussing the office space.	(C) Nodding OK regarding the solution offered.
	1802	1957	155	(A) Introducing an option to the present design.	(C) Summing up and agreeing on next step focusing on options.
			511	**Total lines of co-evolution identified in Meeting A2**	

Table 1. Analysis scheme of design meetings showing co-evolution episodes.

in order to get them resolved. The client was also given room to bring in issues they were concerned with. In that sense it was an open dialogue with input from both parties.

Research question 4: how are the issues finally resolved within a co-evolution episode? Table 1 shows that episodes seem to end when some sort of shared satisfactory situation is reached. Satisfactory situations are: the problem is 'solved' satisfactorily, a consensus about a vision is reached, or there is agreement on a further course of action. Where further design activities are needed there seems to be agreement about more abstract issues, sometimes made concrete by naming an analogous situation ('Le Corbusier like' windows, built in seating like at the city centre church, etc.). Where a resolution occurs the whole subject seems to disappear from the discussion (for example the issue of signage).

3.2 *Co-evolution in two episodes*

The two co-evolution episodes we selected for further analysis were A1 (327–548) which concerned the size of the sanctuary, and A1 (1130–1207) which concerned the stepping stones. The criteria for selecting these episodes were: the clarity of the episode, real client involvement in the solution, and real engagement of the architect with the problem.

Research question 5: what are the detailed patterns of interaction between 'problem space' and 'solution space' during the co-evolution episodes? To answer this question, we first discuss the model of co-evolution in a multi-actor design situation and then go on to reflect in general on the conversations.

3.2.1 Co-evolution model in a multi-actor situation

Concerning the building blocks of the model, we see Anna and Charles both as clients. In general, it was unclear whether the utterances of the actors should be coded as PROBLEMS or as SOLUTIONS. It was often not clear whether the actors talked about the problem or about a possible solution. This holds for the architect as well as the client. Extract 1 shows one example.

Extract 1, Referring to the problem or the solution.

Adam well [*begins to sketch*] there's a couple of things I can do I can make the whole thing bigger for a start to help you make this work I can also splay the opening a little bit wider as well so you've got an even better view in if that helps and that would certainly enable you to get a wider catafalque inside there that might be suitable for two people the architectural idea here is to have like a cylinder which will be top lit at the top perhaps a glass pyramid something like that so you can get top light

The text not underlined might be considered to be related to the solution, whereas the underlined text seems to refer to the problem.

The participants seem to discuss something more like their image of the use of the current building and the future operation of the new building. Use seems to lie between problem and solution. For the architect, use was more closely linked to the solution, whereas for the client (Anna especially), use was very close to the problem. Extracts 2 and 3 show examples of this.

Extract 2, Use for the architect is related to the solution.

Adam OK [*begins to point*] so having got this far everyone is now under cover at this point erm + the way it's designed at the moment the roof edge is actually on this line here + so that's the bit that covers you OK so that is a length of about nine nine metres or so OK + from this point you go through a lobby into the chapel [*turn continues*]

Extract 3, Use for the client is related to the problem.

Anna I don't know I think I would say it might just I mean at the moment they can just they can just go in side by side but it's difficult to squeeze in to put the coffins on at the moment even because you've also you've got the two catafalques in side by side and you need to have four routes for people to go either you need the one in the middle for both people to go and the ones at the end for them to drop the coffins off erm but even two catafalques isn't always enough we've had three or we've had car accidents you know we've had three coffins and we've not been able to accommodate all the you know I mean if we can do two [*turn continues*]

These observations raise the question of whether it is possible at any time in a design project to make a strong separation between 'problem' and 'solution', even if theoretically one might be able to distinguish them. The only exception to this is when there is a clear problem without a solution or a clear solution without a problem. For example extra features, like the stepping stones in A1 (1153–1155), are often introduced by an architect, who has his own goals, ambitions, and feeling for architectural quality; strictly speaking they are outside the specifications[8]. (For a more detailed treatment of this phenomenon see Glock's[9] fine-grained analysis of part of meeting A1, in which he accounts for what he finds design practice to be by distinguishing goal-orientation from goal-direction.)

Turning now to the relationships detailed in our original model (Figure 1), we were able to distinguish each type of relationship in the data, but it was difficult to indicate the separate building blocks of the model. We were also not able to count the number of alternations between problem and solution, as was proposed in

8.
Harfield, S. (2007) On Design 'Problematization': Theorising Differences in Designed Outcomes, *Design Studies*, 28, pp. 159–173.

9.
Glock, F. (Chapter 16) Aspects of Language Use in Design Conversation.

the method, or to quantify the number of relationships of a certain type in order to discuss how prevalent they were in the data. An example from the data of each type of relation is given below.

Relationship 1: The client knowing the 'problem' and the architect aiming at understanding the problem. Architect: "my question for you is how wide would it need to be for two coffins" (A1, 368). But the client knowing the problem doesn't know how to convey that message to the architect stating: "we'll have a measure up on that" (A1, 370).

Relationship 2: The architect knowing the solution and the client aiming at understanding the solution. For example the architect says: "so if you'd like me to increase the size of that space I certainly can do" (A1, 389). However, the client needs additional information to evaluate the solution and asks the architect: "how many people would you get on the seats" (A1, 391). Other examples are when the architect proposes to: "...chalk out a three point one diameter circle on the ground somewhere" (A1, 438), and when the client proposed: "I think if we could put a bridge or something that looks like a bridge" (A1, 1148), which the architect follows with: "you can have that as a bridge if you wanted not necessarily stepping stones" (A1, 1149).

Relationship 3: The client knowing the problem and the architect knowing the solution. The client tries to match the use of the building with the solution of the architect. For instance, after having investigated the solution the client wonders: "I'm just trying to think if people are at the end of those seating they also need to be able to see the coffin..." (A1, 397). To which the architect responds: "I didn't see it as being a doored off space I mean rather like an antechapel in a cathedral or whatever I just saw it as a space that you could walk in to..." (A1, 411). And then it's clear to the client who responds: "yes that's fine" (A1, 414). When the architect does not want to change the design, he tries to change the client's understanding of the problem, the discussion about the stepping stones as an extra architectural feature, for example (A1, 1153). During the discussion the architect mentions a number of times that there are more routes to the entrance by which he aims to resolve the concerns of the client. In the end the client decides to keep the idea and proposes a meeting with future users (A1, 1204).

Relationship 4: The architect focusing on the problem and the client focusing on the solution. Within the two meetings there are several occasions when it is possible to see a slight reversal of roles happen. During an episode where the architect and client discuss the light reflecting off the pond into the circular antechapel the architect mentions: "I saw the pond initially as being quite still but there's no reason why it couldn't have a fountain in it or something" (A1, 511). Elsewhere, during the discussion of the audio-visual room, the architect mentions a possible problem: "that's a difficult one because if they can see out then people would be able to see in" (A1, 583). The client then introduces the solution of using one-way mirrored glass (A1, 585).

The mechanism by which the client and architect co-evolve is by talking about the future use of the building. Use, here, mediates between the problem and the solution, explicating their coherence to each other. We come back to this in the discussion of research question 6 and in Section 4.3.

3.2.2 Reflections on the design conversation

In looking at the empirical data, discussion has been the object of study. Discussion and collaboration are needed because on one side the designer needs the detailed 'use' knowledge of the client (problem side) in order to develop and improve the solution, and on the other side the client needs to acquire better knowledge of the proposed solution in order to agree with it. Many aspects of the discussion that aims at increasing these knowledge levels can be recognised in the data. Utterances in the episodes can be characterised as, for example, announcement, answer, appointment, arguing, concern, confirmation, defence, detailing, elaboration, preference, postponement, question, recognition, reference, repetition, summarising, and translation. Ideas are being taken on and new ideas brought in; ambitions are increased or discarded and compromises made. All these forms of communication result at the end of each episode in increased individual knowledge levels as well as increased levels of sharedness regarding the subjects discussed.

Research question 6: is it possible to identify during the co-evolution episodes individual strategies that aim to clarify the problem and solution, for example by making explicit the implicit knowledge the different actors might have? First, actors do aim to explicate their implicit knowledge structures during the co-evolution episodes. The strategies for doing so that we identified are: the application of use scenarios and use plans, introducing precedents, drawings, and role-switching behaviour.

Scenarios of use are especially helpful to the client, who would never be able to create a design brief that contains all the information required. Sometimes it seems as if the client is reporting a film that shows the behaviour and the thoughts of the users both of the current building and future users of the new building. Such a narrative seems a very effective way of making explicit the implicit knowledge structures of the client that help the architect to better understand the future use of the building. For instance, the client mentions the funeral arrangements for babies and young children (A1, 416) which need a special area that can somehow be 'curtained off' (A1, 456).

What is also interesting to note is that the architect applies a form of use to externalise his knowledge. By giving descriptions of certain activities in and around the building, the architect aims to explain to the client the reasoning behind the design. These descriptions of behaviour are by no means as rich as the use scenarios given by the client. They are more like what Roozenburg and Eekels[10] and Houkes et al.[11] call 'use plans'. The use plans of the architect and the use scenarios of the client need to coincide in order to prevent problematic use situations surfacing in the future. The architect enriches his use plans by taking over elements of the scenarios of the client, although his concluding remarks are still always made in the terms of his own 'thought world'. The application of use plans and use scenarios seems to be the most important bridging activity among these actors. McDonnell[12], in her analysis of the same meeting data, makes similar observations and discusses in some detail the way scenarios are exchanged and elaborated between the architect and client as they negotiate how the design is to be developed.

10.
Roozenburg, N., and Eekels, J. (1995) *Product Design: Fundamentals and Methods*, Wiley.

11.
Houkes, W., Vermaas, P.E., Dorst, K., and de Vries, M.J. (2002) Design and Use as Plans: An Action-Theoretical Account, *Design Studies*, 23, pp. 303–320.

12.
McDonnell, J. (Chapter 14), *op. cit.*

Another strategy that actors apply to transfer their implicit knowledge structures is to refer to precedents as commonly available knowledge (for example the Ronchamp chapel of Le Corbusier (A1, 493)) and in the form of drawings, e.g. where the architect shows some conceptual diagrams that illustrate where his design is coming from (A1, 660). These reference points sometimes help elicit implicit knowledge while in other situations they serve to convey to other actors the meaning or origin of a complex solution without ever being able to explicitly describe that solution.

The final strategy that we have identified is the role-switching behaviour of the actors which happens when actors forget their own 'formal' role either as client or as designer. An example of this is where the client takes the architectural role when she suggests swapping two rooms in order to hide the technicians from the visitors and to provide a better view for them of the service (A1, 572). Similarly, when the client introduces the solution of one-way mirrored glass for the technicians' room to prevent visitors seeing the operators inside tapping on keyboards (A1, 585). This idea is later incorporated in the more integrated solution of the architect (A1, 724), but again later questioned by the client because of reflections that the visitors might see of themselves in the mirrors (A1, 833). At these points the actors act in a very trusting way and there are no distinct roles for a certain period of time. Approaching the resolution of the matter, the actors again resume their roles; the architect asks the client to write a new brief (A1, 971). This strategy would appear to be related to their aim of better understanding the design solution, on the one hand, and the design problem, on the other.

From the analysis above it is clear that co-evolution in a multi-actor setting supports the externalisation of implicit knowledge structures that either reside within the cognitive system related to the architect's solution space or to the client's problem space.

4 DISCUSSION

After a short discussion about the coding and analysis process, we reflect on the co-evolution model and propose alternative modelling ideas.

4.1 *The coding of PROBLEM and SOLUTION*

In general, there was a very close match between the two coders. Concerning the coding of the design meetings, we can remark that sometimes there was disagreement about the start of a new subject; the start is often just an introduction, reiterating what was known before. Coder 2 did not always include the summary at the end of the discussion in their coding. For the coding of the two episodes, there was no agreement about whether an utterance concerned the problem or the solution. Coder 1 labelled many utterances as SOLUTION where Coder 2 labelled them as PROBLEM. What we found was that, when discussing these differences, the concept of 'use' became extremely important.

4.2 *Reflection on the co-evolution model*

From the analysis of the data we can conclude that the codes PROBLEM and SOLUTION are problematic, and that therefore modelling co-evolution in terms of problem and solution is difficult. Conversation in an area 'between' problem and solution such as 'use' seems to more accurately describe exactly how the actors reach agreement. Given the kind of data analysed – design meetings between designer and client – we only looked at co-evolution taking place when the designer or client contributed to the problem or solution. As we stated earlier, however, co-evolution can also take place in the cognition of a single actor.

In the co-evolution episodes themselves all actors are equal with formal roles becoming less important for these short periods of time. These episodes are playful and have their own dynamics. Outside of co-evolution episodes we see many factual statements (e.g. informative, explanatory).

4.3 *Alternative modelling ideas*

In order to explain how co-evolution takes place in the identified co-evolution episodes, we suggest two alternative directions; learning in conversations, and boundary objects within 'thought worlds'.

4.3.1 Learning in conversations

Co-evolution is a natural part of the design dialogue, part of the dialectic process of design deliberation. The architect has to allow co-evolution, has to be open to the possibility of the design changing as a result of discussion. This is demonstrated in the way he asks his questions, summarises his answers, and in how client and architect negotiate to an agreed solution. They are, in effect, operating as a reflective team. Progress comes through a dialogue. By discussing the design problem and solution, they change both the design and their own points of view. In essence they learn.

Learning is a very important part of co-evolution, though it is not defined by personal learning: it is defined by de facto changes in the design problem or the design solution. But learning must take place. In the data, the client and architect have several points where they clearly get a new insight into the subject matter, making it possible to reframe their position in the discussion. Learning was important to discriminate between co-evolution and non co-evolution episodes. But the co-evolution model does not explicitly describe learning. Analysing the conversation from a learning perspective might help in studying co-evolution in more detail.

4.3.2 Boundary objects and thought worlds

From our data analysis we have found that it is difficult to distinguish the concepts of 'problem' and 'solution' and that 'use', mediating in between problem and solution, seems to be what is discussed. We found that the concept of use could be just such a bridging concept. Use corresponds with the concept of 'use plan' developed by Houkes et al.[13] It also corresponds with the concept of behaviour as defined by Gero and Kannengiesser[14] who position behaviour between function (corresponding with problem, the domain of the client) and structure (corresponding with solution, the domain of the architect). One could say that the 'battlefield' of designing is about behaviour. Architect and client talk, in their dialogue, about the use of the existing and new building, and get insight through this: the client designs the future use of the building (to anticipate problems) by presenting rich use scenarios, while the architect needs to build up an understanding of that desired use because many details simply cannot be captured in a brief. Other bridging concepts that we identified were precedents, drawings, metaphors and analogies. This would seem an important way forward in studying co-evolution: the use of a bridging or mediating concept, also referred to as 'boundary objects' in the literature[15,16,17].

Alternative directions could be in: studying the synchronisation of different thought worlds[18,19]; the individual 'object worlds' described by Bucciarelli[20], or the team mental models of Mohammed and

13.
Houkes, Vermass, Dorst and de Vries (2002), op. cit.

14.
Gero, J.S., and Kannengiesser, U. (2004) The Situated Function–Behavior–Structure Framework, *Design Studies*, 25, pp. 373–391.

15.
Star, S.L., and Griesemer, J.R. (1989) Institutional Ecology, 'Translations' and Boundary Objects: Amateurs and Professionals in Berkeley's Museum of Vertebrate Zoology, 1907–39, *Social Studies of Science*, 19(3), pp. 387–420.

16.
Bucciarelli, L.L. (2002) Between Thought and Object in Engineering Design, *Design Studies*, 23, pp. 219–231.

17.
Smulders, F.E. (2007) Team Mental Models in Innovation: Means and Ends. *CoDesign*, 3, pp. 51–58.

18.
Smulders, F.E. (2006) Get Synchronized! Bridging the Gap between Design and Volume Production, Ph.D Thesis, Delft University of Technology, The Netherlands.

19.
Dougherty, D. (1992) Interpretative Barriers to Successful Product Innovation in Large Firms, *Organization Science*, 3, pp. 192–202.

20.
Bucciarelli, L.L. (1988) An Ethnographic Perspective on Engineering Design, *Design Studies*, 9, pp. 159–168.

Dumville[21] that contain implicit and explicit knowledge structures[22]. As indicated above, designers and clients proceed by storytelling although meanings often don't correspond between different actors. Krippendorff[23] gives the example of surgeons who have no difficulty speaking about what their instruments are for, but have much more difficulty in saying how they handle them during an operation. This, of course, is precisely what a designer wants to know[24].

21. Mohammed, S., and Dumville, B.C. (2001) Team Mental Models in a Team Knowledge Framework: Expanding Theory and Measurements across Disciplinary Boundaries, *Journal of Organizational Behavior*, 22, pp. 89–106.

22. Kim D.H. (1993) The Link between Individual and Organizational Learning, *Sloan Management Review*, 35, pp. 37–50.

23. Krippendorff, K. (2006) *The Semantic Turn, A New Foundation for Design*, Taylor and Francis, New York; p. 224.

24. See also for this topic the discussion of expertise in **McDonnell (Chapter 14), *op. cit.***

5 CONCLUSION

In this chapter we have tried to further develop the notion of co-evolution in design. We have used the DTRS7 data to do this. By looking in detail at a real-world setting we have been able to get a much richer picture of co-evolution. To operationalise the concept of co-evolution for this particular dataset – a conversation between an architect and his client, in the latter stages of the design project – we first developed a simple descriptive model of what co-evolution under these circumstances might look like, which made possible the analysis of the data in terms of problems, solutions, and their relationships.

Taking the discussion between the actors as the subject of study, a descriptive model with which to analyse the data has allowed us to look at co-evolution as it was made explicit by these actors. This is a step forward from earlier studies of co-evolution solely focused on individual designers.

What we observed was that the proposed model doesn't go far in explaining how co-evolution takes place in two co-evolution episodes selected from the data. Two alternative ways to look at co-evolution have been briefly touched on; first, the concept of learning in conversations, and second, concepts of boundary objects and thought worlds. Our conclusion is that co-evolution cannot be explained by referring directly to utterances, a layer of interpretation needs to be added.

The limitations of this research concern the type of architectural project, with infrequent mention of resources like time and money (money was only mentioned at three points: A1, 1418, A1, 1467 and A1, 2033), the stage of the project with the design almost finished, and most conceptual issues decided, the type of architect, who is open for dialogue and stakeholder participation: he asks questions and offers the client space to put forward her ideas and solutions. The clients have an equal position to the architect, which makes it safe for them to make suggestions. A final limitation is the extent of the data analysed, only two meetings of a larger design process. The fact that these were just two meetings within the context of a single project, and participants that clearly have a personal style in their approach to it, does limit the scope of our conclusions however many episodes and utterances within the meetings have been studied. Although the dataset doesn't contain the most intense moments of the conceptual design phase in the project, we think that the (small) steps made in these multi-actor sessions are a vital part of the design activities adding a social dimension to the design process. It would be interesting to study whether the distinction between collaborative negotiation and collaborative design as proposed by McDonnell[25] corresponds to 'normal' and co-evolution episodes.

25.
McDonnell (Chapter 14), ***op. cit.***

For future research we propose studying the same data with another, more fine-grained theoretical framework in mind, and a more precise coding scheme[26], to also look at the engineering meetings to see if co-evolution can be identified, and to gather more data from different design projects in different fields, dealing with different phases of the design process. The study of co-evolution in other fields such as organisation design might also offer useful insights[27]. The aim of these future studies would be to find patterns in conversations on the basis of which we could start to discern different kinds of co-evolution (perhaps at different levels – detailed and at more general levels, for example – or more normative – better or worse – types of co-evolution and the criteria to distinguish them), perhaps working towards a typology. This could help to stimulate and improve co-evolution processes in design education and in professional design practice.

26.
The coding schemes used for the data analysis in this volume could be useful here, for example the behavioural categories of Le Dantec and Do, the design criterion and design features of Akin, the problem definition and idea generation of Atman et al.
Le Dantec, C.A., and Do, E.Y. The Mechanisms of Value Transfer in Design Meetings (Chapter 6).
Akın, O. Variants and Invariants of Design Cognition (Chapter 10).
Atman, C.J., Borgford-Parnell, J., Deibel, K., Kang, A., Ng, W.H., Kilgore, D. and Turns, J. Matters of Context in Design (Chapter 22).

27.
Lewin, A.Y., Volberda, H.W. (1999) Prolegomena on Coevolution: A framework for research on strategy and new organizational forms, *Organization Science*, 10, pp. 519–534.

What we have shown in this chapter is that co-evolution which was initially introduced to contrast with 'normal' problem solving theory in an effort to account better for design is clearly one of the thought patterns, and one of the patterns discussion takes, within the design arena – and possibly a very important one. We hope, with this chapter, to have opened up the notion of co-evolution for further research. Now it is time to develop the concept and to further inform, define and scope it.

ACKNOWLEDGEMENTS

The authors would like to thank the participants of the DTRS workshop in London in 2007 for their comments on our presentation. Their many reflections and comments served to improve the chapter. We would also like to thank Peter Lloyd for his review of this chapter and his many editorial improvements.

Part 2

Values in Designing

5
Ethical Imagination and Design*

Peter Lloyd

Using the DTRS7 data of both architectural and engineering design meetings this chapter shows how the fields of ethics and design interrelate, especially in the area of creative imagination. The chapter first draws on Medway and Clark's[1] concept of 'the virtual building' to show how essential aspects of designerly thinking can apply to ethics. It then goes on to show how, in the process of designing, designers engage explicitly and implicitly with ethical issues. The chapter discusses four extended examples from the data – two involving the design of the crematorium and two involving the design of the digital pen – before suggesting that by addressing ethical subjects without framing them in explicitly ethical ways, the design process allows us to 'imaginatively trace out the implications of our metaphors, prototypes and narratives' a key element of ethical decision-making according to Johnson[2].

Mark Johnson, in taking a cognitive view of the individual in his book *Moral Imagination*, concludes that: "we must cultivate moral imagination by sharpening our powers of discrimination, exercising our capacity for envisioning new possibilities, and imaginatively tracing out the implications of our metaphors, prototypes, and narratives"[3]. Although Johnson doesn't mention it, put in this generic way, this description sounds strikingly like the activity of designing.

The aim of this chapter, in exploring the relationship between design thinking and ethical thinking, is twofold. First, to look at the way that designers talk about, and resolve, ethical issues – implicit and explicit – in the process of design. Second, to consider how the process of design, the *social* process of design, and the ways of imagining that it entails, functions as a way of thinking about and resolving ethical problems. Underlying these two areas of inquiry is the idea of moral imagination which defines what we take 'ethics' to be in the chapter. There are three aspects to this ethics. First, the aim of acting in the world to change or influence behaviour. Second, the imagination of alternative actions and their consequences. Third, the evaluation of those consequences in terms of good or bad. Importantly, it is the relationship between aesthetics and experience, first identified by John Dewey[4], which points to *designing* as a profitable way forward when considering moral imagination.

A number of chapters in this book are relevant to the present inquiry. Le Dantec and Do[5] describe the ways in which values, and here

1. Medway, P. and Clark, B. (2003) Imagining the Building: Architectural Design as Semiotic Construction, *Design Studies*, 24, pp. 255–273.

2. Johnson, M. (1993) *Moral Imagination: Implications of Cognitive Science for Ethics*, University of Chicago Press, Chicago.

3. *ibid*, p. 198.

4. Dewey, J. (2002) *Human Nature and Conduct*, Prometheus Books, New York (original publication 1922).

5. Le Dantec, C. and Do, E.Y. (Chapter 6) The Mechanisms of Value Transfer in Design Meetings.

*Reprinted from: Design Studies Vol. 30 No 2 (March 2009), pp. 154–168, DOI: 10.1016/j.destud.2008.12.004

they include ethical value, are transferred between participants in a design process. Dong et al.[6] analyse the differences between what they term 'affect-in-cognition' and more logical types of thinking, concluding that "affect helps us to conditionally and unconditionally value situations with respect to value codes". What both chapters try to show is how feelings about what is valuable are brought out into the open during the process of design and that idea, applied to ethical thinking, is very much in keeping with the approach taken here. Atman et al.[7] also study the more general contexts, including ethics, that designers both draw on and design for.

6.
Dong, A., Kleinsmann, M. and Valkenburg, R. (Chapter 7) Affect-in-Cognition through the Language of Appraisals.

7.
Atman, C., Borgford-Parnell, J., Deibel, K., Kang, A., Ho Ng, W., Kilgore, D. and Turns, J. (Chapter 22) Matters of Context in Design.

How and why do designers imagine their designs? Ball and Christensen[8] find that mental simulation strategies help to reduce uncertainty in the design process, citing the example of Frank Lloyd Wright who was able to "conceive the building in the imagination, not on paper but in my mind…". Medway and Clark[9], along similar lines, develop the concept of 'the virtual building'. This term doesn't indicate virtual reality as we have come to know it – simulation using computers – Medway and Clark talk in more abstract terms about the ontology of buildings yet to be built: "this imagined building, despite being unreal in a physical sense, is *a solid social fact*, something known, often in great detail, to participants, both inside and outside the office, in the activities that cause the building to get conceived, financed, approved and built"[10]. The virtual building is a fluid entity, an agglomeration of thoughts, agreements, and partial visualisations; Medway and Clark describe it as a palimpsest. What is not in doubt is the existence of the virtual building, since this is what is talked about during the process of design. The concept is insightful in that it reveals the social nature of imagination in design; the virtual building is, more than anything, a provisional agreement. That a building (or a pen for that matter) is imagined socially also has ethical consequences since responsibility for the design cannot reasonably be said to fall on a single designer.

8.
Ball, L. and Christensen, B. (Chapter 8) Analogical Reasoning and Mental Simulation in Design: Two Strategies Linked to Uncertainty Resolution.

9.
op. cit., Medway and Clark.

10.
op. cit., Medway and Clark, p. 256, italics added.

The virtual building and the virtual pen – if we extend the idea to other types of design – are constructs that imagination can work within, and importantly where imagined action can have consequences. Such a construct fits very well with John Dewey's idea of how deliberation ('dramatic rehearsal (in imagination)')[11] forms an important component of moral action. "Deliberation is an experiment in finding out what the various lines of possible action are really like", Dewey writes, "the trial is in imagination, not in overt fact". As with Johnson the resonance to designing is striking.

11.
op. cit., Dewey, p. 190.

With these concepts in mind it is the purpose of this chapter to look at how conceptions of ethics intertwine with the unfolding process of design. The chapter argues two things. First, that the social process of designing, and the kind of imagination that designers engage in, can function as a good way of making ethical decisions. Second, that the process of any designing contains within it episodes that can be construed as ethical in nature. To do this, the chapter focuses in detail on four discussions from the DTRS7 data, two from the architectural meetings, and two from the engineering meetings. We show how it is the nature of the discussion, not the specifics of the particular design products, where one must look for the ethical issues in design.

1 METHOD

The method used in analysing the data has been rather simple. All DTRS7 videos were viewed in full; then the transcripts to all meetings were read. In this reading a number of keywords have triggered the identification of episodes that seem to reveal more generic aspects of designing (particularly 'meta' explanations of activity "what I'm trying to do is..."), or a kind of tacit discussion about ethics (words such as 'right', 'correct', and 'good' are indicators here, though some concepts, such as 'intimidation' or 'violence', have obvious ethical content). From the reading about 10 episodes or themes were identified from which five were analysed in detail. This analysis took the form of piecing together the various elements of the conversation into a coherent sub-narrative, before drawing out the more generic and underlying issues. Four main themes were selected for the chapter together with four examples from the transcripts: *relative value in design*, looking at the process of making a designerly judgment between competing aspects of the problem; *prescribing correct use*, looking at how designers discuss their designs as if they were passing legislation; *how form constrains behaviour*, looking at how aesthetic or 'formal' decisions can have ethical consequences; and *the ethics of creative imagination*, how association in thought can operate over ethical boundaries.

2 RELATIVE VALUE IN DESIGN: CAR-PARKING AND THE ENVIRONMENT

Ethical decision making, in design and elsewhere, is often characterised *post hoc* by points at which 'big' life or death decisions have to be made. Yet trying to identify where in the process this happened is often very difficult. Even a sophisticated cost-benefit model of design simply doesn't explain the nature of design activity because it misses the key component of *imagination*. Scruton[12] has been one of the few philosophers to look seriously at designing

12. Scruton, R. (1979) Architecture and Design, in Scruton, R. *The Aesthetics of Architecture*, Methuen, pp. 23–36.

with the observation that, in advance of seeing the designed thing, we find it difficult to make judgements of relative value:

> "In every serious task there are factors which, while of the greatest importance, cannot be assigned a relative value – not because their value is absolute, but because a man [sic] may not be able to judge in advance just when he is prepared to tolerate their remaining unsatisfied."[13]

13.
ibid, p. 25.

How can one gauge, in advance of seeing say, a new road layout, that the value derived from addressing environmental considerations – through routing and planting – should be reduced in order to increase the value of saving lives – through lighting and safety barriers? Scruton argues that the point at which we would tolerate losing life in order to improve the environment is impossible to determine 'rationally', using design methods. Instead he argues for the application of good judgement to arrive at 'appropriate' solutions. Putting aside the question of what 'appropriate' might mean here, judgements of relative value do seem to be at the core of any design process. But how, in reality are these kind of judgements made?

The fine balance between competing design issues is illustrated in the discussion about car-parking at the new crematorium and highlights a trade-off between two putative discourses in design theory, 'usability' and 'improvement'. The usability discourse assumes that a new design should support existing modes of behaviour and hence be based around people's practice. The improvement discourse, however, is about changing people's behaviour, to make it better in some respect[14]. Charles starts off the discussion by reporting a concern that has been voiced that:

14.
Nigel Cross quotes the architect Dennis Lasdun: "Our job is to give the client, on time and cost, not what he wants, but what he never dreamed he wanted; and when he gets it he recognises it as something he wanted all the time…". which suggests that designers are somehow able to meet the unconscious needs of their clients. If this is true it is easy to imagine designers quietly educating and 'improving' people without them realising. Cross, N. (1990) The Nature and Nurture of Design Ability, *Design Studies*, 11, pp. 127–140.

> "people will go into the wrong chapel because as soon as they come in they'll park up and then wander down to this one and then find out the (wrong place they're in) and then walk [to the other chapel] so yes you've got to do it with signage" (A1, 1758–1761).

The solution that has been suggested is one of providing signs to let people know in which direction to go, but Anna suggests that this will not be enough:

> "[you're really suggesting that] if you're here for MR SMITH's funeral go left if you're here for MRS JONES' funeral go right really aren't you? Because … they won't go the way that they're [signposted to] even if you put different chapels … [the] separation has got to be at the entrance" (A1, 1778–1783).

Adam, the architect, although seeming to concur with Anna, actually suggests something that she has already said will not work:

> "it can't be that difficult having signage here that says 'east and west chapel car parks this way' and you have a sign here that says 'west chapel car park straight ahead' and you keep having signs up 'west chapel car park straight ahead' until they get there." (A1, 1807–8110).

Anna mentions that they've tried suggesting a separate entrance for the new chapel, but, because of planning issues, this has not been allowed. Charles replies that he thinks the idea of a 'car park corridor' is a good one, and Adam agrees, suggesting that the 'flowing system' that he's proposing simply extends the 'system' that is already in place, and adding that even if people *were* to miss the first (east) chapel: "it wouldn't be the end of the world" (A1, 1848–1849) if they had to park around a one-way circulation system and walk back. Charles suggests the idea of a cut through road for people lost in the road system but Adam doesn't like it: "It's messy + and Colin Cook wouldn't like his (ancient) hedge messed up" (A1, 1851–1852). This is the first sign of a trade-off between two areas of the design problem/solution: 'usability' on the one had against 'environment' on the other. The design solution suggested for the possibility of people getting lost is being weighed against the consequence of breaking the 'ancient' hedge. Charles comments later that:

> "The reason I mentioned [the problem] was because of this ancient hedge here, I mean if you're going to drive another road through it..." (A1, 1885–1886).

Anna notes that there is already a gap in the hedge, and that other parts of the environment elsewhere, though much more recent, could be developed as an alternative. So there is a judgment of value here, crudely characterised as the value of preserving an ancient environment against the value of making the new thing usable. The trade-off that is made in the discussion is that the road shouldn't disrupt the hedge (for reasons of messiness according to Adam) and that if people get lost they can park elsewhere and can walk back. Whether people will or do actually get lost is an empirical point. In the discussion the idea of people getting lost in the virtual design has been imagined, examined, and added to the palimpsest.

The environment and signage issue briefly re-emerges in the second architectural meeting and it is interesting to see how the previous discussion some months earlier, and undoubtedly forgotten, is recast. The discussion is about the colour and materials that will be used for the roof of the new crematorium. Tony has a preference for copper and lists a couple of options: "you can get it pre-patinated or you can have natural orange copper that sort of changes with time" (A2, 1377–1378). Anna asks about the colour and Tony responds: "you can get [it] in ... turquoise green, or you can get it in green and orange, which is sealed in to stay slightly orange." (A2, 1389) Anna pushes further: "why do you think copper's best then" (A2, 1402) to which Tony responds: "I just think that the turquoise colour of the roof would really suit the colour of the remainder of the building" (A2, 1404–1405). Anna then asks if Tony likes it because it will be a contrast to the existing crematorium's roof: "or just because it would be nicer" (A2, 1409). Tony acknowledges that it would be a contrast, but also that people wouldn't necessarily link the two buildings, however Anna is thinking in more dynamic terms: "people drive past one, it's just nice to have a separate sort of … you could point people out and say 'the one with the green roof'" (A2, 1415–1416).

Anna has, inadvertently or not, hit on a solution here to the original problem of signage. Clearly distinguishing the two buildings in terms of roof colour would be one way of communicating which crematorium to go towards when parking. What's interesting is that there is also an environmental element to this solution, Tony comments: "I think it'll speak for itself in terms of relationship to the trees in the background". Here we have an integrative design solution, able to encompass the relative values usability and environment. Rather than making an explicit trade-off, as Scruton suggests designers must do, design thinking has produced a solution that brings both 'values' together in a single form. The material of copper, chosen by Tony for aesthetic reasons functions both as a sign, by contrasting with the roof colour of the existing crematorium, and as a way for the new building to blend in with the surrounding environment. Perhaps this distinction between buildings will be enough to preserve the ancient hedge, and indeed enhance the existing natural environment.

How design affects the environment is becoming of increasing concern and is often explicitly addressed in design briefs. Yet here we have an example of a solution, slowly worked out through discussion, which brings together disparate problems in a single elegant solution that combines usability, aesthetics, and environment. It is this type of thinking that characterises designing and can clearly apply to ethical situations where there is usually some sort of conflict or resource issue.

3 PRESCRIBING 'CORRECT' USE: WRITING WITH THE PEN

The ways in which design can be used to prescribe behaviour in trying to train users to: "[do] a good job" (E1, 601) can be found in the first engineering meeting transcript (E1, 550–611). This episode comes after a discussion about various methods of controlling the pen, and particularly the way in which the head of the pen can be made to stay in contact with the paper as a user writes with it.

Reflecting on the various methods that have been discussed for controlling the pen head Tommy frames the debate by outlining two options, either: "you take all opportunities for control away from people ... and make the pen do everything" (E1, 552–553) or you: "give some feedback so that people know when they're doing the right thing" (E1, 555–556). The first option essentially gives the user no choice but to use the pen in a prescribed way, while the second option suggests a learning process for the user, training them, through the feedback that is given by the pen, to use the pen in the 'correct' way. Tommy then adds a third option that: "forces you to hold [the pen] the right way" (E1, 559–560) suggesting that the form of the pen could afford the use of the pen in the correct manner.

Alan echoes Tommy's third option that they: "design the shape of the thing, so [the use of the pen] can only be done in one way and that's the correct way" (E1, 572–573). He also subtly discounts Tommy's second option by suggesting that option three would involve: "less sort of learning to be done by the user" (E1, 573) because "if they don't get the results pretty quickly they'll just give up on it" (E1, 574–575). Chad expands this line of exploration by suggesting a mechanism by which lines produced with the pen could be made 'heavier' or 'lighter', concluding that sensors in the pen: "attend to what the child's doing and ... [respond] accordingly" (E1, 590). Todd agrees, commenting: "so you can trick the child into doing..." (E1, 592) The product, then, should give the appearance of control to the user, whilst at the same time limiting the extent of what they can do. Chad, seeming to diverge from Tommy's early view that users should only be allowed to do one thing, suggests that a range of responses from the pen might be possible, depending on the abilities of the child: "even if the child's really uncoordinated basically you get a reasonable result, but if the child's very coordinated you can actually do some quite sophisticated things" (E1, 596–598). Alan picks up on this by suggesting that good performance with the pen should be rewarded: "you could have ... some feedback in terms of colour LEDs on the pen saying that he's done a good job or she's done a good job" (E1, 600–601).

A little later on the designers return to the theme. This time Alan suggests an LED that lights when the pen is being held at the correct angle for use. Robert disagrees with this approach, acknowledging that it might be a training option, but also that: "it's kind of patronising to have this sort of device" (E1, 697) to tell you what to do. A little later, prompted by a new product he is shown by Alan, Todd states that: "there are pens that you do have to hold in a certain orientation" (E1, 715) to which Sandra responds: "yes ... it's really hard because I don't hold my pen in the way that they expect you to... I scrunch all my fingers up at the top, whereas that one expects that you do more + [*demonstrates a conventional position*] that + which I don't do" (E1, 716–720). Alan jokingly notes that: "if you had … you would write in the right way" (E1, 723) with Chad adding, again jokingly, "you would do it properly" (E1, 724).

The discussion continues but it is clearly an issue that is of concern to the designers as they return to it on a number of occasions. Jack suggests that: "you design it in such a way that that's [*demonstrates*] a reasonably comfortable position and if you don't put it in that position it doesn't work" (E1, 813–815), while Todd suggests that a possible function of the pen might be to: "teach people to hold a pen properly" (E1, 867–868). Later on Robert remarks that he's: "seen a pen … where it's actually got this kind of loop in it so that if you hold it with your index finger … it supports the rest of your hand" (E1, 974–980). Alan asks: "so you almost wear it like a glove?" (E1, 983), to which Robert answers: "yeah, or sort of like a ring on your finger with a pen attached to it, but *it forces you to hold it in a very specific way*" (E1, 984–985, italics added).

Wrapped up in this example are fundamental issues about what design seeks to achieve in prescribing behaviour. Indeed the example also contains, in a number of judgments, the designers' own realisation of this; Todd thinks the idea of reinforcing correct behaviour is 'patronising' while Sandra notes that her own way of holding a pen would not be possible with such a device, even invoking a 'they', distancing herself from the designers of previous devices.

15.
Verbeek, P.-P. (2006) Materializing Morality: Design Ethics and Technological Mediation, *Science, Technology, and Human Values*, 31, pp. 361–380.

What the designers are doing is writing a kind of script[15], a sequence of things that the users of the pen must do if they are to use the pen 'correctly'. What we see is the consideration of the user's interests: Should we force them to do this? Should we give them a choice? Should we help them learn? Or should we just leave it to them? This consideration, again essential to the design process, is a fundamental ethical thing to be doing, working out what the consequences are for any given design decision. And this is all carried out in this imaginative social space that we are terming 'the virtual pen'. The underlying agreement here, though, is that formal properties have controlling effects. Indeed near the beginning of the whole episode Alan explicitly links the 'shape' with 'doing the right thing'. Doing the right thing here, of course, means getting the pen to work properly as it was intended by the designers. This is not an ethical 'right thing', but a technical 'right thing'.

16.
Winner, L. (1980) Do Artifacts Have Politics?, *Daedalus*, 109, pp. 121–136.

The references to right and wrong, however, are interesting because we normally associate right and wrong with some kind of law or principle. What the designers are doing is creating legislation in the form of an artefact; constructing what is 'right' and 'wrong'. That artefacts have a political dimension is not a new idea, Winner[16] notes various examples and Grint and Woolgar[17] discuss in detail the script written into a personal computer. However, to see how this inscription is negotiated in practice during the design process is striking.

17.
Grint, K. and Woolgar, S. (1997) Configuring the User: Inventing New Technologies in Grint, K. and Woolgar, S. (eds) *The Machine at Work: Technology, Work and Organization*, Polity Press, Cambridge, UK, pp. 65–94.

4 CONSTRAINING BEHAVIOUR: WAITING AND FIGHTING

The previous example showed, in general, the way that designing is used to construct laws, and this has obvious application for ethical thinking in areas where the 'right' and 'wrong' *is* ethical. Where designers know in advance that undesired outcomes are possible, design can be used to prevent bad situations occurring. The following example shows how design decisions can influence (ethical) behaviour.

Early on in the first architectural meeting the waiting area in the new building is discussed and we see directly how architectural form can affect human behaviour. As services at the crematorium take place throughout the day, often one after another, there needs to be a waiting area where people can gather before their service begins, and while the previous service ends. The size of the waiting area that Anna has specified is of concern to Adam, who thinks it looks: "kind of small" (A1, 93). Initially Anna is unconcerned, drawing on her experience of the existing crematorium she concludes that the: "eight to ten, maximum twelve seats" (A1, 114) that Adam suggests are maybe too many. However, after Adam suggests increasing the space Anna nuances her position, acknowledging that while the space would be adequate for the majority of the time she could envisage situations that would require more room particularly when: "people ... arrive and ... keep in their little groups, they don't want to mix with other people" (A1, 134–135). A larger space would allow people to keep themselves separate from one another. Anna suggests that the space should have a: "feeling of keeping [people] segregated" (A1, 135).

Before we go further with this it is interesting to briefly note a number of points here, chiefly concerning the tacit responsibilities of the architect. First it seems obvious to point out that, although we normally think of 'space' – particularly 'public space' – as something that brings people together, 'space', in buildings, segregates as much as it integrates. The requirement that has become evident here, through discussion, is that the space of the waiting area should help to keep people apart if they wish to be apart. In this respect the architect is being asked to reinforce what Anna has observed to be 'normal' behaviour; the fact that people who come to a funeral service might not know each other, and might not want to know each other. A waiting space where everyone could see everyone else might be advantageous in a building such as a theatre but for the new crematorium, Anna is emphasising, such an 'open' space would be inappropriate.

Although an uncontentious design issue here – Adam is clearly willing to respond to Anna's request – this is nevertheless a decision point where, safely back in the office, an architect might intervene for a number of reasons. Having a space where people could hide from one another might go against the aesthetic vision of the architect[18] or the architect might feel it their job to 'improve' behaviour through design: that people *should* see and talk to each other. The architect, then, has to decide whether to support existing behaviours with space – behaviours that they might personally think strange or reactionary – or whether to try and use design to change or modify the behaviour of building users.

18.
Adam is clearly, as Oak also points out, an architect whose aesthetic vision has consequences for the nature of the space: "I felt that *architecturally* it needed to be a great deal more bold" (A1, 337–338); "I think we'd have to review this whole area and the chances are we couldn't retain the *symmetry*" (A2, 685–686); "I might still be able to integrate it into the *architectural form* [as] I've got other *bold forms* on the *axis*" (A1, 725–726); "I'm trying to keep the spaces *pure*" (A1, 819); italics added in all quotes.
Oak, A. (Chapter 17) Performing Architecture: Talking 'Architect' and 'Client' into Being.

The consequences of not providing a space with a 'feeling of keeping people segregated' are made explicit, as Anna continues, countering her own intuition that: "you'd think [a funeral] would bring [families] together" (A1, 140–141) with the comment that not keeping people separated: "can escalate to sort of violence at times" (A1, 149). This is extreme behaviour, but Anna is implicitly making a link here between the nature of a space and its potential to create, in this case, violence. What is interesting again to consider is the responsibility of the architect, and indeed the client that agrees to the new scheme. Say, for example, the architect did disagree with the client in this instance, feeling that a space with a feeling of segregation would unduly compromise their aesthetic vision. And say that, a couple of years after completion of the building, a fight did indeed break out, which ended with someone being badly injured. How would we then weigh the responsibilities of the individuals concerned in the fight with those of the architect, particularly as the issue had been discussed? Perhaps it is interesting just to note the ethical consequences of design decisions here; how easily an aesthetic decision can lead to having moral consequences. A building is not an ethically neutral thing in this sense. One could even push the point further here and suggest that an aesthetic decision is also an ethical decision since form and behaviour *are* so closely linked.

5 THE ETHICS OF CREATIVE IMAGINATION IN DESIGN: 'BAD' USES FOR THE PEN

To pick up on Dewey's phrase mentioned in the introduction, one feature of the virtual product is that it leads to 'trials of the imagination', not action 'in overt fact'. The virtual product allows the designers to 'play' with the object, to see where it might take them. In this play of imagination they cross easily over ethical lines which might be unacceptable in 'overt fact'. The process of design, in fact, allows them – even forces them – to think in ways which we might consider unethical (if we allow the possibility that thinking could be unethical), and this is necessary since it allows the designers to think about, and work out, the possible bad consequences of their design. This is not done from within any clear moral framework, following any guidelines or checklists, if anything it is done with an intuitive sense of right and wrong.

An example of this behaviour occurs in the second engineering meeting. Extract 1 below explores what kind of representations can be stored in the (virtual) digital pen:

Extract 1, E2, Associative thinking leads to unethical consequences.

Jack	can it store things like digital signatures
Tommy	erm store ... what in the pen
Jack	yeah
Tommy	so
Jack	store your own signature
Tommy	you could + it would be quite narrow, but you could print anything you
	like yeah erm if you could download to it then you could er download
	any image you like ( ) small number of dots this might have thirty
	forty dots erm that's the sort of thing that erm
Jack	you need you need to remember you know either unique things or
	long numbers
Stuart	or forge signatures with it
Jack	mmm or forge a cheque

This extract is interesting because it shows first, how one concept associatively slides into another (signature > any image > unique things > long numbers > forging signatures > forging cheques) and second, how the concept that is slid into might be what we'd consider to be somehow unethical. Jack first asks if the pen can store digital signatures, before qualifying by saying 'your own' signature. Tommy confirms that it can, before going into the technical limitations of the pen if it were to do this. The word 'signature' has obviously set off a possible use for Stuart, who suggests quite naturally that you "could forge signatures with it". Jack, concurring and taking this line of reasoning forward, suggests that being able to print signatures would allow the possibility of forging cheques.

The discussion moves on, and no-one comments on the possibility of the pen being used to forge cheques, which is interesting in itself. No judgement is made about it being right or wrong, or the pen being used for other malign purposes. Indeed what is striking is the way in which the discussion *does* move effortlessly on. The 'forging cheque' scenario is left unresolved, perhaps because to make a definitive judgment about its ethical nature would be to put a brake on the associative process driving the underlying technical discussion. The discussion, as Tommy's interjection (E1, 304–307) shows, is about exploring how the creative possibilities of the pen can be matched with current technical possibilities. The fact that what we might consider an unethical possibility arises during the course of the discussion about the use of the pen is almost a by-product. In fact, a detailed analysis about how the pen *could* be used for forgery might have been another possible pathway to shed light on this underlying technical discussion.

These few lines show how ethical issues are closely intertwined with other issues during the creative process of design. But the exploration of ethical issues does not occur as Dewey would have it. The designers are not using the virtual product to explore ethical consequences, indeed they show no signs of even being aware that this might be an ethical matter at all. They have simply happened upon a particular use for the pen stepping easily, as we mentioned earlier, over a putative ethical line.

It is instructive to think out a line of consequences here. Suppose the pen that is designed and produced is able to store handwriting, and suppose it is found to be an exceptionally good means of forging signatures, and suppose that several million pounds are stolen using it. Where would that leave the designers if we return to their forgery discussion? Would they be responsible in any way for this crime, having previously envisaged its possibility?

There are other examples of this kind of thinking. In the first engineering meeting the subject of bullying arises. Tommy initially mentions the idea of kids playing with the pen in the back of a car and Patrick follows with: "yeah, but you'd get lots of burn holes in your car seats" (E2, 971). The idea of burning triggers another thought as Tommy suggests that: "if you were a school thug you could pin people to the ground..." (E2, 974). There is laughter, then Tommy realises that what he has suggested could also be done with a simple pencil, which in turn allows Patrick to conclude that: "we probably ought to go through this list and look at which ones you can actually do with a pencil" (E2, 981–982). In this quick exchange of comments we again see not only the fluidity of association but how, in imagination, designers alight upon situations that, if they were real, would be considered in ethical terms. For the designers, the school bully situation described by Tommy is simply a means of getting to somewhere else, not a scenario to explicitly consider in itself. Nevertheless, as with the forgery example, the question of the designer's responsibility *should* the pen turn out to be a useful implement for bullying is interesting, especially as they have imagined that it *could* be used for bullying.

The heat generated by the pen's print head is an issue that reappears at various points in the two engineering meetings. The safety consequences of having a hot surface are mentioned a couple of times. Tommy remarks that: "there's a potential for a safety issue" (E1, 1444) and Roman suggests the idea of having a barcode to control the heat (E2, 1134). The idea of heat triggers other associations as well. An interesting example comes in the second engineering meeting when Rodney suggests that brail could be a possible application area for the pen: "isn't there a big demand for being able to print brail?" (E2, 846) he asks. The heat association prompts Stuart to comment: "you could make blisters in the paper … from the heat" (E2, 858–860). The idea of heat at the tip of the pen, then, has been imagined in a number of what we might consider ethical contexts. The pen might be used for intimidation, or it might be used to print brail for blind people – two ends of an ethical spectrum. The discussion, based around the virtual pen, could thus be construed ethically; 'what can we do that's *bad* with this pen?' versus 'what can we do that's *good* with this pen?'. Yet these remain ethically neutral concepts to the designers. We might assume a tendency on the part of the designers to seek 'good' applications for their virtual product, but the data would suggest otherwise.

6 DISCUSSION

This chapter set out to explore the relationship between design thinking and ethical thinking in two ways. First, by looking at the ways in which designers talk about ethical issues. Second, by considering the role that imagination plays in the process of design. Ethics was defined in three ways: acting in the world to change behaviour, the imagination of alternative actions and their consequences, and the evaluation of those consequences in terms of good or bad. The chapter has shown how designers engage with, and resolve, ethical issues in the process of design (although they are often not aware that they are *ethical* issues) as well as how a 'virtual design' is socially created, allowing the consequences of the design to be explored in the careful construction of the design 'script'.

Any discussion of ethics and design naturally centres around questions of what is valuable, but it is the means by which designers are able to discuss what is valuable (and right) that has been of interest. Johnson[19] places imagination at the centre of ethical decision-making and it is clear that the designers studied in the DTRS7 data have a well-developed ability to construct, virtually, a vehicle with which to think out the consequences of making certain choices; a vehicle that also allows participants to share their expertise in reaching common agreement. This is a very focused form of imagining, and here lies the relevance of design to ethical thinking. Designers are not just being creative for the sake of it,

19. *op. cit.* Johnson.

generating endless concepts, but are very consciously exploring a range of options, all the while looking for ways to integrate competing values into a solution ("is there anything you get for free" (E2, 1742) as Jack asks in the second engineering meeting). It is simply this ability, to hold an incomplete solution in mind and work through different consequences of actions, which makes the process of designing prototypical of the kind of ethical thinking that Johnson describes.

Moreover, the data has shown how, even where ethics is not explicitly mentioned, there are episodes that can be construed as being about ethical issues; spaces that avoid conflict, for example, or malign uses of the digital pen. The generic aspects of design – that it legislates and changes behaviour – ensures that ethical issues are never far from the surface, even if the products under discussion aren't obviously 'ethical' products that might, for example, be explicitly to do with safety or sustainability.

One issue that has been referred to at several points during the chapter are the responsibilities that designers might have as a result of being able to imagine the bad consequences of their designs should those bad consequences come to pass (and with the designer having done nothing to prevent them in the design process). Is it enough to be able to envision consequences without necessarily acting on them? This is perhaps an area where design and ethics diverge and where designers, as well as people more generally, need an explicit ethical framework within which to make decisions. Without an explicit ethical framework upon which to draw, decisions about whether to act or not act – in 'overt fact' or imagination – tend to be made with 'ethical intuition'[20]. The responsibilities of designers here would be no less than anyone else making an informed decision about a future state of affairs affecting people's behaviour.

20. Lloyd, P. and Busby, J. (2003) 'Things That Went Well – No Serious Accidents or Deaths' : Ethical Reasoning in a Normal Engineering Design Process, *Science and Engineering Ethics*, 9, pp. 503–516.

From the data ethics in design would also appear to be about what one might consider the extreme consequences of design – fighting breaking out in waiting areas before a funeral, for example – and the use of humour in the meetings often arose from the discussion of these types of consequence. Jokes tended to be made when imagination had led to strange or unexpected consequences, away from 'normal' behaviours, where people's response was often simply to laugh. What is interesting is that part of ethics in design *is* about extreme behaviours and unexpected consequences, so there is some kind of link here between negotiating ethical subjects and the use of humour. This might also, of course, be a mechanism to avoid explicitly talking about ethical consequences.

Nevertheless, the issue of responsibility once a possible ethical consequence has been imagined is important to resolve, and one way of doing this might be to argue that the process of designing is *intrinsically* responsible, forcing extreme consequences to be considered and 'designed for'.

A final remark concerns the research methodology used in the chapter. The qualitative methodology, sketched in Section 2, identified episodes from the data that could be construed as addressing ethical concerns. While the episodes were obviously based on direct quotation the analysis of those episodes often posed the question, as the designers themselves did in the meetings, 'what if…?' This way of looking at the possible consequences of situations that aren't, in fact, discussed (and that the data could never support) but are nevertheless plausible, is something that deserves comment. It is an approach that would seem particularly well suited to empirical design research – and indeed is employed to good effect by Matthews[21] – in that it allows the researcher to, in effect, retrospectively join the design process, pointing out the ways in which the process could have gone. In a sense this involves creativity on the part of the researcher, imagining possible consequences, and exploring them from a particular theoretical perspective. That perspective was an ethical one in this chapter, but it could equally have been one focussed on creativity, for example, or decision-making, or even product alternatives. This could be construed as a designerly approach to research and it is an approach that could be used more often in the field of design research; the use of actual design processes to investigate questions about the possibilities of design.

21.
Matthews, B. (Chapter 2) Intersections of Brainstorming Rules and Social Order.

6
The Mechanisms of Value Transfer in Design Meetings*

Christopher Le Dantec & Ellen Yi-Luen Do

Values play an integral role in design: they inform the kinds of trade-offs the designer makes when considering different solutions; they create a basis for the client to assess how a particular artefact may fit into their lives; and they are an important part of negotiating a common understanding in collaborative design settings. In this chapter, we present a grounded analysis of a collaborative architectural design process. We examine the interactions between architect and client to better understand how different values are brought into the design discourse. By analysing the verbal content and non-verbal communication between the architect and client, we identify patterns of discourse that imbue design problem-solving with the language and concepts that express values. From this analysis we develop a theory of value transfer and describe the social mechanism that facilitates this transfer during design negotiation. Our theory provides an observational basis for understanding value transfer in the context of collaborative design and is relevant to design domains beyond architecture.

Values play an important role in design; tracing back to the Roman architect Vitruvius, the values of "firmitas, utilitas, venustas" – stability, utility and beauty – were imbedded in the early codification of architectural practice[1]. Looking beyond architecture, design practice, from industrial design to interaction design, is deeply steeped with questions of values. It is through the process of design that values are exposed and negotiated in the search for potential solutions. The presence of different values in turn affects the adoption, use, and social impact of a particular designed artefact.

The kinds of design inquiries that encourage a broader consideration of values in design have only emerged in the last decade. Going back to the early 1990's, Lawson[2]. and Rowe[3], for example, have each contributed to work focusing on design process and the act of designing. These works largely considered design as an individual activity performed by a designer and as such, do not provide a good perch for considering the collaborative nature of design. But by the middle of the 1990's the notion of design as a collaborative activity had become a topic of study. During this period, Brereton et al.[4], Cross and Cross[5], and Radcliffe[6] all studied the affects and mechanisms of team-based design, yet in these studies the focus was still on small teams of designers and not on the interaction designers have with clients or other stakeholders.

1. Pollio, V. (1914) *Vitruvius, the Ten Books on Architecture*, Harvard University Press.

2. Lawson, B. (1990) *How Designers Think*, Butterworth, London.

3. Rowe, P.G. (1995) *Design Thinking*, MIT Press.

4. Brereton, M.F., Cannon, D.M., Mabogunje, A., and Leifer, L. (1996) Collaboration in Design Teams: How Social Interaction Shapes the Product in Cross, N., Christiaans, H. and Dorst, K. (eds), *Analysing Design Activity*, John Wiley and Sons, pp. 319–341.

5. Cross, N. and Cross, A.C. (1996) Observations of Teamwork and Social Processes in Design in Cross, N., Christiaans, H. and Dorst, K. (eds), *Analysing Design Activity*, John Wiley and Sons, pp. 291–318.

6. Radcliffe, D.F. (1996) Concurrency of Actions, Ideas and Knowledge Displays within a Design Team in Cross, N., Christiaans, H. and Dorst, K. (eds), *Analysing Design Activity*, John Wiley and Sons, pp. 343–364.

*Reprinted from: Design Studies Vol. 30 No 2 (March 2009), pp. 119–137, DOI: 10.1016/j.destud.2008.12.002

Looking specifically at visual design, Frascara[7] and Tyler[8] each addressed the role of values and audience. Frascara considered graphic design primarily as an activity of persuasion and asserted that graphic artefacts should be considered on more grounds than aesthetics alone. Tyler echoed a similar view in the scope of visual communication design, and further discussed how the values of the audience influence their interpretation of the design and the persuasive power it possesses. Both of these views are borne of a semiotic understanding of visual design and the rhetoric of the image[9]. By considering design in this manner, Frascara and Tyler elevated the discussion of design to include the human values expressed by the designer and interpreted by the consumer.

Despite these moves to acknowledge the human values imbedded in a designed artefact, the fact that these values are an integral part of the process continued to be overlooked. In Brereton et al's Delft Protocol analysis[10], designers were noted to make appeals to values, for example Kerry, one of the designers working on the design of a bicycle rack, made an appeal to elegance. This appeal to a design value was bold, yet at the time, there was no deeper analysis of how appeals to such values contributed to the design. Similarly, Cultural Probes have received much attention in the interaction design community for their ability to generate inspirational responses from a user population, yet there has not been much investigation into how the results of the probes are incorporated into the design process[11]. In both cases, the presence of different types of values is tacitly understood, but the role those values play during the act of designing have not been thoroughly investigated.

Much of the work considering human values comes at a time when the broader field of design is in the middle of an evolution: the consumer is becoming a 'co-designer'[12]. This change indicates a shift in how values are reflected in the design process. Where a consumer once took what was given, the co-designer is empowered to accept and reject design choices much earlier in the process, thus exercising an increased influence on the shape of the final product[13].

In the domain of Human-Computer Interaction (HCI), more researchers are considering 'value' as an integral part of design and evaluation. The conversation about values in design started with Suchman's seminal article *Do Categories Have Politics?*[14] in which she lays the foundation for discussing how values are built into software systems. Suchman's work is rooted in the participatory design tradition that emerged in the 1960's and 1970's[15]. This same tradition provides the underpinning Sanders identifies as motivating design disciplines toward co-design. The relevance these works have to exploring values in design is the collaborative

7.
Frascara, J. (1995) Graphic Design: Fine Art or Social Science in Margolin, V. and Buchanan, R. (eds), *The Idea of Design*, MIT Press, pp. 44–55.

8.
Tyler, A.C. (1995) Shaping Belief: The Role of Audience in Visual Communication in Margolin, V. and Buchanan, R. (eds), *The Idea of Design*, MIT Press, pp. 104–112.

9.
Barthes, R. (1977) *Rhetoric of the Image: Image, Music, Text*, Fontana, Glasgow, pp. 32–51.

10.
Brereton, Cannon, Mabogunge and Leifer (1996), *op. cit.*

11.
Gaver, W., Dunne, T. and Pacenti, E. (1999) Design: Cultural Probes, *Interactions*, 6, pp. 21–29.

12.
Sanders, E.B.N. (2005) Information, Inspiration and Co-creation in *Proceedings of the 6th International Conference of the European Academy of Design*, University of Arts, Bremen, Germany.

13.
Sanders, E.B.N. (2006) Design Research in 2006, *Design Research Quarterly*, 1, pp. 1–8.

14.
Suchman, L. (1997) Do Categories have Politics? The Language/Action Perspective Reconsidered, *Human Values and the Design of Computer Technology*, Center for the Study of Language and Information, Stanford University, pp. 91–106.

15.
Sanoff, H. (1973) *Integrating User Needs in Environmental Design*, National Institute for Mental Health, Raleigh North Carolina State University.

nature of participatory design; it is through these roots that asking questions about values becomes easier since the collaboration between designer and client (or user) is explicitly recognised as a goal of the process.

As a result, work in HCI has produced different approaches for coping with values in the design and evaluation of software systems. Friedman's Value Sensitive Design (VSD)[16] is a methodology that proposes engaging design with conceptual, empirical, and technical investigations to identify and address values in software systems. Another framework, from Blythe and Monk[17], suggests that technology designed for the home – a nascent context of inquiry for HCI – should be analysed using three scales: enjoyability, inclusivity and recodification, which stand in contrast to the traditional scales used in HCI of efficiency and productivity. What these efforts demonstrate is a recognition of human values as crucial to the experience of using technology and a concerted effort to account for them across different contexts of use.

16. Friedman, B. (1996) Value-sensitive Design, *Interactions*, 3, pp. 16–23.

17. Blythe, M. and Monk, A. (2002) Notes towards an ethnography of domestic technology, *Proceedings of the Conference on Designing Interactive Systems: Processes, Practices, Methods, and Techniques*, ACM Press, London, pp. 277–281.

Looking outside HCI, social researchers have examined how technologies emerge in society. Social Shaping of Technology (SST), a theory put forward by Williams and Edge[18], asserts that technology is developed through the negotiation of social, technical and economic factors. In this regard, VSD and SST are similar as they both emphasise the interplay between the development of a technology and the social context that gave rise to, and eventually adopts that technology[19]. We consider technology, here, as any intentionally designed artefact and do not limit the definition to computational devices. The compelling argument in theories like SST is the light they shine on the intersection of human values and designed artefacts. It is a move away from technological-determinism towards a more nuanced understanding of how society shapes design as much as design shapes society.

18. Williams, R. and Edge, D. (1996) The Social Shaping of Technology, *Research Policy*, 25, pp. 865–899.

19. Friedman, B. and Kahn, P.H. (Jr.) (2003) Human Values, Ethics, and Design, *The Human-Computer Interaction Handbook: Fundamentals, Evolving Technologies and Emerging Applications*, Lawrence Erlbaum Associates, pp. 1177–1201.

In looking at the DTRS7 architectural design meetings, our goal was to establish a better understanding of how architectural practice incorporates values into the design process. By analysing one specific design activity, we sought to create an understanding of value transfer that can be applied to other design domains.

1 DEFINITIONS

In this chapter, we examine the role values play in design. Before presenting our analysis, and after already having used the word 'value' extensively, it is important to acknowledge the breadth of meaning encompassed by the term. For some, values are ethical considerations; in Lloyd's account of values in the design process[20], he calls attention to judgments that are made in relation to

20. Lloyd, P. (Chapter 5) Ethical Imagination and Design.

ethical considerations. Other considerations of value may involve economic factors and whether something is, or is not, a good value. We are considering a set of motivations that may be ethical in nature, as well as those that could be construed as a value-add (for example a professional skill or experience that might be sought for inclusion on a development team).

Broadly, we define values as the principles, standards, and qualities that guide actions. These may be personal, cultural, or professional. For example, the decision to avoid disrupting the habitat of local floral and fauna during the design of the crematorium crosses both ethical and professional value lines: the ethics of preserving the natural environment and the professional judgment of how to integrate the design within the physical constraints[21]. Regardless of where the emphasis is placed, values motivate the decision and guide the actions of the designer and client. Ultimately, values serve as the basis for how designer and client assess the design.

21. ***ibid.***

The kinds of instances that we are associating with the communication of values include: assertions of form or aesthetics, descriptions of how people are to use the space and anecdotes that illustrate the human condition behind the function. We are referring to communication about these aspects of the design as 'design discourse'.

Finally, we use the word 'client' to refer to the person who conveys the needs of end users and owners. The client represents stakeholders and communicates concerns of value assessment or judgment with the designer.

In looking at value transfer, we need to understand the different types of values contributed by designers and clients. The values the designer brings to the design meeting include professional expertise, knowledge of the design domain, and the personal values that make up their individual character. Likewise, the client comes to the design meeting with notions about how the artefact will be used and how it will fit into their lives. Some of the client's values will correspond with those of the designer, while some values will be foreign to the designer. It is our assertion that in order for the designer and the client to come to agreement on a suitable solution, each must begin to understand the other's values. As we will see, this exchange of values occurs more vigorously during analysis and synthesis phases of design. As the design evolves toward completion, these values are used in the design meetings to further define, validate, and assess the proposed design solutions.

It is instructive to consider our notion of value transfer in the light of Nelson and Stolterman's taxonomy of design judgments[22]. Many of the values that we are pulling out of the design discourse

22. Nelson, H.G. and Stolterman, E. (2002) *The Design Way: Intentional Change in an Unpredictable World: Foundations and Fundamentals of Design Competence*, Educational Technology Publications, New Jersey.

may appear to be various forms of design judgments – especially those that designers would recognise as the result of training and experience. We believe, however, that these values are the underpinning for design judgments; they complement the Nelson and Stolterman taxonomy and present a way of understanding how design judgments develop in the context of exchange that occurs between designer and client.

2 DATA AND METHOD OF ANALYSIS

We chose to focus on the two architectural meetings because they consisted of direct contact between the lead designer, Adam, and the clients, Anna and Charles. Over the course of the two meetings we were particularly interested in identifying how each party talked about values. In order to begin to understand the type of social transactions that enable value transfer, we undertook an approach based on Grounded Theory. Grounded Theory is a systematic methodology of qualitative data analysis where the analysis begins without any pre-supposition of what results will be found in the data. Instead, patterns that exist in the data are brought forward through rigorous iterative coding. The goal of Grounded Theory is to end up with one central code, the theory, which relates all observed behaviours[23,24]. In applying this approach to the protocol data, we examined the transcripts iteratively, applying open coding where we identified and categorised phenomena observed in the transcripts, and axial coding which focused on identifying causal relationships between the set of open codes.

23.
Miles, M.B. and Huberman, A.M. (1994) *Qualitative Data Analysis: An Expanded Sourcebook*, Sage.

24.
Strauss, A. and Corbin, J. (1998) *Basics of Qualitative Research: Techniques and Procedures for Developing Grounded Theory*, Sage, London.

We had some idea of a hypothesis – that values are an important part of design discourse – and we were looking for events in the data that might support that; therefore, we diverged from adhering to Grounded Theory in the strictest sense because we were interested in paying specific attention to the following events:

- verbal exchanges that explicitly revealed values to be reflected in the final design,
- verbal exchanges that were implicitly about values and their relation to the design,
- verbal cues that indicated one or the other party understood a particular value.

Phrases were the key unit of analysis. In some cases, a concise phrase communicated a value clearly, as in Extract 1, where Adam's assertion of 'design purity' is clear on its own.

Extract 1, A1, Example of DESIGN VALUE, PURITY.

817	Adam	well it's not as pure a summation as I was looking for but I mean

In other cases, our comprehension of the significance of the coded phrase benefits from considering a larger section of verbal exchange to provide context or clarity. Extract 2 is an example of such a situation where Anna's description of the human values involved during a ceremony start with a short phrase (A1, 140), but benefit from considering the discourse that follows to further illuminate the details of the human values involved.

Extract 2, A1, Example of HUMAN VALUE, JEALOUSY.

140	Anna	police attendants quite often you know you'd think it would bring
141		them together but it actually makes it worse
142	Adam	really gosh
143	Anna	yeah and they sit separately in the chapel as well it's all to do with
144		money and you know they've left someone something wonderful
145		that's most of the time what it is or the other family are cross because
146		one family has arranged it and they used they never visited her while she
147		was alive and how dare they get involved with this and it all escalates

We used the video recordings of the design meetings to clarify ambiguous statements in the transcripts. Through this iterative process, we categorised the social transactions into a set of codes describing the types of exchanges of interest. These open codes were in turn refined into a set of five axial codes (see Table 1) that enabled us to clearly delineate subjects of discourse.

The first axial code is labelled DESIGN VALUES and includes the open codes AESTHETIC, UNIQUENESS, PURITY, FORM, SOLITUDE and MATERIAL. These codes describe parts of the discourse that touch on

Axial codes	Open codes	Description
DESIGN VALUES	FORM, MATERIAL, AESTHETIC, UNIQUENESS, PURITY, SOLITUDE	Applies to comments about architectural purity or vision, to form and material, as well as perceptual awareness.
HUMAN VALUES	SPIRITUALITY, RESPECT, JEALOUSY, FAMILY, RELIGION, MOURNING, COMFORT, TRADITION	Identifies phenomenological experience and symbolic meaning comments that may or may not directly result from the designed space.
REQUIREMENTS	ACTIVITY, SPATIAL, PHYSICAL, REVIEW	Reserved for comments that addressed functional needs or activities that take place in the designed space.
NARRATIVE	DIRECT SUPPORT, INDIRECT SUPPORT, PROCESS DETAIL, JUSTIFICATION, TANGENT	Used to identify anecdotes that either designer or client engaged in during the discourse.
PROCESS	COMMUNICATION, PROBLEM-SOLVING	Delineates meeting activities concerning meeting mechanics or when additional research would be needed.

Table 1. Axial and open codes.

values primarily originating from the designer. FORM and MATERIAL specifically address physical characteristics of the building. AESTHETIC, UNIQUENESS, and PURITY address values of how the building relates to its surroundings. SOLITUDE was used to capture the phenomenological experience of the crematorium and represents an aggregation of values like privacy and reclusiveness.

The HUMAN VALUES axial code includes codes for SPIRITUALITY, RESPECT, JEALOUSY, FAMILY, RELIGION, MOURNING, COMFORT, and TRADITION. Each of these codes were used to represent either the desired phenomenological experience of the designed space, or a description of how the human condition impacts the activities that take place in the building. In Extract 2 above, Anna identified the emotional tenor of the waiting area. Her description of tension between family members exposes some of the values that accompany mourning – in this case, JEALOUSY over an inheritance and inequity in care-giving during illness. Adam, as the designer, must consider how these values will relate to the built space. What is revealed here, as Lloyd points out in his analysis[25], is an alignment with a particular understanding, or valuing, of space that enables privacy.

25.
Lloyd (Chapter 5), ***op. cit.***

The axial code REQUIREMENTS contains codes for ACTIVITY, SPATIAL, PHYSICAL and REVIEW. These codes were used when the design discourse touched on the basic functional requirements of the design and were noted as being the target for value-laden statements. ACTIVITY and SPATIAL requirements captured, for example, the flow of pedestrian and vehicle traffic, and the spatial requirements that would enable that flow. The PHYSICAL code captured requirements like needing a re-usable space to display religious objects (see Extract 3).

Extract 3, A2, Example of REQUIREMENTS, PHYSICAL.

Charles	yeah what about religious ( ) religious symbols
Anna	yeah I mean we'll be inviting the inter-faith groups and we've just had the Sikhs donate err- a symbol to us as well er and so it's just trying to think about how we would allow a symbol to be shown that would be removable in a sense or something like a cross because it can't be + one faith
Adam	well there's a couple of ways of doing it you could add the symbol on the plasma TV screens

NARRATIVE codes include DIRECT SUPPORT, INDIRECT SUPPORT, PROCESS DETAIL, JUSTIFICATION, and TANGENT. These codes identify pieces of text in the transcripts that support functional requirements for either of the axial code groups for values. They represent the anecdotes and justifications offered in support of a particular idea.

The final axial code, PROCESS, contains the codes COMMUNICATION and PROBLEM-SOLVING. The code COMMUNICATION was used to identify instances when either client or designer referred to communication with stakeholders who were not present at the meeting. PROBLEM-SOLVING was used to identify design discourse that centred on defining functional requirements. The codes in this grouping do not share a particularly strong affinity and indicate the need to consider a more comprehensive study of collaborative design, particularly one that includes all designer-client interactions, from project start to completion.

These codes were refined through four repeat processes that analysed the data from scratch. Each iteration occurred after spending several weeks away from the data and had a high degree of test-retest.

3 RESULTS

In developing a theory of value transfer during design meetings, we found it useful to examine trends of code occurrence across the two meetings. By looking at these trends, we were able to identify large-scale phenomenon and relate it to the specific design discourse that indicated value transfer. The summary of events for each axial code for A1 and A2 can be found in Tables 2 and 3 respectively. These tables display the number of behaviour codes contributed by each participant and the percentage of their total contribution; e.g. in Table 2, Adam contributed 53 instances of codes in the DESIGN VALUES code which is 45.7% of his total contribution to the coded discourse. The right-most columns in Tables 2 and 3 show the distribution of all codes between the participants and give an idea of where the action took place during the design discourse.

Both meetings exhibited roughly the same pattern. Across the two meetings about 30% of the coded events were DESIGN VALUES, 10% HUMAN VALUES, 30% NARRATIVE, 20% REQUIREMENTS and around 10% PROCESS. The total contribution of coded events was roughly even between architect and client as seen in Tables 2 and 3. A closer examination of Tables 2 and 3 shows that in meeting A2 there was a decrease in the number of coded DESIGN VALUES (30% down from 31.1%), HUMAN VALUES (6.9% down from 10.7%), and REQUIREMENTS (16.9% down from 20.9%), and an increase in events coded as NARRATIVE and PROCESS (from 26.7% to 33.1% and from 10.7% to 13.1% respectively). Broadly, these numbers show that by the second meeting there was a decrease in discourse about requirements and values.

3.1 *Indications of value transfer*

Another, perhaps better, indication of the content of the meetings can be found by breaking down contributions by axial code.

Tables 4 and 5 show how each participant contributed to the content of the meeting. The percentages in these tables are derived from the data in Tables 2 and 3; for example from Table 2, Adam's 53 DESIGN VALUE codes are 75.7% of the total 70 DESIGN VALUE codes recorded in A1.

	DESIGN VALUES	HUMAN VALUES	NARRATIVE	REQUIREMENT	PROCESS	Total
Adam	53 (45.7%)	9 (7.8%)	26 (22.4%)	13 (11.2%)	15 (12.9%)	116 (51.6%)
Anna	16 (17.4%)	15 (16.3%)	32 (34.8%)	24 (26.1%)	5 (5.4%)	92 (40.9%)
Charles	1 (5.9%)	–	2 (11.8%)	10 (58.8%)	4 (23.5%)	17(7.6%)
Cat. Total	70 (31.1%)	24 (10.7%)	60 (26.7%)	47 (20.9%)	24 (10.7%)	225 (100%)

Table 2. Axial code summary for A1.

	DESIGN VALUES	HUMAN VALUES	NARRATIVE	REQUIREMENT	PROCESS	Total
Adam	31 (41.3%)	–	22 (29.3%)	6 (8.0%)	16 (21.3%)	75 (46.9%)
Anna	16 (20.5%)	11 (14.1%)	30 (38.5%)	18 (23.1%)	3 (3.8%)	78 (48.8%)
Charles	1 (14.3%)	–	1 (14.3%)	3 (42.9%)	2 (28.6%)	7 (4.4%)
Cat. Total	40 (30.0%)	11 (6.9%)	53 (33.1%)	27 (16.9%)	21 (13.1%)	160 (100%)

Table 3. Axial code summary for A2.

	Adam	Anna	Charles
DESIGN VALUES	53 (75.7%)	16 (22.9%)	1 (1.4%)
HUMAN VALUES	9 (37.5%)	15 (62.5%)	–
NARRATIVE	26 (43.3%)	32 (53.3%)	2 (3.3%)
REQUIREMENTS	13 (27.7%)	24 (51.1%)	10 (21.3%)
PROCESS	15 (62.5%)	5 (20.8%)	4 (16.7%)

Table 4. A1, Axial code contribution.

In Table 4, 75.7% of the coded DESIGN VALUES came from Adam, 22.9% from Anna and 1.4% from Charles. The HUMAN VALUES in A1 came primarily from Anna at 62.5%. Adam contributed 37.5% of the HUMAN VALUES while Charles contributed no events coded as HUMAN VALUES.

Coded discourse labelled as NARRATIVE was fairly evenly split between designer and client with Anna and Charles accounting for 56.6% together, and Adam claiming the remaining 43.3%.

	Adam	Anna	Charles
DESIGN VALUES	31 (64.6%)	16 (33.3%)	1 (2.1%)
HUMAN VALUES	–	11 (100%)	–
NARRATIVE	22 (41.5%)	30 (56.6%)	1 (1.9%)
REQUIREMENTS	6 (22.2%)	18 (66.7%)	3 (11.1%)
PROCESS	16 (76.2%)	3 (14.3%)	2 (9.5%)

Table 5. A2, Axial code contribution.

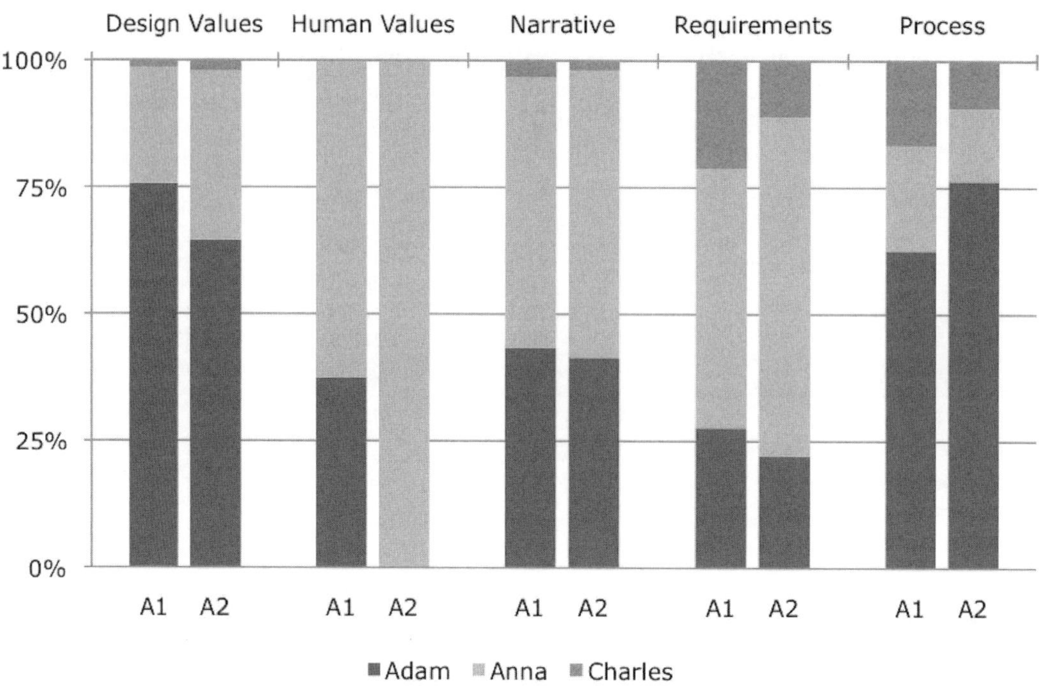

Figure 1. Comparison of category contributions in A1 and A2.

REQUIREMENTS were mostly driven by Anna (51.1%) and Charles (21.3%). The business of running the meeting, noted by PROCESS, fell primarily to Adam (62.5%).

Table 5 shows the contributions for meeting A2. Adam's contribution to DESIGN VALUES decreased to 64.6% while Anna's increased to 33.3%. The contributions to HUMAN VALUES were one-sided with Anna contributing all events coded for HUMAN VALUES.

NARRATIVE events were again fairly evenly split with Adam contributing 41.5% and Anna 56.6%. In combination, Anna and Charles contributed 77.8% of events coded as REQUIREMENTS, while Adam clearly drove the business of the meeting with 76.2% of the events coded as PROCESS.

To better understand these trends, it is important to consider the context of each design meeting. During A1, the details of the design were still being refined. Adam spent the meeting 'walking' Anna and Charles through the design, and at each step clarified requirements and suggested modifications. The discourse throughout A1 was a volley of values between architect and client where each asserted, listened, and responded to statements of values from the other.

In examining the change from A1 to A2, we suggest that the transfer of values can be seen as a process of osmosis where higher concentrations of each type of value begin to permeate a lower concentration of those values. This progression can be seen in A2 where Anna contributed more discourse to the DESIGN VALUE codes (33.3% in A2 up from 22.9% in A1).

Adam's contribution, however, presents a problem. During A2, Adam did not contribute a single HUMAN VALUE coded event during the discourse. This result frustrates our attempt at creating a coherent theory but may be explained by examining where, in the design process, meeting A2 took place. The amount of time that passed between A1 and A2 was significant at seven months, and by A2 the design process had progressed considerably. The lack of HUMAN VALUE statements from Adam could be because his focus had moved to advancing the design towards a planning application (in fact, this intention of moving the design on to the planning phase was repeated several times by Adam during A2). What this situation suggests is that the big design problems, with a few exceptions, had been solved and Adam no longer needed to synthesise new information about the design space.

3.2 *Value transfer and problem solving*

In examining the transfer of values between client and architect, we have said little about the contributions made by Charles, the second 'client' present at the meetings. Charles's role in the meetings was slightly different from Anna's. Through both meetings he typically let Anna lead the conversation, commenting only sparingly. His contributions came primarily in the form of REQUIREMENTS or PROCESS. He was specifically engaged in discussions about building features that were less well defined. One such instance occurred in A1 during a long discussion about the audio-visual system. Throughout this part of the design discourse, Charles presented functional requirements and engaged in problem solving with Adam:

Extract 4, A1, Example of REQUIREMENT, ACTIVITY.

628	Charles	and the other bonus of them not being actually sitting in there was that
629		they could communicate then outside the other issue we looked at was
630		because this person erm [*begins to point*] also is monitoring in the ideal
631		world what's happening out here and what's happening out here so ah
632		they're not only dealing with this the current funeral but the previous one
633		and the one to come
634	Anna	see when they're arriving
635	Adam	[*begins to sketch*] the answer is then to have a door there
636	Charles	a door
637	Adam	maybe a window
638	Charles	a window
639	Adam	and they can
640	Charles	and a window this way

The pattern of discourse between Charles and Adam was different in that a specific remedy satisfying the requirement was not immediately apparent. As a result, the discussion did not touch HUMAN or DESIGN VALUES much at all. This suggests that before the discussion of values occurs, the functional requirements of the building must be met, at least in part, so that the proposed solution can be judged against those values. The absence of statements about values during problem-solving is consistent with the findings of Luck and McDonnell[26], in their investigation of architect and user interaction, discussions that occurred early in the process did not touch on phenomenological experiences but focused on the functional and structural needs of the design.

26.
Luck, R. and McDonnell, J. (2006) Architect and User Interaction: The Spoken Representation of Form and Functional Meaning in Early Design Conversations, *Design Studies*, 27, pp. 141–166.

3.3 *The mechanics of value transfer*

Throughout both A1 and A2 a consistent discourse pattern emerged around value transfer. The pattern begins with a REQUIREMENT introduced by either client or designer. A VALUE concept is then associated with the REQUIREMENT, and finally, NARRATIVE elements support and further expound the VALUE. At the end of this exchange, there is often some kind of affirmation of understanding. This mechanism sits at the centre of our theory of value transfer in design. It is an iterative interaction that enables either party to negotiate aspects of the design based on their values and provides a framework for understanding how those values are exposed and responded to within the design activity.

In characterising the types of responses we observed, we found that Adam's affirmative response usually came in the form of a specific change to the design, which is consistent with ideas about how architects effectively communicate through drawing[27]. Anna's response often came as a restatement of the idea, as seen in Extract 5.

27.
Robbins, E. (1994) *Why Architects Draw*, MIT Press.

Extract 5, A1, Example of value transfer discourse.

Adam	well last time we spoke you thought it was comfortable to have a space
	like this because you said that there might be large families visiting
	wanting to arrange a funeral and if you couldn't get them into the office
	for that purpose you could bring them in here
Anna	yes
Adam	so just like this space this space here would double up as a kitchenette
	staff room and meeting room for large meetings
Anna	you'll be able to see like we can we can see the cremators from here
	at the moment which is always
Adam	no you can't see them from here
Anna	no you can't see them so that's not a that's an issue yeah that's well
	some people you know
Adam	don't want to see them
Anna	don't want to see it you know

(continued)

(continued)

1053	Adam	yes I can understand that
1054	Anna	they also see that there's been you know we're sitting here chatting
1055		having tea coffee and lunch and that's so that's quite nice that you
1056		don't actually see it although you're near to it
1057	Adam	just like this room you get a view out over the gardens in this
1058		direction OK
1059	Anna	OK
1060	Adam	and er the staff wing this area if you like has all the staff (support)
1061		accommodation they have their own disabled loo cleaners' store
1062		shower changing area at the end here you have a (coat) store couple of
1063		ordinary loos and on the front of house the really posh bit you get
1064		lovely views from both the vestry and the office over the pond and you
1065		get a formal entrance lobby on this axis the vestry has its own WC so
1066		that the clergyman or priest whoever's taking the service can change
1067		and so on and so forth

Extract 5 begins with Adam reasserting a requirement communicated to him previously (A1, 1039). Anna and Adam then negotiated and agreed on the HUMAN VALUES present in the staff room (A1, 1046–1054). In this exchange both Anna and Adam were synchronised in their understanding of the phenomenological experience of the space and they traded comments that support and validated the

Extract 6, A2, Example of problem-solving discourse.

448	Charles	and I think you need an out-and-out office here
464	Anna	here ++ I mean we've got the bigger waiting room but the vestry we
465		we felt had to be this size for some reason we just felt it was rather
466		than coming all the way through here +
469	Anna	they'll say first thing they'll say when we get the consultation is they
470		don't want to be over there walking across the water or coming in
471		through this way they would probably prefer to be around this edge
472		where the the ordinary people are so they can mingle with the people
473		sort of here before the service starts
501	Anna	but I j- I just have a feeling that they will not they will feel although
502		there's the reasons why tha- that's quite a good idea I think they will
503		feel too far away from the arrival of the cortege and the people
504		milling around I think that would be one of the things they will say
505		++++ I would think they would feel that they were sort of out of the
506		way a bit and they'd like to sort of be hanging around here especially
507		if this is now covered and especially if they're sort of sitting in there
508		they can see that's such a nice idea they don't have to move +++
529	Anna	I'm not too sure that I wanted it over there and I don't think they
530		would perhaps want it over there either but down here that's quite a
531		nice idea I quite like that if that's possible
532	Adam	yeah that would make it very similar to the existing building

shared understanding. This can be seen in how Anna repeated or restated what Adam said (A1, 1049; A1, 1052; A1, 1054).

Extract 6 highlights segments from a longer section of conversation regarding the need for office space and its relationship to the vestry (A2, 448). The functional requirement was followed by a number of comments, mostly from Anna, describing the needs of the minister or officiating person and how they would feel in the space. Anna also discussed how to help these officials provide the best support for arriving mourners waiting for the service to start (A2, 464; A2, 469; A2, 502). Anna's comments and narrative describe the human elements of the activity, adding necessary details so Adam can accurately judge what an appropriate solution might be. Adam closed this segment by agreeing to the change and asserting a DESIGN VALUE of form, which was formulated as a comparison to the current building to help Anna and Charles judge the appropriateness of the change.

Going back to Extract 2, the discussion of the tensions in the waiting room (A1, 140) motivates the way Adam responds to the requests for changes in the waiting room.

Extract 7, A1, Example of value transfer discourse.

158	Anna	so I'd like it a little bit bigger I think + not hugely because there is it is
159		a wasted space most of the day really
160	Adam	yeah well I would have thought another couple of metres on there [*writes*
161		*on drawing*] would do the trick so shall we agree a two metre extension
162		yes or thereabouts hmm
163	Anna	I mean the other suggestion that perhaps I could make at this stage
164		would be perhaps for a small amount of outside seating because people
165		like to smoke at funerals they like to have a and the seat that we've got
166		out by the car park at the moment the half seat even if it's cold and not
167		very nice is actually people feel more happier out there then they do
168		sometimes in the waiting room
169	Adam	yes well we can certainly add some outdoor seating out here if you
170		wish we have got some outdoor seating here we've got a number of
171		benches there erm but we can add-

The REQUIREMENT for the change to the waiting room size is associated with the VALUE statement that started in Extract 2. This association develops in Extract 7 to motivate Adam suggesting how to enlarge the waiting room (A1, 160). Anna then responds with another REQUIREMENT for outside seating that has connections to the HUMAN VALUE of JEALOUSY expressed in Extract 2 (i.e. the need for facilitating personal space) as well as an expression of additional HUMAN VALUES, RESPECT and MOURNING (motivating the desire to provide comfort to family members at the crematorium) (A1, 163). Adam then responds to the outdoor seating REQUIREMENT by connecting it to attributes in the design that are already present (A1, 169).

The examples provided here show the intricate nature of value transfer in a design meeting. Although in Extracts 5 and 6 the progression is somewhat linear, Extract 7 shows how the interaction around value transfer goes both forward and backward as the values in play (JEALOUSY, MOURNING, RESPECT) are connected to a larger set of REQUIREMENTS and a NARRATIVE that develops the complex social interactions and tensions present in the waiting area of a crematorium. It is only through understanding these factors that Adam is able to develop a response that both he and Anna will be able to recognise as appropriate.

3.4 *Evidence of mutual understanding*

The mechanics of value transfer described above facilitates the generation of mutual understanding between the designer and the client. As evidence of this, we looked for occurrences where either the designer or the client demonstrated increased comfort when discussing aspects of the domain that were initially the purview of the other.

Starting with the first meeting, when Anna was contributing DESIGN VALUES, she typically spoke in deference to Adam. Her concerns were about the uniqueness of the project and specifically, the PURITY of the final form. While these were her goals, she deferred to Adam's judgment as to how those goals could be met and what the right design decisions might be.

Extract 8, A1, Example of design value – Anna's deference to Adam.

816	Anna	OK is that too heartbreaking for you [*all laugh*]
817	Adam	well it's not as pure a summation as I was looking for but I mean
818		maybe there's another way of doing it maybe if I keep my thinking cap
819		on because you can see I'm trying to keep the spaces pure the
820		purer the space the more spiritual I think it will be the more you mess
821		around with it

In Extract 8, Anna was concerned by the impact a required change would have on the overall design (A1, 816). The joke, and nervous laughter (A1, 816), belie her desire for a coherent design even as she is unsure how to achieve it.

Looking at A2, Anna asserted DESIGN VALUES in a more confident manner, indicating her comfort with those values.

The conversation of Extract 9 shows Anna expressing AESTHETIC remarks (DESIGN VALUES). She began by expressing a goal for creating a certain phenomenological experience (A2, 803; A2, 806) and she presented a specific idea of how the design could meet that goal (A2, 825; A2, 829).

Extract 9, A2, Example of DESIGN VALUE – Anna's stronger expression of form.

803	Anna	will that be coloured or will it be-
804	Adam	could be if you wanted it I hadn't thought of that but if that was
805		something you you'd er be interested in us looking at we could do that
806	Anna	mood lighting I think they call it don't they ++
813	Anna	as well I was looking at something for stained glass or
814	Adam	yes no
815	Anna	something that was sort of
816	Adam	we're with you one hundred percent I think /we we\-
817	Anna	/the sun\ comes up this way and sets sets this way so it would be sort
818		of erm that would be you know quite nice to do but then I mean that
819		obviously adds more expense
825	Anna	so we're of thinking something like COVENTRY CATHEDRAL
826	Adam	yes
827	Anna	you know with that sort of effect in a way more
828	Adam	yes
829	Anna	and EDINBURGH's got sort of quite similar erm sort of ss- ss- streaks of
830		light coming through erm and that was the sort of- not that- this is sort
831		of slightly bigger but you know something + in a sense that has some
832		sort of feel of sort big- of something attractive I mean thinking of that
833		but obviously that would add extra expense

Our quantitative analysis of coded occurrences shows that Anna contributed more DESIGN VALUE discourse events in the second meeting. The shift in speaking more often about DESIGN VALUES was accompanied by a qualitative change marked by the ability to speak more fluently about those DESIGN VALUES. Taken together, these two changes indicate to us that DESIGN VALUES had been internalised by Anna, and that a transfer of values from Adam, the designer, took place by way of their interaction in the design activity.

4 CONCLUSION

We are encouraged that our findings are congruous with other analyses presented in this volume. While several chapters in this volume have examined the social aspects of the design process, two specific analyses exhibit traits similar to our understanding of the social interaction that facilitates value transfer Luck[28] and McDonnell[29]. What Luck refers to as 'design in talk' incorporates both the kinds of social interaction we are associating with value transfer, and an analogous outcome that leads to a more comprehensive understanding of the design space. McDonnell's analysis of negotiation during the design process brings to light additional characteristics of the fluid exchange between designer and client; in particular, the blurring of established boundaries of expertise and

28.
Luck, R. (Chapter 13) "Does this Compromise Your Design?" Socially Producing a Design Concept in Talk-in-Interaction.

29.
McDonnell, J. (Chapter 14) Collaborative Negotiation in Design: A Study of Design Conversations between Architect and Building Users.

identification with certain design problems is consistent with the idea of value transfer. These analyses, taken together, complement each other and present a rich description of the depth of exchange that takes place during collaborative design. Our analysis adds to them by developing an explanation of value transfer as an underpinning that motivates both 'design in talk' as well as the dynamism in expertise and problem identification. But more than describing the motivational factors for actions taken in a design interaction, our analysis provides an explicit way to talk about a variety of external influences that both designer and client bring to bear during the course of collaborative design.

By focusing on how values are transferred in design discourse, we are able to understand more about the significance of designer-client interaction. The core components of this transfer are the presence of a REQUIREMENT, the expression of VALUES that relate to the REQUIREMENT, and finally, a supporting NARRATIVE that helps convey how the REQUIREMENT and VALUES are situated together. Through the interplay of these elements, participants consider and exchange information about the design space and the users who will inhabit it. It is during this exchange that value transfer takes place. Within the data that we analysed, the transformation is apparent from meeting A1 to meeting A2. In A1, while the design was still under revision and the details supporting each requirement were still unclear, the transfer of values was in full swing. The designer contributed DESIGN VALUES to which the client responded and the client contributed HUMAN VALUES to which the designer responded. During A2, the client contributed more expressions of DESIGN VALUES; moreover, the client expressed those values in a more fluent manner. This transformation demonstrates that the client was able to internalise new information in the form of DESIGN VALUES.

In looking for similar phenomena from the designer, we are left only to speculate about what might have happened earlier in the design process. We did not observe a similar increase or mastering of HUMAN VALUES expressed by the designer from meeting A1 to meeting A2. This may be a characteristic of where in the design process each meeting took place – it is possible that the transfer of HUMAN VALUES to the designer started at an earlier point in the design process that we did not have access to. This hypothesis is supported by the fact that by A2 the building design was largely finished and the goal of the meeting from the designer's point of view was to advance onward to planning. Regardless, the lack of strong evidence from the designer encourages us to further investigate this kind of design interaction and include in our data meetings that take place earlier in the process to clarify our theory of value transfer.

As designers of all disciplines continue on the path toward co-design, it is important to examine how different domains accommodate values in the design process so that those same values may be present in the final artefact. It was with this in mind that we began our investigation of the architecture meetings. Our grounded analysis of the two architecture design meetings identified an important pattern during the communication of values between designer and client and sets a foundation for understanding how values are woven into design discourse.

7
Affect-in-Cognition through the Language of Appraisals*

Andy Dong, Maaike Kleinsmann & Rianne Valkenburg

The premise of this chapter is that affect is a basis for rationality and that affective processing is a constituent component of design thinking. The chapter focuses on the influence of the valence of affective judgments on design thinking. The transcripts of design meetings are coded according to a formal, linguistic analysis of the semantic resources for appraisals, the display of sentiment and subjectivity in language. The research indicates that the appetitive or aversive orientation of appraisals have design thinking consequences on knowledge integration and generation. During knowledge integration, negative appraisals accompany periods of technical analysis and engineering new design solutions; conversely, positive appraisals accompany a reliance on general knowledge and background experience. During knowledge generation, positive appraisals are associated with the creation of knowledge while negative appraisals are associated with 'being stuck'.

In this chapter we present a study of the linguistic display of affect in all four of the DTRS7 design meetings using a formal method of analysis for understanding affect in the language of design; the language of describing and doing design. Specifically, the analysis focuses on the valence of affective judgements. Valence is an interesting aspect of affect to study in relation to design thinking since prior research has indicated that positive and negative affective states influence cognitive processes related to design, including generative and creative thinking[1].

The research draws on hypotheses put forward in neurobiological and psychological theories of affect and the appraisal theory of emotions. Affect is the neurobiological state incorporating emotion, feelings and other affective phenomena such as mood and temperament. Neuroscientists model affective judgments as a mechanism for organisms to unconditionally and conditionally valuate situations[2] as the basis for action. Neurobiologists define the affective processor as 'components of the nervous system (conceptual and neurophysiological) involved in appetitive (positive) and aversive (negative) information processing'[3]. For neuroscientists, emotions prepare an organism for action because they are an 'episode of coordinated brain, autonomic, and behavioural changes that facilitate a response to an external or internal event of significance for the organism'[4]. While not all affective conditions are emotions,

1. Fiedler, K. (2000) Toward and Integrative Account of Affect and Cognition Phenomena Using the BIAS Computer Algorithm in Forgas, J.P. (ed) *Feeling and Thinking: The Role of Affect in Social Cognition*, Maison des Sciences de l'Homme and Cambridge University Press, pp. 223–252.

2. Burgdorf, J. and Panksepp, J. (2006) The neurobiology of positive affect, *Neuroscience & Biobehavioral Reviews*, 30(2), pp. 173–187.

3. Cacioppo, J.T. and Bernston, G.G. (1999) The Affect System: Architecture and Operating Characteristics, *Current Directions in Psychological Science*, 8(5), pp. 133–137.

4. Davidson, R.J., Scherer, K.R., Goldsmith, H.H. (2003) *Handbook of affective sciences [electronic resource]*, Series in Affective Science, Oxford University Press.

*Reprinted from: Design Studies Vol. 30 No 2 (March 2009), pp. 138–153,
DOI: 10.1016/j.destud.2008.12.003

affective judgments on the state of the situation, how the situation is believed to come about, and the implications of the situation, all precede emotions, according to the appraisal theory of emotions[5].

5.
Ellsworth, P.C. and Scherer, K.R. (2003) Appraisal Processes in Emotion in Davidson, R.J., Scherer, K.R. and Goldsmith, H.H. (eds) *Handbook of affective sciences [electronic resource]*, Oxford University Press.

The chapter presents a rigorous coding mechanism of linguistic appraisals as a means to identify realisations of affect through language. The coding is used to impute the valence of affective judgments from the orientation of the linguistic appraisal. In turn, the valence is expected to steer the design dialogue, exercising regulatory effects consequential to design thinking; these consequential findings are reported in Section 6.

1 LINGUISTIC CODING STRATEGY

Understanding how language is used to construe emotion has been theorised within the tradition of Halliday's theory of systemic-functional linguistics (SFL)[6]. According to Halliday, two linguistic features evoke appraisals: semantic meaning and grammar. We used this connection to develop a coding strategy for appraisals. One researcher coded all appraisals using the method of functional grammar. A second researcher coded the appraisals interpretively. The differences were discussed, resulting in a set of rules and heuristics which assisted in the application of the coding scheme to the design meeting transcripts, described in Sections 3–5.

6.
Halliday, M.A.K. (2004) *An introduction to functional grammar*, Arnold, London.

The coding scheme is based on the analysis technique of functional grammar, which prescribes an analytical, criterion-based method for the functional-semantic analysis of the grammar and the participants in the grammar. The coding scheme is necessarily detailed and systematic. The nefarious nature of affect motivates us to apply rigorous linguistic analysis to characterise affect rather than to rely on interpretive schemes, which are limited in their application by other researchers to other data sets. Further, our interest is how the language of appraisal in design functions as an analysis tool itself to reveal the nature and number of the semantic resources of appraisal. The coding process is 'bottom-up' and relies on accounting for all of the possible semantic resources for appraisal to ensure that the identification of appraisals is not ad hoc. Empirical research on the relations between linguistic appraisal and affect is not yet sufficiently mature to propose and test specific hypotheses based on a limited set of semantic resources for appraisal.

A full explanation of functional grammar analysis is provided elsewhere[7]. In Figure 1, we provide a flavour of functional grammar analysis in order to highlight the relatively high level of objectivity. The representative appraisal clause is taken from the first architectural meeting, A1: "yes well we can certainly add some outdoor seating" (A1, 169).

7.
For example: Eggins, S. (2004) *An Introduction to Systemic Functional Linguistics*, Continuum International Publishing Group, London, pp. 206–253.

1. Identify the Participants and PROCESS. Using the rules of the TRANSITIVITY system in SFL, decide the appropriate Process type: mental (thinking), material (doing), relational (having, being), existential (existing) or behavioural (behaving). Code the Participant(s) according to their respective categories based on the PROCESS type.

Yes well	**we**	{can} certainly {add}	**some outdoor seating**
	Participant: **Actor**	Process: Material	Participant: **Goal**

2. Identify semantic resources for APPRAISAL, which are indicated by square brackets [ ]. Interpret whether the orientation of the appraisal is positive or negative. The orientation provides the valence of the affective judgment.

Yes well	we	{can} [**certainly**] {add}	[**some**] outdoor seating
	Participant: Actor	Process: Material **Graduation: Force** **Orientation: Positive**	Participant: Goal **Graduation: Force**

3. Classify the clause as being about the design product, the process of designing, or the people doing the design. These three categories, Product, Process and People, describe the design situations (states) and events which may be the stimuli of affective judgment and therefore the subject of the linguistic appraisal.

Figure 1. Steps in functional grammar analysis summarised.

Methods to identify the semantic resources for appraisals (Step 2 in Figure 1) and to categorise clauses (Step 3) using the PROCESS and Participant types are covered in Sections 3–5.

2 FRAMEWORK FOR THE LANGUAGE OF APPRAISAL IN DESIGN

Within the system of APPRAISAL, linguists define five high-level semantic resources for conveying appraisals: *Attitude, Engagement, Graduation, Orientation* and *Polarity*[8]. The most important resource *Attitude*, has to do with ways of taking evaluative stances through affect, appreciation or judgment. *Engagement* is often considered an appraisal of the appraisal. It deals with the grading of the speaker's commitment to what is said by either promoting or demoting the possibility of negotiation with the speaker (monogloss) or with the reader (heterogloss). *Graduation* deals with the strength of the evaluation and can be made by increasing the size of the appraisal (force) or its specificity (focus) on the subject. *Orientation* relates to whether the appraisal is positive or negative. *Polarity* is labelled as marked or unmarked, depending upon whether the appraisal is scoped. Each of the semantic resources is used to identify appraisals. To illustrate the explicit registration of each of the semantic resources for appraisal, consider the following clauses:

8. Martin, J.R. and White, P.R.R. (2005) *The Language of Evaluation: Appraisal in English*, Palgrave Macmillan.

1. This is the architectural concept.
2. This is not a good architectural concept.

3. This is a terrible architectural concept.
4. It seems that this is a terrible architectural concept.

The first clause is not an appraisal as it does not negotiate an attitude toward the *architectural* concept; it is an existential clause. The second clause is an appraisal which expresses an attitude (appreciation) with negative orientation, and a marked polarity through the use of the word 'not'. The third clause invokes a larger negative orientation through the force of the word 'terrible' as opposed to merely 'not good'. The final clause 'uncommits' from the appraisal slightly by diminished engagement through the words 'It seems that' which makes use of heteroglossia (leaving open the potential to negotiate with the reader).

3 CODING APPRAISALS OF PRODUCT, PROCESS AND PEOPLE

In the next three sections we further detail the coding scheme. The discussion of the scheme is limited to what is needed for the reader to conduct an independent analysis of the data. Further details about the rationale of the scheme are described elsewhere[9].

9. Dong, A. (2006) How am I doing? The language of appraisal in design in Gero, J.S. (ed) *Design Computing and Cognition '06 (DCC06)*, Kluwer, Dordrecht, pp. 385–404.

All linguistic labels are indicated by the use of SMALL-CAPS to distinguish from the design category of Process and where there might be ambiguity. Appraisal clauses are indicated in *italic text* or underlined with brackets [ ] indicating a semantic resource for appraisal. Curly braces { } indicate the PROCESS. Where the PROCESS type is also a resource for appraisal, as with MENTAL, both roles are indicated, with the PROCESS type taking hierarchical precedence.

Appraisals of *Product* are directed towards the design work, including its requirements and goals and the data informing the construal of the design brief. Appraisals of *Product* can justify (provide rationale) decisions taken during the design process. That is, appraisals of *Product* can explain how the designers' *feelings* toward the design work influenced the designing of the work. In the appraisal of *Product*, the designer may rely on semantic resources that apply an external, normative *judgment* or a personal, subjective *appreciation*. To distinguish *Product* appraisals, we applied the criteria shown in Table 1.

Taking stances towards tangible tasks and actions performed during designing identifies the appraisal of *Process*. Appraisal of *Process* is generally associated with concrete actions. In all of the process-oriented appraisal clauses, a tangible action is being evaluated. The evaluation associates a position toward the state of being of the action. The criteria shown in Table 2 guide the identification of appraisals towards design tasks and activities.

Criterion	Example
The appraisal deals with the concept of the work	"**I** (Senser) [***quite***] (Graduation: force) {[***like***]} (Process: mental) (Attitude: affect) **the exam th- th- practice test idea** (Phenomenon) **because you could take that right through and have it so the whole answer**" (E2, 454–455).
The appraisal deals with the structure of the work	"**...a deep wall *it*** (Participant) '{*s*} (Process: relational, attributive) **a** [***very deep***] (Attitude: judgment) **wall** (Attribute) **walls at least three hundred millimetres thick there and it would be there would be the opportunity to have the**..." (A1, 500–501).
The appraisal deals with the behaviour of the work	"**well and that system is going to be up and running in our chapel so the spaghetti and all the other software** (Carrier) {**is**} (Process: relational, attributive) [***quite***] (Graduation: force) [***vast***] (Attitude: judgment) (Attribute) **and** {***is***} (Process: material) [***at the moment***] (Graduation: force) [***sort of***] (Engagement: heterogloss) [{***spoiling***}] (Process: material) (Attitude: appreciation) **the appearance of our chapel** (Goal) + [***at the moment***] (Graduation: force)" (A1, 597–599).
The appraisal deals with the design brief and its influence on the work	"**oh** [***definitely***] (Engagement: monogloss) [***no***] (Polarity: marked) **I mean** [***any***] (Graduation: focus) **sort of thoughts from** [***anybody***] (Graduation: focus) **that** (Carrier) '{**s**} (Process: Relational) **outside of the process** (Attribute) – {**is**} (Process: relational) [***really***] (Graduation: force) [***helpful***] (Attribute)" (A2, 2028–2031).
The appraisal deals with the context where the work will be used and their reciprocal influences	"**...area stand around there with chairs and candles and sit there so the the dimensions of the area that we've got in our existing chapel** (Carrier) [***at the moment***] (Graduation: force) {**is**} (Process: relational) [***more than***] (Graduation: force) [***enough***] (Attribute) (Attitude: judgment) **for what they do** (Circumstance) **then so probably...**" (A1, 459–461).
The appraisal deals with ideas influencing the work[10]	"***the ALHAMBRA*** (Carrier) (**the relief there**) **oh** [***fantastic***] (Attribute) (Attitude: appreciation)" (A1, 2048).
The appraisal deals with secondary effects of the work in its operational context	"**it** (Carrier) '{**s going to be**} (Process: relational) [***very***] (Graduation: force) [***difficult***] (Attitude: appreciation) **for them** (Carrier) ***to survive*** (Attribute) **and wildlife survival when we develop over...**" (A1, 2060–2061).
The appraisal deals with the social or cultural significance of the work	"**there** (Carrier) '{**s**} (Process: existential) **an area** (Existent) **that** [{***cares***}] (Attitude: affect) **and** [***doesn't***] (Polarity: marked) [{***worry***}] (Process: mental) (Attitude: affect-cognitive) **about that it** (Carrier) '{**s**} (Process: relational) **an area of** [***sort of***] (Graduation: force) [***spirituality heaven***] (Attribute) (Attitude: appreciation) (Phenomenon) **you know that sort of thing as they think**" (A1, 2134–2135).

Table 1. Criteria for coding appraisals of *Product*.

Appraisals of *People* express evaluations of a person's (a stakeholder in the design process) cognitive and physical states of being. Appraisals of *People* are generally associated with the MENTAL and BEHAVIOURAL PROCESS or the RELATIONAL PROCESS where the Carrier is a sentient being and the second participant is Attributive or Identifying. One of the major challenges is coding what counts as appraisals of *People*, since all first-person and third-person descriptions of *People* could be construed as advancing an opinion. Descriptions of *People* tend to take on an air of normative evaluation about how people should and should not be or behave. We applied Iedema's[11] criteria of social esteem (normality, capacity or tenacity) and social sanction (truth and ethics) to identify appraisals as judgments of *People*, shown in Table 3.

10. Solovyova, I. (2003) Conjecture and Emotion: An Investigation of the Relationship Between Design Thinking and Emotional Content in Cross, N., Edmonds, E. (eds) *Expertise in Design: Design Thinking Research Symposium 6*, Sydney, Creativity and Cognition Studios Press.

11. Iedema, R., Feez, S. and White, P.R.R. (1994) *Media Literacy*, Disadvantaged Schools Program, NSW Department of School Education, Sydney.

Criterion	Example
The appraisal is taken toward a specific task or action	"**yes we** (Actor) {**could**} [***certainly certainly***] (Graduation: force) {**think**} (Process: mental) **about that** (Phenomenon)" (A2, 807).
The appraisal is commenting on the need for an action	"**we** (Actor)**'ll** [***have to***] (Graduation: force) {**describe**} (Material: process) **the materials** [***in detail***] (Graduation: force) **on the planning drawings**" (A1, 1306–1307).
The appraisal is taken towards generic design processes	"**brainstorms erm the** [***most***] (Graduation: force) [***important***] (Attitude: appreciation) **one** [***really***] (Graduation: focus) **is being that we should**[***n't***] (Polarity: marked) {**criticise**} (Process: material) **ideas of other people or** [***even***] (Graduation: force) [***your own***] (Graduation: focus) **ideas**" (E1, 7–8).

Table 2. Criteria for coding appraisals of *Process*.

Criterion	Example
The appraisal judges social esteem (normality)	"**I** (Carrier)**'m** (Process: relational) [***just***] (Graduation: focus) [***very***] (Graduation: force) [***gobby***] (Attitude: cognitive-behavioural)" (A2, 1695).
The appraisal judges social esteem (tenacity)	"***I*** (Actor)'{**ve been** [***trying***]} (Process: material) **my** [***hardest***] (Attitude: judgment)" (A1, 667).
The appraisal judges social sanction (ethics)	"**we** (Actor) **did**[***n't***] (Polarity: marked) {**have**} (Process: relational) that [***forward thinking***] (Attitude: judgment) [***back in eighty nineteen eighty two***] (Engagement: force)" (A2, 701–702).

Table 3. Criteria for coding appraisals of *People*.

4 METHODS OF APPRAISAL BEYOND THE CLAUSE

Appraisals often occur beyond a single clause and do not follow the grammatical forms of the system of APPRAISAL described above. We have identified two linguistic techniques beyond the clause that designers use to appraise: *enumeration* and *comparison*.

In *enumeration*, designers list a set of characteristics, typically involving the structure of the work and how it behaves, meant to evoke an attitude about the object being appraised. Extract 1 shows an appraisal of carpet (the appraising characteristics are underlined).

Extract 1, A2, An example of enumeration.

Anna that would be the issue but + it frays I've seen it fray dirty trip
hazard ooh er cleaning constantly need to keep it clean
it needs to match up with other things like the kneelers and everything else and
they'd look faded compared to the- so everywhere that I've been

In the case of *comparison* the designer will appraise by comparing one subject to another. Extract 2 shows a dialogue between Tommy, Patrick and Roman, in which they discuss the desirability of rechargeable batteries. Rechargeable batteries are more desirable because they could be recharged using a charger or from the computer's USB port. The comparison is signalled by the word 'whereas'.

Extract 2, E2, An example of comparison.

1435	Patrick	but the batteries won't last very long + whereas if you use rechargeables
1436	Roman	yeah
1437	Patrick	to charge it you could supply a charger with it a mains charger that's
1438		expensive or you plug it into your USB which is free essentially

Both *enumeration* and *comparison* are combined when the designer really needs to 'drive the point home'. In (A2, 797–802), the repetition of "it allows" and "so that" enumerate the benefits of using the diffuser ('it').

5 WHEN SUBJECTIVITY IS NOT AN APPRAISAL

One of the most significant challenges in identifying appraisals is distinguishing between an appraisal (= an evocation of an attitudinal stance) and subjective content that does not valuate. In identifying appraisals, the following three questions are useful.

Is the appraisal expressed by taking an external viewpoint? Extract 3 shows that the engineering design team is brainstorming possible uses for the pen. They offer up many possibilities. One possibility is to use the pen as a way for people to take note of high scores in a video game.

Extract 3, E2, Expression by taking an external viewpoint.

910	Tommy	you'd probably use mobile phones to photo photograph the screen
911		these days wouldn't you
912	Jack	yeah probably
913	Tommy	[*laughs*] right

Tommy's evaluation (at 910) is evaluative by acknowledging alternative positions as heteroglossic diversity and by increasing the importance of the statement of fact about mobile phone use through the use of current time ("these days") as a resource for graduation. He uses "you" as a resource of engagement as heterogloss to mark alternative positions. An alternative appraisal using the resource of engagement as monogloss might have been 'I think mobile phones are better suited for capturing video game scores'.

Is the expression descriptive but not evaluative? In these cases, the subjective content stands in for potentially unknown objective data, for example: "it may be a bit lower than that" (A2, 749). This clause uses modifier terms to describe the height of the chapel. There is no attitudinal stance in the statement. It could have been stated as 'it may be a centimetre or two lower than that'. Compare the above clause to the following one: "there's [*vast*] (Graduation: force) amounts of heat to be recovered" (A2, 886). The aggrandising of the 'amounts of heat' by the use of 'vast' indicates that this is a significant problem in the design context that needs to be addressed. A more descriptive expression could be 'there's heat to be recovered'.

Can the clause be rewritten to express a neutral sentiment? If the clause can be rewritten to express a neutral sentiment, then it is an appraisal, e.g. "the [*whole*] idea was we [{*wanted*}] to make [*sure*] the architectural concept worked [*through*]" (A2, 771–772). This clause could be more neutrally expressed as 'the idea is to make the architectural concept work'.

6 FINDINGS

Table 4 reports quantitative data from the coding of appraisals in the design meetings about what was the object (i.e., what is the stimulus for the affective judgment) of the appraisals and how many appraisals were made (i.e., how much affect is explicitly linguistically embedded). The design Product was the focus of majority of appraisals. There were fewer appraisals in the engineering design meetings than the architectural meetings. This may be due to motivation. In the video-recording of an interview with the architect Adam (supplied as part of the dataset), he expressed very positive attitudes toward the crematorium architectural design project. The architect was "frankly gobsmacked" when awarded the project, considering the project "fabulous" and his "greatest opportunity" so far. The architect remarked that the clients were enthusiastic and engaged. The positive motivation and engagement of the architect and the clients toward the project likely influenced the appearance of appraisals. The prior enthusiasm is unknown for the engineering meetings.

A qualitative analysis of the data led to the two main findings of the study. These are discussed below.

Appraisal Type	A1	A2	E1	E2
PD+	221	133	129	117
PD–	148	73	74	66
Total PD	369	206	203	183
PR+	133	77	24	7
PR–	54	12	10	6
Total PR	187	89	34	13
PP+	51	11	2	1
PP–	41	10	6	0
Total PP	92	21	8	1
Total Appraisals	648	316	245	197
Transcript Lines	2342	2124	2019	1867
Appraisal Rate	28%	15%	12%	11%

Key: PD Appraisal of *Product*; PR Appraisal of *Process*; PP Appraisal of *People*; +/– positive/negative orientation.

Table 4. Frequency and type of appraisals in DTRS7 data set.

6.1 *Negative appraisals are accompanied by a focus on analysis of situational data whereas positive affective content displays a reliance on background experience*

We observed that negative appraisals are accompanied by periods of technical analysis of the situation and 'engineering' new solutions. Psychologists have theorised that the affective state may influence the degree to which people rely on general knowledge rather than focusing on specific data presented from the current situation[12]. In this data set, we find that negative affective states, indicated by negative appraisals, influence decision-making, where negative affect causes the designers to focus more on technical data and analysis before deciding, whereas positive affect allows them to rely on prior knowledge (that might not be expressed verbally) in order to proceed onward. Let us examine the cases of negative affect first. Before Alan took control of the brainstorming session, as shown in Extract 4, Tommy began the conversation with a series of negative appraisals of methods for heating up the media (E1, 74–81).

12. Bless, H. (2000) The Interplay of Affect and Cognition in Forgas, J.P. (ed) *Feeling and Thinking: The Role of Affect in Social Cognition*, Maison des Sciences de l'Homme and Cambridge University Press, pp. 201–222.

Extract 4, E1, Example of consequence of negative appraisals.

76	Tommy	now erm technically that's more demanding (PD–) because energising the entire ( )
77		printout takes time hammers the batteries (PD–) and it's quite a demanding
78		thing to do (PD–) so technically the stuff where you've got a low percentage fill
79		sort of faces and text and things like that is a lot more realisable (PD+) but if we

Tommy and Chad join the discussion with other ideas before Alan intervenes and sets the direction of the brainstorming. Later, Jack proposes another solution (stabilisers): "well I guess the easiest way to keep the pen at a right angle would be to have a set of stabilisers on it" (E1, 154–155). His positive appraisal is followed by Alan's positive appraisal of this idea: "yeah that's a good idea"

(E1, 158). Then, the group moves onto the next idea: "I was thinking that a sort of maybe like a flat base with a sort of universal joint like windsurf mast" (E1, 160–161). Tommy appraises the shape of: "the size of the thing in contact with the paper" as "it needs to be quite narrow" (E1, 176). This negative appraisal sets the group off into a technical discussion of the shape despite Alan trying to steer the group: "not be too preoccupied with the shape" (E1, 180–181). As the group continues to discuss this issue and engineer solutions, Tommy appraises the problem that the angle of contact of the pen with the media as something: "that's going to limit us for the time being" (E1, 260). This negative appraisal sets the group off onto doing more engineering in relation to this problem (E1, 266–327). Here, negative affect is indicated by negative linguistic appraisals. The extent of affective influence seems greater than in other episodes of the dialogue due to the elaboration of negative appraisals, which appear over about 300 transcribed lines of continuous design engineering.

Conversely, positive affective content allows the group to rely on general knowledge and background experience. Extract 5 shows some conversation where the group discusses the design of a staff room which could double as a meeting space for large families organising a funeral. This section is marked by a series of positive appraisals of the space.

Extract 5, A1, Example of consequence of positive appraisals.

1039	Adam	well last time we spoke you thought it was comfortable to have a space
1040		like this (PD+) because you said that there might be large families visiting
1049	Anna	no you can't see them so that's not a that's an issue (PD+) yeah that's well
1055	Anna	having tea coffee and lunch and that's so that's quite nice (PD+) that you
1063	Adam	ordinary loos and on the front of house the really posh bit (PD+) you get
1064		get lovely views (PD+) from both the vestry and the office over the pond and you

There really is not much designing or objective evaluation of the space as there is a set of subjective exchanges on the virtues of the space. This dialogue exhibits another pattern, which is that positive affective appraisals co-occur with a focus on global aspects of the space such as its views and its 'feel' (e.g., it's 'nice'). As discussed above, the special status of the crematorium project to the architect Adam meant that he felt motivated to add a feeling to the building beyond its functional needs. In Extract 5, Adam discusses the feeling created through the architecture with Anna. Anna as a client is likely to be more interested in the feeling of the architecture rather than the architectonic details.

Both examples show consistency with the levels-of-focus hypothesis[13], which states that positive affect promotes attention to global information and negative affect to local information. In the dialogue shown in Extract 5 it is almost as if Adam and Anna were accessing their own personal global schemas about spaces to interpret and construct an experience about the current space rather than examining specific details of this space. From the interview with the architect, the architect described the communication with the clients as referring to the feeling that they wanted to create in the building. Anna defined this feeling by referring to the procedures executed during funerals, while Adam created this feeling through his architectural concept. The concept for the crematorium itself was based on the feeling of Louis Kahn's Kimbell Museum mapped into the crematorium design as a division between 'served and servant space'.

13. Clore, G. L. and Gasper, K. (2000) Feeling is Believing: Some Affective Influences on Belief in Frijda, N.H., Manstead, A.S.R. and Bem, S. (eds) *Emotions and beliefs: how feelings influence thoughts*, Cambridge University Press, pp. 10–44.

This reliance on positive affective appraisal as a source of information for the interpretation of information, or what might be called 'a hunch', is portrayed in Extract 6, a dialogue between Charles, Anna and Adam about a footpath over a stream. Given the positive reception to the stepping stone idea, the group decides to pursue this design concept, without really considering why Anna likes the idea or how workable the solution is, for example, if local flooding would cover the stepping stones, if they would be a slip hazard, if mobility impaired people could cross the stream, etc. The series of positive appraisals may indicate a mild form of groupthink, a finding supported by Adam's explanation in the interview that there were no major disagreements with the clients.

Extract 6, A1, Dialogue about a footpath over a stream.

1178	Charles	I think something that looks like a bridge would be well or something
1179		that's solid would be the unfortunately although I can see the benefit
1180		of having I mean I quite like the idea of stepping stones (PD+)
1181	Anna	yes I do I like the idea (PD+)
1182	Adam	well if you like it why don't we run with it (PR+) until somebody says you
1183		don't want to do that

6.2 *Positive affect has a congruency effect with knowledge generation and negative affect hampers or inhibits knowledge generation*

During the meeting E1, the engineering group discusses ways to keep the print head on, whilst in use, and off otherwise (E1, 1689–1857). The elaboration of the switch-based idea is an unusual section of the transcript since it contains relatively few negative appraisals. Tommy expresses this positive affect as: "yeah I think there's going to be loads of ways to low cost switch" (E1, 1711).

In engineering the switch-based idea, the group runs into the following technical issues: "floating the head" (E1, 1719), "the head springing forward" (E1, 1751), and "holding the heads flat" (E1, 1780). They do not experience any blocks in finding solutions to these problems. This is not just a positive affective appraisal of various print head designs; the group's positive affective state seems to assist the group in generating further print head designs.

This contrasts with the group being 'stuck' on the problem of the weight of the pen in order to hold the sheath down properly:

> "actually five hundred grams that means this sort of thing isn't going to work oh no it means this has got to be more than five hundred grams to push that back if you had a sheath you have to press that has to be more than fi[ve]" (E1, 1937–1966).

The group realises that 500 grams may not be sufficient to hold the sheath down. Chad, being somewhat optimistic, suggests that users have just: "got to press harder" (E1, 1950) and that: "five hundred grams isn't very much" (E1, 1957). Yet, Tommy and Alan feel that 500 grams: "is a lot is a lot for a pen" (E1, 1964) and: "and then you've got all the batteries" (E1, 1967). The group comes to no agreement about the weight, and does not spend much effort planning other solutions or planning ways to engineer a solution, possibly because the group also realises that the session is ending.

7 IMPLICATIONS FROM OTHER RESEARCH IN THIS VOLUME

The method of analysis we have taken is based on systemic-functional theory which emphasises the systematic relationship between language and its social context. While the participants may not realise it, they are performing complex linguistic manoeuvring by construing interpersonal relations to relay design knowledge. As it is not possible to verify with the participants what influence their appraisals of the design situations *actually* had on their design thinking, it becomes important to consider the background and social context of the affective judgments, in line with SFL theory. It is this issue which is taken up by two other chapters in this volume.

14. LeDantec, C. and Do, E. (Chapter 6) The Mechanisms of Value Transfer in Design Meetings.

Le Dantec and Do[14] draw on the notion of value-based design to discuss the ways in which principles, standards and qualities guide designers' actions. In the cases of their examples of design values and human values, it is interesting to note that the statements of design values are stated as appraisals, for which we have coded a few representative examples in Table 5. Note that the third example

Type of value	Example
(A1, 819–820) Example of design value (Purity)	"**the** {***purer***} (Attitude: judgment) **the space the** {***more***} (Graduation: force) {***spiritual***} (Attitude: appreciation)"
(A1, 141) Example of human value (Jealousy)	"**it** (Actor) [***actually***] (Graduation: force) {**makes**} (Process: material) **it** [***worse***] (Attitude: judgment)"
(A2, 701–702) Example of design value (Aesthetic)	"**we** (Actor) **did**[***n't***] (Polarity: marked) {**have**} (Process: relational) **that** (Attribute) [***forward thinking***] (Attitude: judgment) [***back in eighty nineteen eighty two***] (Graduation: force)"

Table 5. Examples of values (cf **LeDantec and Do (Chapter 6)** ***op. cit.***)[15]

in Table 5 is an appraisal of People as a means to express a design value.

In the first two examples, facts about the space are constructed through subjective values. Agreements on the values that can shape these facts have apparently been backgrounded as there is no discussion on the rationale for these values. Nonetheless, which values are stated, sharpened or softened, are facts that are constructed. These facts incline the participants to think in one way or another, or to think about one thing or another. This principle is discussed in relation to the transfer of values between the client and the architect[16], which emphasises one of the principle tenets of the appraisal theory of emotions – the way that Adam believes the world to be (how people should communicate) has cognitive consequences that are mediated by his subjective experience of affect. Positive affect likely resulted from Adam receiving positive feedback on the design which is consistent with his design values. In experiencing positive affect, he seems more amenable to knowledge integration leading to shared understanding with the clients. This shared understanding is not a blanket proposition, however. The possibility of arriving at a shared understanding of the design work when there is a disagreement on matters of apparently emotional investment is only possible when one party acquiesces. This is evident when Anna agrees to the design requested by Adam[17], made obvious by Anna's perception of Adam's emotional investment in the proposed solution. Such data suggest that the consequences of the valence of affect on design thinking depend on prior belief structures (i.e., design values). This principle is further observed by Luck[18].

Luck focuses on the exchange of talk as driving the development of the concept, focusing on patterns of relations between the architect and the client. One key aspect Luck examines is the agency of the design concept and its alternation between the architect,

15.
ibid.

16.
***ibid.*, Extract 4, A1.**

17.
***ibid.*, Extract 8, A1.**

18.
Luck, R. (Chapter 13) 'Does this compromise your design?' Socially Producing a Design Concept in Talk-in-Interaction.

Adam, and the clients, Anna and Charles. In one of the eponymous passages on compromise, Adam, Anna and Charles negotiate the design concept agency. Luck[19] infers that Adam is open to modifications to the concept, letting go of the agency of the concept. This is done through an appraisal:

19. *ibid.*

> "**it** (Carrier)'{**s**} (Process: relational) {**not**} (Orientation: positive; Polarity: marked) {***necessarily***} (Graduation: force) **a** {***compromise***} (Attribute) (Attitude: appreciation)" (A1, 1153).

By diminishing the force of the appraisal, Adam marks alternative positions. In the second meeting, they hit an impasse. The indication of the impasse is marked by an appraisal:

> "**well this** (Carrier) {**is**} (Process: relational) {***fairly***} (Graduation: force) {***fundamental***} (Attribute) (Attitude: judgment) … **deciding the number of cremators**" (A2, 618)

How to get out of the impasse is also marked with an appraisal:

> "**I** (Senser) {***think***} (Process: mental) (Attitude: affect-cognitive) **we** (Senser) {***need***} (Process: mental) (Attitude: affect-cognitive) **to** {***just***} {***clear***} ++ {***clear***} ***direction*** (Phenomenon) **from you**" (A2, 687–688).

In these utterances, there is no diminishment of the evaluative position. By promoting or demoting an appraisal through the resource of engagement, the importance of evaluative statements is made explicit.

20. *ibid.*

Luck's[20] analysis illustrates an interesting phenomenon, which is the difference in sensitivity to affect between beliefs and objective knowledge. In the second exchange, the discrepancy between what Adam thinks should be the best number of cremators based on a *belief* in symmetry and what Anna and Charles *objectively know* as the best number given the functional requirements based on historical evidence is not so relevant. What is most interesting is Adam's belief in symmetry is more affect-sensitive; he tries to argue his case, through appraisals, and is not entirely amenable to hand over agency of the concept, only to get 'better direction' from Anna and Charles. In the first exchange, he *is* willing to let go of agency 'if there's a good reason' based on some objective knowledge (about slippery stones). This implies that beliefs are more likely than objective knowledge to be sensitive to affect. The occurrence of affective appraisals when the subject of the appraisal is closely tied to the person's beliefs rather than objective knowledge suggests that beliefs are not easily adjusted to be compatible with internal evidence in the form of affect.

8 CONCLUSION

Our analysis focused on the influence of the valence of the affective appraisals, as the orientation of the linguistic appraisals, on knowledge integration and generation. Although the designers in these studies might not be aware of their functional use of appraisals, (e.g., the architect claims in the interview to "prefer to sell an idea on the base [sic] of logical reasons. It is easier to sell with logic instead of having discussions about appearance and so on…"), the analysis shows that affective judgments expressed though language do actually influence design thinking. Counting the frequency of occurrence of the options for expressing attitude in appraisals could further elaborate this chapter's findings.

We conjecture that appraisals might serve more important meta-functions in design than the affective judgment of Product, Process or People. In the spirit of Halliday's meta-functions of language, we hypothesise that the language of appraisal serves three meta-functions in design: firstly, to regulate design thinking where affect-in-cognition is part of a highly coupled regulatory network on rationality; secondly, to signal and control the pacing and sequencing of design actions; and thirdly, to place a value on design knowledge. Each of these meta-functions is mediated by the valence of affect. We predict that the valence of affect, which is often publicly displayed in language, is most consequential to design thinking.

ACKNOWLEDGEMENTS

This research was supported under Australian Research Council's *Discovery Projects* funding scheme (project DP0557346).

Part 3

Aspects of Design Cognition

8
Analogical Reasoning and Mental Simulation in Design: Two Strategies Linked to Uncertainty Resolution*

Linden J. Ball & Bo T. Christensen

This chapter aims to further our understanding of the nature and function of analogising and mental simulation in design through an analysis of the DTRS7 engineering data. Analogies were coded for 'purpose' and in terms of whether they were within-domain or between-domain. Mental simulations were coded for 'focus': technical/functional or end-user. All expressions of uncertainty were also identified. Analogies were found to be typically between-domain (indicative of innovative reasoning) and were evenly distributed across solution generation, function finding and explanation. Mental simulations were predominantly technical/functional. Our most striking observation was that analogies and mental simulations were associated with conditions of uncertainty. We propose that analogising and mental simulation are strategies deployed to resolve uncertainty – a claim that is supported by the fact that uncertainty levels returned to baseline values at the end of analogising and simulation episodes.

Analogical reasoning involves accessing and transferring previously acquired knowledge of objects, attributes and relations to support current problem solving and decision making activities. Such reasoning has long been viewed as central to intelligent thought and creative cognition[1,2,3], with recent studies confirming the importance of analogising for scientific discovery[4], organisational management[5], and innovative product development[6,7,8,9]. A recent study by Christensen and Schunn[10] has further clarified the key characteristics of analogising in design via a detailed 'in vivo' examination of design meetings in an international R&D company specialising in the design of medical plastics. Around nine hours of discussion by a core design team was analysed, with transcripts spanning the first five months of a design project. Christensen and Schunn's[11] analysis indicated that analogising was a frequent strategy deployed by the team, with analogies serving three primary functions or purposes: problem identification, problem solving, and explaining[12]. In addition to their function-oriented characterisation of design analogies, Christensen and Schunn[13] also coded analogies in terms of their 'distance' from the domain of medical plastics. Using a binary classification system of within-domain versus between-domain they revealed that *within-domain* analogies prevailed during problem identification, whilst *between-domain* analogies prevailed during explanation. In contrast, solution

1.
Gentner, D. and Stevens, A.L. (1983) (eds) *Mental Models*, Erlbaum, Hillsdale, NJ.

2.
Holyoak, K.J. and Thagard, P. (1995) *Mental Leaps: Analogy in Creative Thought*, MIT Press, Cambridge, MA.

3.
Schank, R.C. (1999) *Dynamic Memory Revisited.* Cambridge University Press, Cambridge, UK.

4.
Dunbar, K. and Blanchette, I. (2001) The In Vivo/In Vitro Approach to Cognition: The Case of Analogy, *Trends in Cognitive Sciences*, 5, pp. 334–339.

5.
Bearman, C.R., Ball, L.J. and Ormerod, T.C. (2007) The Structure and Function of Spontaneous Analogising in Domain-Based Problem Solving, *Thinking and Reasoning*, 13, pp. 273–294.

6.
Ball, L.J., Ormerod, T.C. and Morley, N.J. (2004) Spontaneous Analogising in Engineering Design: A Comparative Analysis of Experts and Novices, *Design Studies*, 25, pp. 495–508.

7.
Casakin, H. (2004). Visual Analogy as a Cognitive Strategy in the Design Process: Expert Versus Novice Performance, *Journal of Design Research*, 4.

8.
Casakin, H. and Goldschmidt, G. (1999) Expertise and the Visual Use

*Reprinted from: Design Studies Vol. 30 No 2 (March 2009), pp. 169–186, DOI: 10.1016/j.destud.2008.12.005

generation was characterised by an equal distribution of within-domain and between-domain analogies.

The DTRS7 dataset afforded an opportunity to delve further into issues surrounding analogising in design. By focusing on the engineering meetings (E1 and E2) we aimed to replicate and extend Christensen and Schunn's[14] evidence that analogies are used for different purposes in innovative design situations. Likewise, we wished to validate the links between analogical purpose and analogical distance. Finally, we aimed to examine the relation between analogising and 'epistemic uncertainty'. The latter concept refers to situations where people have metacognitive awareness of the limitations of their current knowledge or understanding[15,16]. Epistemic uncertainty is integral to non-routine design contexts[17], where the complex, multi-facetted and ill-defined nature of problems means that designers are continually working at the extremity of their current knowledge. One idea that we wanted to examine was whether analogising may be a strategy that is deployed under conditions of uncertainty in order to reduce or resolve such uncertainty. For example, designers may use analogies to enhance their grasp of poorly understood design requirements and constraints, to clarify the nature of ill-defined problems, to inform the completion of partially developed solution concepts, or to augment the communication of vague ideas.

In addition to inspecting the engineering transcripts for instances of analogising we also aimed to examine them for evidence of another cognitive strategy, 'mental simulation', where a sequence of interdependent events is consciously enacted or 'run' in a dynamic *mental model* to determine cause-effect relationships and to predict likely outcomes[18,19]. Model-based mental simulation appears to be primarily an *evaluative* strategy, where the designer's imagination is used to test out ideas and validate solution concepts. There are certainly striking personal accounts in the literature signifying the importance of mental simulation in design and architecture. For example, Frank Lloyd Wright's famous anecdote of the design for Fallingwater (one of the USA's most acclaimed residential buildings) indicates that before committing ideas to the drafting board he was able to: "Conceive the building in the imagination, not on paper but in the mind, thoroughly…"[20].

The mental models that underpin the process of mental simulation are assumed to involve qualitative rather than quantitative reasoning, relying, for example, on ordinal relationships and relative judgements[21]. Thus when running mental models people neither estimate precise values and quantities nor carry out mathematical computations in predicting device behaviour. Despite lacking a focus on exact quantifications, however, such mental simulations can be very powerful, having the great advantage of facilitating the process of reasoning on the basis of incomplete knowledge. In mechanical domains

of Analogy: Implications for Design Education, *Design Studies*, 20, pp. 153–175.

9.
Visser, W. (1996) Use of Episodic Knowledge and Information in Design Problem Solving in Cross, N., Christiaans, H. and Dorst, K. (eds.) *Analysing Design Activity*, Wiley, Chichester, UK, pp. 271–289.

10.
Christensen, B.T. and Schunn, C.D. (2007) The Relationship of Analogical Distance to Analogical Function and Pre-Inventive Structure: The Case of Engineering Design, *Memory and Cognition*, 35, pp. 29–38.

11.
ibid.

12.
See for example Bearman, Ball, and Ormerod's (2007) evidence for two primary functions of analogising in management decision making: problem solving and illustration, *op. cit.*

13.
Christensen and Schunn (2007), *op. cit.*

14.
ibid.

15.
Davidson, J.E., Deuser, R. and Sternberg, R.J. (1996) The Role of Metacognition in Problem Solving in Metcalfe J. and Shimamura, A.P. (eds) *Metacognition: Knowing about Knowing*, MIT Press, pp. 207–226.

16.
Quayle, J.D. and Ball, L.J. (2000) Working Memory, Metacognitive Uncertainty, and Belief Bias in Syllogistic Reasoning, *Quarterly Journal of Experimental Psychology*, 53A, pp. 1202–1223.

17.
Schlosser, J. and Paredis, C.J.J. (2007) *Managing Multiple Sources of Epistemic Uncertainty in Engineering Decision Making*, SAE World Congress and Exhibition, Detroit, MI, USA, April 2007.

there is also evidence that the inferential processes associated with mental models are modal and analogous to the physical properties of the systems and processes being simulated[22,23,24]. In essence, then, mental simulation provides a relatively quick and cognitively economical way for an individual to test out the behaviour of a physical system, including how a system might function under changed circumstances or with altered features.

In strategic terms mental simulation would seem to be especially useful in creative domains such as science and design, where tasks involve constructing novel solutions within a large space of possibilities. Previous studies using verbal protocol analysis have confirmed the important role of mental simulation in both domains. For example, Trickett et al.[25] located the presence of mental model 'runs' in the protocols of scientists conducting data analysis, whilst Christensen and Schunn[26] identified key instances of mental simulation in design protocols. The latter research tested three core assumptions of mental simulation theories: that mental simulations are run under situations associated with subjective uncertainty; that mental simulations of possibilities inform reality through inferences that reduce uncertainty; and that the role of mental simulations is approximate and inexact. Christensen and Schunn[27] successfully demonstrated support for all three assumptions: initial representations in simulations had higher than baseline levels of uncertainty; uncertainty was reduced after the simulation run; and resulting representations contained more 'approximate' references than either baseline data or initial representations. The DTRS7 dataset presented an opportunity to examine further the links between mental simulation and uncertainty in an attempt to generalise and extend Christensen and Schunn's[28] findings.

In summary, our aims in the work presented here were to pursue a detailed protocol analysis of the transcripts of the two engineering meetings in order to further an understanding of how mental simulation and analogising may be linked to epistemic uncertainty. Although the analyses that we present below follow on closely from previous research we note that our approach departs from earlier studies in terms of both the intended *breadth* of interest on the three concepts of analogising, mental simulation and uncertainty, and the intended *depth* of analysis in relation to the nature and function of different 'types' of analogising and mental simulation.

1 TRANSCRIPT CODING

In order to divide up the data into discrete units of spoken discourse we decided to use the line-based segmentation scheme already present in the meeting transcripts. Our analysis, therefore, involved a total of 3886 line-segments of data (henceforth simply referred to as 'segments'). Below we describe the approach that we

18.
Gentner, D. (2002) Psychology of Mental Models, in Smelser, N.J. and Bates, P.B. (eds) *International Encyclopedia of the Social and Behavioral Sciences*, Elsevier, Amsterdam, pp. 9683–9787.

19.
Nersessian, N.J. (2002) The Cognitive Basis of Model-Based Reasoning in Science in Carruthers, P. and Stich, S. (eds) *Cognitive Basis of Science*, Cambridge University Press, pp. 133–153.

20.
Tafel, E. (1979) *Years with Frank Lloyd Wright: Apprentice to Genius*. Dover Publications, New York.

21.
Forbus, K.D. (1997). Qualitative Reasoning in Tucker, A.B. (ed) *CRC handbook of Computer Science and Engineering*, CRC Press, pp. 715–733.

22.
Hegarty, M. and Just, M.A. (1993) Constructing Mental Models of Machines from Text and Diagrams, *Journal of Memory and Language*, 32, pp. 717–742.

23.
Hegarty, M., Kriz, S. and Cate, C. (2003) The Role of Mental Animations and External Animations in Understanding Mechanical Systems, *Cognition & Instruction*, 21, pp. 325–360.

24.
Schwartz, D.L. and Black, J.B. (1996) Analog Imagery in Mental Model Reasoning: Depictive Models, *Cognitive Psychology*, 30, pp. 154–219.

25.
Trickett, S.B., Trafton, J.G., Saner, L. and Schunn, C.D. (2005) 'I Don't Know What's Going on There': The Use of Spatial Transformations to Deal with and Resolve Uncertainty in Complex Visualizations in Lovett, M.C. and Shah, P. (eds) *Thinking wih Data*, Erlbaum, pp. 65–86.

adopted to code the protocols for occurrences of analogies, mental simulation and epistemic uncertainty.

26. Christensen, B.T. and Schunn, C.D. (in press (b)) The role and impact of mental simulation in design, *Journal of Applied Cognitive Psychology*.

27. *ibid.*

28. *ibid.*

29. Christensen and Schunn (2007), *op. cit.*

1.1 *Coding of analogies*

The transcripts were coded for presence of analogies by applying Christensen and Schunn's[29] approach. Any time a designer referred to another source of knowledge and attempted to transfer concepts from that source to the target domain then this reference was coded as an analogy. All analogies were also coded for ANALOGICAL DISTANCE using a binary categorisation scheme where *within-domain* analogies involved mappings from sources that related to tools, mechanisms and processes associated with graphical production and printing, whilst *between-domain* analogies involved mappings from more distant sources (see Extracts 1 and 2).

Extract 1, E1, Example of within-domain analogy.

819	Alan	the other thing to to think about is in almost all cases when I look at pens
820		the apart from re-wired sort of micropens the th- tip is actually the
821		narrowest part of the product whereas in what we're looking at it could
822		actually be as wide or wider-

Extract 2, E2, Example of between-domain analogy.

247		a pen that looks like this and you just put water in it um er the best
248		analogy I can think of is er if you like wet t-shirt effect where-
249	All	[*laugh*]
250	Tommy	the top layer the top layer of paper gets wet
251	Sandra	[*laughs*]
252	Tommy	the top layer of material gets wet and reveals what's underneath it
253	Sandra	oh right

We also coded analogies for ANALOGICAL PURPOSE (i.e. the goal or function of the analogy) using a tripartite scheme based on that developed by Christensen and Schunn[30]. This scheme (see Extracts 3 to 5) categorised analogical purpose in terms of

30. Christensen and Schunn (2007), *op. cit.*

- *problem identification* – noticing a possible problem in the emerging design, where the problem was taken from an analogous source domain
- *solution generation* – transferring possible solution concepts from the source domain to the target domain
- *explanation* – using a concept from the source domain to explain some aspect of the target domain to members of the design team.

Extract 3, E1, Example of problem identification analogy.

1026	Alan	in fact in some ways we should think about the fact it isn't even a pen
1027		because a pen you you'll always learn to write from left to right whether
1028		you're left handed or right handed so actually what you end up doing is
1029		left handed people is you smudge over over your work which is a problem
1030		but actually with this you're dragging it you're not pushing it are you
1031		most people will drag it

Extract 4, E1, Example of solution generation analogy.

1291	Tommy	like a garage door type of thing
1292	Todd	yeah push the button then it goes open
1294	Todd	but that's probably overly complicated
1295	Rodney	garage door well it could be a roller
1296	Todd	a roller door

Extract 5, E2, Example of explanation analogy.

172	Tommy	yeah this is a bit like photographic paper in a way where you're erm
173		developing what's on the paper whereas here you're just enabling the bits
174		you need to print so here you're kind of getting in to normal text

During the application of this a priori coding scheme, however, we found that a significant number of analogies were not readily categorisable as oriented toward problem identification, solution generation or explanation. Thus a new analogy purpose emerged from the transcripts relating to situations involving the active mapping of *new functions* to the *design form* currently being developed (i.e. a thermal printing pen). We refer to this new analogy type as *function finding* (see Extract 6). With the addition of this category to the scheme it was possible to code all analogies within the transcripts.

Extract 6, E1, Example of function finding analogy.

986	Todd	um that's intriguing sort of like a like a could be like a finger puppet
987		couldn't it
988	Sandra	yeah cos wearing it like a finger puppet – the feel of it might be fun
989	Todd	exactly so you can make you can make the footprints-
990	Rodney	I I think I think the sort of design not very good at ( )

1.2 *Coding of mental simulations*

The codes for mental simulations were based on those developed by Christensen and Schunn[31], which were, themselves, adapted from Trickett and Trafton[32]. A mental model 'run' is taken to be

31. Christensen and Schunn (in press (b)), *op. cit.*

32.
Trickett, S.B. and Trafton, J.G. (2002) The Instantiation and Use of Conceptual Simulations in Evaluating Hypotheses: Movies-in-the-Mind in Scientific Reasoning in *Proceedings of the Twenty-Fourth Annual Conference of the Cognitive Science Society*, Erlbaum, pp. 878–883.

a mentally constructed model of a situation, phenomenon, object or system of objects than is grounded either in memory or in the mental modification of design artefacts that are currently present. As such, mental simulation enables designers to reason about new possible states of a design artefact in terms of its perceptual qualities, functions, features or attributes, but without the need for physical manipulation of the actual artefact. Mental simulations are not limited merely to technical design properties, but can also relate to envisioning a whole range of changed circumstances, including those arising from end-user interactions with the design artefact. We coded simulations in the present transcripts using a binary scheme (see Extracts 7 and 8) whereby they related either to: technical/functional aspects of the product (e.g. altering its form, function, or features); or end-user behaviour associated with the product (e.g. people's use habits, usability comprehension, or interaction experience).

Extract 7, E1, Example of technical/functional simulation.

1755	Tommy	there's two forces there isn't there [*bangs it*] there's sort of the
1756		momentum of the thing itself
1757	Alan	mmm
1758	Tommy	yeah it's not going to be anything like this heavy is it
1759	Jack	no well as I say you need to shock that down ( )
1760		+++
1761	Tommy	er
1762	Jack	you're smash you're gonna smash the edge of this protective sheath
1763		before this does anything in here
1764	Tommy	yeah also they're not that ( ) made out of ceramic and glass
1765	Jack	mmm I think that's-I think that the other other protective thing is whether
1766		they smash it off the table before momentum

Extract 8, E1, Example of end-user simulation.

691	Rodney	I think if I was in their shoes using this I'd prefer they'd be something where
692		I decide whether it's in the right position or whether I want something
693		lighting up and saying der-derrr-
694	Alan	mmm
695	Rodney	that kind of thing
696	Alan	well there's a bit of training there isn't it
697	Rodney	true but it it's kind of patronising to have these sort of lights

Notwithstanding the different foci of mental simulations, their key feature involves a simulation 'run' that alters mental representations to produce a change of state. What this means is that mental simulations entail a specific sequence of representational changes, beginning with the creation of an initial representation, then involving the running of that representation (where it is modified by spatial transformations where elements or functions may be extended, added to, deleted, etc.), and ending up with a final, altered representation[33]. These three elements (initial representation, simulation run, and changed representation) are not mutually exclusive, but could occur in the same protocol segment in the present transcripts, although typically they extended over several segments.

33. Christensen and Schunn (in press (b)), *op. cit.*

1.3 *Coding of uncertainty*

Epistemic uncertainty was coded using a purely syntactic approach – adapted from Trickett et al.[34] and Christensen and Schunn[35] – which employs 'hedge words' to locate segments displaying uncertainty. In the present analysis these hedge words included terms like 'probably', 'sort of', 'guess', 'maybe', 'possibly', 'don't know', '[don't] think', '[not] certain' and 'believe'. Segments containing these words were located and were coded as *uncertainty present* – but only if it was also clear that the hedge words were not simply being stated as politeness markers by members of the design team. All segments that were not coded as *uncertainty present* were coded as *uncertainty absent* (see Extract 9 for an example). Segments containing uncertainty made up 13% of the dataset.

34. Trickett, , Trafton, Saner, and Schunn (2005), *op. cit.*

35. Christensen and Schunn (in press (b)), *op. cit.*

Extract 9, E1, Example of an 'uncertainty present' segment (bold typeface) and an 'uncertainty absent' segment (normal typeface).

1592	Todd	my arg what I was trying to say before is erm you could do it with one
1593		switch if if the casework say the casework is comes out f-further out errm
1594		**+++ it may or may not work erm [*clears throat*] if you what it does is you've**
1595		got one switch that if it's but if it's on and back i-it works but then if you pivot
1596		pivot this can pivot a bit but it but it pivots more on the c- but the case
1597		if you pivot a little bit on the c-the case comes in contact and it starts to
1598		come away

1.4 *Inter-coder reliability checks*

The second author acted as primary coder. To assess coding consistency an individual unassociated with this research coded one hour of data (segments 500–1771 from E1). This independent coder received general training in protocol analysis and was also given some familiarisation and practice with the present coding categories using 'spare' data from the transcripts. All coding categories reached acceptable levels of reliability (i.e. greater than .70), with near perfect reliability for analogy type. Kappa reliability coefficients are reported in Table 1.

Coding category	Kappa coefficient
UNCERTAINTY	.88
ANALOGY	.77
ANALOGY PURPOSE	.85
ANALOGICAL DISTANCE	.99
SIMULATION	.75
SIMULATION TYPE	.71

Table 1. Kappa coefficients for inter-coder reliability.

36.
Christensen, B.T. and Schunn, C.D. (in press (a)) 'Putting Blinkers on a Blind Man': Providing Cognitive Support for Creative Processes with Environmental Cues in Wood, K. and Markman, A.B. (eds.) *Tools for Innovation*, Oxford University Press.

37.
Dahl, D.W. and Moreau, P. (2002) The Influence and Value of Analogical Thinking During New Product Ideation, *Journal of Marketing Research*, 39, pp. 47–60.

38.
Christensen and Schunn (2007), *op. cit.*

39.
ibid.

40.
ibid.

2 RESULTS AND DISCUSSION

2.1 *Analogies*

Across the two transcripts we identified 147 unique analogies, which ranged from 1 to 20 segments, averaging 3.5 segments per analogy. Analogies thus made up 13% of the segments across E1 and E2 and are clearly used frequently by the present designers during their product development meetings.

2.2 *Analogical distance*

Of the 147 analogies produced the vast majority (84%) were *between-domain*, with 16% *within-domain*. Previous evidence[36,37] suggests that distant analogies have a positive effect on the estimated originality of resulting product designs. Thus our observation of very high levels of distant analogising in the present context may indicate that an elevated level of innovative design was taking place. Our findings here contrast with Christensen and Schunn's[38] results, which revealed that *within-domain* and *between-domain* analogies were equally distributed across team design meetings in the area of medical plastics. Such discrepant findings may reflect domain differences between engineering sub-disciplines; alternatively, this discrepancy may have a basis in the different goals of the meetings in the two studies. In the present case the designers were primarily engaged in brainstorming activities aimed at solving problems associated with the print head mounting and pen format. Such brainstorming – with its emphasis on the creative exploration of the design space – may well encourage a focus on distant analogies rather than close ones. In addition, the present designers had been specifically requested prior to E1 to do some 'homework' so that they could come prepared with 'a product (or a picture of a product) that has to glide smoothly over contours'. This request would be likely to focus the designers on distant analogies rather than within-domain ones. In Christensen and Schunn's[39] study, in contrast, the designers were developing improvements to an existing product; this may have encouraged more within-domain analogising linked to aspects of similar products, rather than far-flung between-domain analogising.

2.3 *Analogical purpose*

Analogies were fairly evenly distributed in terms of their purpose across the categories of *solution generation* (37%), *function finding* (33%) and *explanation* (27%), with only a minority being directed toward *problem identification* (3%). The level of analogy-based problem identification in the present study is markedly lower than that observed by Christensen and Schunn[40], and, we wonder if this may again be a consequence of discrepancies in the broader goals of the project meetings in the two studies. In particular, since Christensen and Schunn's[41] designers were striving to refine an existing product, then it makes perfect sense that problem identification would have been a key activity that could have been bootstrapped through analogising. In the present study, however, problem identification was arguably not a priority since for both of the engineering meetings a number of 'problems' had already been presented to the design team as givens. For example, in relation to E1 a set of four problems had been stipulated in advance as a focus for the meeting, namely, keeping the print head level; protecting the print head; activating the print head; and constraining the print head angle. It is unlikely, then, that these pre-identified problems would have fuelled any further analogy-based problem identification. Instead, analogy-based solution generation and explanation would have been the more likely outcome of having been presented with problems to solve as part of the design brief (which is precisely what was observed).

41. *ibid.*

Despite the plausibility of these proposals we note that the high prevalence of *function finding* as a form of analogising in the present protocols is curious given that the purpose of such analogising has little to do with actually solving pre-identified problems. This type of analogising – which we believe has not previously been discovered in the literature – was observed when designers actively searched for novel ways in which the thermal printing pen might operate, and involved the mapping of a potential *function* from a source domain onto the existing design *form* in the target domain. Such analogising, whilst not specifically related to problem solving, may well have been a consequence of the remit of these meetings as being to engage in brainstorming, the very point of which is to pursue a creative exploration of the design space. Although much of the brainstorming in the present meetings was focused on the required set of 'problems' that needed to be solved, it appears that the team's brainstorming also transitioned (either strategically or inadvertently) into phases of exploratory concept generation fuelled by function-finding analogies.

This latter *form-before-function* activity involved designers taking the novel mechanical form and reflecting on what could be done with it in terms of functions. This is a very different process to 'design-as-usual', which progresses in a *function-before-form* manner, where functional requirements need to be realised as a blueprint for an implementable artefact. To examine further the relation between different types of analogising and different types of design activity we coded transcripts in terms of the overall *design question* being pursued. In this way we were able to identify two primary types of design episode: a single form-before-function episode and three function-before-form episodes.

Table 2. Distribution of analogy types across 'function-before-form' and 'form-before-function' episodes.

	Function-before-form episodes	Form-before-function episodes
problem identification	5	0
solution generation	52	1
explanation	30	10
function finding	6	43

(NB: Two other episodes related to 'introductory comments' and were excluded from the analysis.) The distribution of analogies across these two episode types is shown in Table 2. It is apparent that the episode where designers were seeking new functions for the existing form was associated with nearly all of the function finding analogies and relatively few of the other types of analogies, an effect that was highly reliable with a chi-square analysis, $\chi^2(3) = 87.85, p < .001$.

2.4 *The relation between analogical distance and analogical purpose*

It is also of interest to know whether the analogical distance parameter (i.e. the extent to which analogies are *within-domain* or *between-domain*) varies in relation to analogical purpose. To examine this issue we conducted a one-way Analysis of Variance (ANOVA), with analogical distance as the dependent variable (1 = within domain; 2 = between domain) and with analogical purpose as the independent variable (see Table 3 for descriptive data). This analysis revealed a reliable effect of analogical purpose, $F(3, 146) = 14.04$, $p < .001$. From Table 3 it is apparent that analogies linked to problem identification were exclusively *within-domain*, whilst all other analogy types were predominantly *between-domain*, with function finding analogies being markedly between-domain in nature. Post hoc analyses using Tukey HSD tests confirmed that *problem identification* analogies were significantly different in analogical distance to all other analogies (*ps* < .001). *Function finding* analogies were also significantly different to *solution generation* analogies ($p = .03$), being more *between-domain* in nature.

2.5 *Analogies and uncertainty*

Our analysis also explored the existence of a possible association between analogising and epistemic uncertainty. We applied the binary scheme described previously to categorise all protocol segments in terms of presence or absence of uncertainty. Our analysis then focused on the number of segments during analogies that showed either presence or absence of uncertainty, and compared

	Mean	Standard deviation
problem identification	1.00	.00
solution generation	1.80	.41
explanation	1.83	.38
function finding	1.98	.14

Table 3. Mean analogical distance scores (plus standard deviations) for analogy types.

	uncertainty absent	*uncertainty present*
Baseline segments	2325	325
Five segments before analogy	309	49
Segments during analogy	421	93
Five segments after analogy	313	51

Table 4. Number of segments revealing presence versus absence of uncertainty before, during and after analogies as well as for baseline segments.

these values with those for the 5 segments before the analogy, the 5 segments after the analogy, and all remaining segments (the latter providing a baseline measure). Table 4 indicates that only analogies had uncertainty levels that were elevated above baseline, an effect that was reliable with a chi-square analysis, $\chi^2(3) = 12.89$, $p < .005$. This very tight temporal coupling between analogising and uncertainty suggests that analogies are instantiated *coincident* with situations of design uncertainty whilst also facilitating the *resolution* of such uncertainty (since post-analogy segments resume baseline uncertainty levels). This makes a lot of sense in that analogising involves exploratory reasoning about concepts and mappings that are synchronised to deal with immediate design uncertainties, with these concepts and mappings typically being helpful in removing doubts and improving understanding.

2.6 *Mental simulation*

Across the two engineering transcripts we located 130 unique simulations (83 technical/functional; 47 end-user). These ranged from 3 to 25 segments, averaging at 8.4 segments per simulation. Simulations thus made up 28% of the segments within the transcripts. Of the 130 simulations, 124 contained identifiable segments for all three simulation parts: initial representation, simulation run, resulting representation.

2.7 *Mental simulation and uncertainty*

To examine whether mental simulations are associated with epistemic uncertainty we compared the coded segments that had been categorised in terms of the presence or absence of uncertainty (see Section 2.5. above) with the segments where simulation was present or where simulation was absent. The data from this analysis are presented in Table 5. It is evident that segments where simulation was present show a higher association with uncertainty compared to segments where simulation was absent. This effect was highly reliable with a chi-square analysis, $\chi^2(1) = 24.75, p < .001$.

Having demonstrated an association between mental simulation and uncertainty our next focus was on examining the temporal relationship between simulation and uncertainty in terms of changes in uncertainty over the three stages of simulations: initial representation, simulation run, and resulting representation. To pursue this analysis we first calculated the proportion of uncertainty segments in each stage to provide mean uncertainty scores for the initial representation, the simulation run, and the resulting representation. A paired-samples t-test revealed that initial representations had significantly higher uncertainty scores (Mean = 26%) than resulting representations (Mean = 15%), $t(125) = 2.58$, $p = .011$, two-tailed. Likewise, simulation runs had significantly higher uncertainty scores (Mean = 23%) than resulting representations (Mean = 15%), $t(124) = 2.43$, $p = .016$, two-tailed. Whilst the difference in uncertainty between initial representations and simulation runs was not reliable, $t(126) = 0.71$, $p = .48$, two-tailed, the essential pattern of results clearly indicates that mental simulations are serving to reduce uncertainty over time.

It is also worth noting that the baseline measure for uncertainty across all segments where simulation was absent was 13%. One-sample t-tests comparing each simulation stage against this baseline revealed that both initial representations and simulation runs were significantly above baseline levels of uncertainty: $t(128) = 3.81$, $p < .001$, two-tailed, and $t(127) = 3.25$, $p = .002$, two-tailed, respectively. The difference between resulting representations and baseline uncertainty was not reliable, $t(126) = 0.72$, $p = .47$, two-tailed, suggesting that by the end of the simulation uncertainty had diminished to baseline levels. Overall, then, our analysis of the temporal associations between mental simulation and uncertainty replicated Christensen and Schunn's[42] observations. Our data thus appear to validate the hypothesis that mental simulation is a strategic aspect of design cognition that functions to reduce epistemic uncertainty.

42. Christensen and Schunn (in press (b)), *op. cit.*

2.8 *The association between mental simulation and analogising*

One final issue of considerable interest concerns the possible existence of meaningful associations between mental simulation and

Table 5. Number of *simulation present* and *simulation absent* segments revealing presence versus absence of uncertainty.

	uncertainty absent	*uncertainty present*
simulation present	892	202
simulation absent	2476	316

analogising. To examine this issue we first determined whether any analogies were embedded or partly embedded within mental simulations in the sense that the analogies 'started' within some part of the simulation, whether in the initial representation, the simulation run, or the resulting representation. It transpired that 48 analogies showed a direct association with mental simulations (see Extract 10 for examples of analogising arising during mental simulation).

It was also possible to examine the way in which analogies having different 'purposes' were linked to mental simulations. Table 6 shows a breakdown of different types of analogies as a function of their actual starting point within the simulation. The distribution of analogies reveals some intriguing results. First, *solution generation* and *function finding* analogies appear early in the mental simulation, seemingly being 'generative' in nature (arguably producing novel variations that are explored in the subsequent run). Second, explanatory analogies tend to appear later in the mental simulation, seemingly arising in order to explain the simulation run or to explain the resulting representation.

Extract 10, E1, Analogising (bold typeface) occurring in association with mental simulation.

1341	Alan	is there is there an issue anyway erm guys with there being or having to be
1342		or a benefit by having like a stand-by mode so it's either completely
1343		switched off with or without the cap on it or there's like a stand-by mode
1344		where it's sort of semi-warm but it's ready for action quickly
1345	Tommy	it has a home so a **docking station**
1346	Alan	yeah nice one
1347	Tommy	we could charge it in there as well plus it might be over budget but let's not
1348		worry about that for now
1349	Alan	yeah so a docking station what would the docking station look like
1350		(charge)
1351	Tommy	well it would just be a **cradle** it would just be somewhere for it to live
1352		when you're not using it like a little protector
1353		like an **inkwell**
1354	Alan	yeah
1355	Todd	[*laugh*] like a **quill**
1356	Rodney	a **quill** that means it's not something that you could put in your pencil case

It was not appropriate to apply a chi-square analysis to the data in Table 6 since 6 cells within the contingency table had expected counts of less than 5. We therefore combined the simulation run and resulting representation categories into a single 'late stage' category. A comparison of *solution generation* analogies versus *explanation* analogies confirmed the contrasting distribution of these analogy types across the 'early' versus 'late' stages of the mental simulation process, $p < .002$, two-tailed, Fisher's Exact test.

Table 6. The distribution of analogies as a function of their starting point within the three stages of mental simulations.

	Analogical purpose		
	solution generation	*explanation*	*function finding*
Initial representation	16	4	9
Simulation run	3	4	3
Resulting representation	1	8	0

A similar comparison of *function finding* versus *explanation* analogies also confirmed the contrasting distribution of these analogy types across the early versus late stages of the mental simulation, $p < .02$, two-tailed, Fisher's Exact test.

3 GENERAL DISCUSSION

Our aim in this work was to use the two engineering transcripts to pursue a detailed analysis of analogising and simulation strategies in design, with a close eye on the potential role of these strategies in dealing with the epistemic uncertainties that typically arise in design contexts[43]. Our previous research has indicated that epistemic uncertainty is critically related to:

- switches between breadth-first and depth-first modes of design development[44,45]
- strategic recourse to 'satisficing' heuristics in design evaluation[46]
- transitions from structural to functional modes of representation during the sketching of design objects[47,48].

In this study we wished to examine possible associations between epistemic uncertainty and strategies based around analogising and mental simulation. Christensen and Schunn[49] have already provided some compelling evidence that mental simulations are run under situations of subjective uncertainty to enable inferences that subsequently reduce such uncertainty. The DTRS7 dataset provided an excellent opportunity to validate this finding with a different design team working in a different design context and tackling a different engineering design task. Likewise, although previous research has informed our understanding of the nature of analogical reasoning in design[50,51,52], we are not aware of studies that have attempted to draw links between analogising and epistemic uncertainty.

In terms of key findings, our analyses revealed that analogising and mental simulation are indeed intimately associated with situations of epistemic uncertainty in design. In the case of analogising, analogies were found to be temporally coupled with situations involving expressions of uncertainty, whereas pre-analogy and post-analogy segments revealed levels of uncertainty that were close to baseline values. Our interpretation of these observations is that analogical reasoning is a core design strategy that

43.
Schlosser and Paredis (2007), *op. cit.*

44.
Ball, L.J., Evans, J.St.B.T., Dennis, I., and Ormerod, T.C. (1997) Problem-Solving Strategies and Expertise in Engineering Design, *Thinking and Reasoning*, 3, pp. 247–270.

45.
Ball, L.J. and Ormerod, T.C. (1995) Structured and Opportunistic Processing in Design: A Critical Discussion, *International Journal of Human-Computer Studies*, 43, pp. 131–151.

46.
Ball, L.J., Maskill, L. and Ormerod, T.C. (1998) Satisficing in Engineering Design: Causes, Consequences and Implications for Design Support, *Automation in Construction*, 7, pp. 213–227.

47.
Kavakli, M., Scrivener, S.A.R. and Ball, L.J. (1998) Structure in Idea-Sketching Behaviour, *Design Studies*, 19, pp. 485–518.

48.
Scrivener, S.A.R., Ball, L.J. and Tseng, W.S-W. (2000) Uncertainty and Sketching Behaviour, *Design Studies*, 21, pp. 465–481.

49.
Christensen and Schunn (in press (b)), *op. cit.*

50.
Ball, Ormerod, and Morley (2004), *op. cit.*

51.
Casakin (2004), *op. cit.*

52.
Christensen and Schunn (2007), *op. cit.*

is instantiated *coincident* with situations of design uncertainty, serving to facilitate the *resolution* of such uncertainty. Turning to mental simulation, our analysis replicated Christensen and Schunn's[53] findings and demonstrated that mental simulations arise concurrent with situations of uncertainty and, moreover, that levels of uncertainty dissipate to baseline values over the course of simulations. A further intriguing aspect of our analysis was that analogies were observed to interleave with mental simulations. Analogies within mental simulations that are aimed at solution generation and function finding appear to have a 'generative' role in design, whereby solution ideas are produced that are then explored and evaluated in subsequent simulation runs. On the other hand, explanatory analogies within mental simulations are mainly invoked to explain the nature of the simulation run or resulting representation; as such, they appear primarily in the later stages of mental simulations.

53. Christensen and Schunn (in press (b)), *op. cit.*

The existence of *function finding* analogies in the present transcripts seems a unique observation that appears not to have been identified previously. We also note that these function finding analogies were associated with a 'form-before-function' mode of design reasoning, with a single, extended episode of such reasoning being apparent in the transcripts. Form-before-function activity is rather different to the more typical 'function-before-form' reasoning seen in many design situations, and its occurrence is suggestive of some distinctive characteristics of the DTRS7 engineering design meetings[54]. In particular, we propose that the focus of the meetings on brainstorming may have shifted the design team into a creatively rich phase of exploratory concept generation fuelled by function-finding analogies. It would be valuable to determine if this observation can be replicated in other brainstorming-oriented design meetings. Our analyses also permitted examination of the nature of simulation types and analogy types in design using previously identified categories (e.g. technical/functional vs. end-user simulations, within-domain versus between-domain analogies, and problem identification versus solution generation analogies). Our data here were broadly consistent with previous findings[55,56], with discrepancies across studies being interpretable in terms of differences in the goals of the design teams (e.g. the brainstorming remit in the present context could account for the increase in between-domain analogies in comparison to the rates observed in earlier research).

54. See Finke, Ward, and Smith (1992) for other evidence – as well as a cognitive model – of a form-before-function approach in creative invention and design. Finke, R. Ward, T.B. and Smith, S.M. (1992) *Creative Cognition: Theory, Research, and Applications*, MIT Press.

55. Christensen and Schunn (2007), *op. cit.*

56. Christensen and Schunn (in press (b)), *op. cit.*

By way of some final comments, we briefly reflect on how our examination of analogy, mental simulation and uncertainty in design relates to ideas presented by other authors within this volume. The chapter that relates most closely to our own analysis is that by Stacey, Eckert and Earl[57], which discusses designers' references to *previous* design objects and the role that such references

57. Stacey, M., Eckert, C. and Earl, C. (Chapter 20) From Ronchamp by Sledge: On the Pragmatics of Object References

play in design thinking, including analogy-based design reasoning. These authors are critical of our own analogy-categorisation scheme (also used by Christensen and Schunn)[58] that codes analogies in terms of 'purposes' such as problem identification, solution generation and explanation. We would argue, however, that our current analysis appears to validate this coding approach since we were readily able to categorise *all* design analogies with high levels of inter-coder reliability. Moreover, the way in which different analogy types were distributed across distinct phases of mental simulations attests to the sensitivity of the coding scheme to interesting nuances in design activity. We are also sceptical about Stacey, Eckert and Earl's[59] attempt to extend an analogy coding scheme through the addition of new categories such as that of 'prepackaged analogies' (solution-generation analogies produced when designers are working alone that are then relayed within the meeting context). We suggest that describing solution-generation analogies as prepackaged adds a confusing *temporal* dimension to a purpose-oriented coding scheme. Prepackaged analogies are clearly *generative* in nature whether they arise prior to or during a meeting. At the same time, we note that Stacey, Eckert and Earl's[60] focus on object references in design is far broader than ours, and we applaud their attempt to extend the analysis of prior object knowledge beyond concerns with analogical mapping. For example, we welcome their identification of object references that are used to set up contrast classes that define what a current design should *not* be like (what they term 'synthesis by exclusion'). We also value their observation that mappings from previous objects can *blend* in complex ways that suggest creative design ideas.

58. Christensen and Schunn (2007), *op. cit.*

59. Stacey, Eckert and Earl (Chapter 20) ***op. cit.***

60. ***ibid.***

4 CONCLUSION

In conclusion, we believe that analysing the engineering transcripts for instances of analogising, mental simulation and uncertainty has provided valuable insights into central aspects of processing in innovative design contexts. In particular, we have replicated and extended previous findings linking epistemic uncertainty to key design strategies. In addition, we have been alerted to novel issues that will be important to examine in the future, such as the role of function finding analogies as a dominant aspect of form-before-function reasoning in creative design.

ACKNOWLEDGEMENTS

We thank Kristoffer Riis Pedersen for undertaking coding reliability checks.

Taking these processes into account we conceptualise the main requirements of teamwork as:

1. Working on the task,
2. Coordination of the process,
3. Coordinating roles and responsibilities within the team,
4. Creating and maintaining team cohesion.

According to these requirements our model (Figure 1) refers to four types of mental models; task, process and two team aspects. This model depicts two different processes related to the development over time: first, the part of creating shared understanding (upper part, solid arrows) and second, the effects of shared understanding on the basis of a team mental model (lower part, dotted arrows). Groups that have achieved shared understanding continue to employ the same cognitive processes, yet the frequency of certain activities decreases once shared knowledge on how to collaborate on the task and within the team has been acquired. The model focuses on the cognitive acts that seem to be the most important in creating shared understanding such as: task-related information exchange, agreement about the approach, agreement about team roles, and signals of confirmation and attentive listening that encourage team members to continue and enhance cohesiveness within the team.

The knowledge gained can be categorised into three different types. Task-specific knowledge refers to the knowledge of the task in hand. Strategic knowledge refers to ways of approaching a design

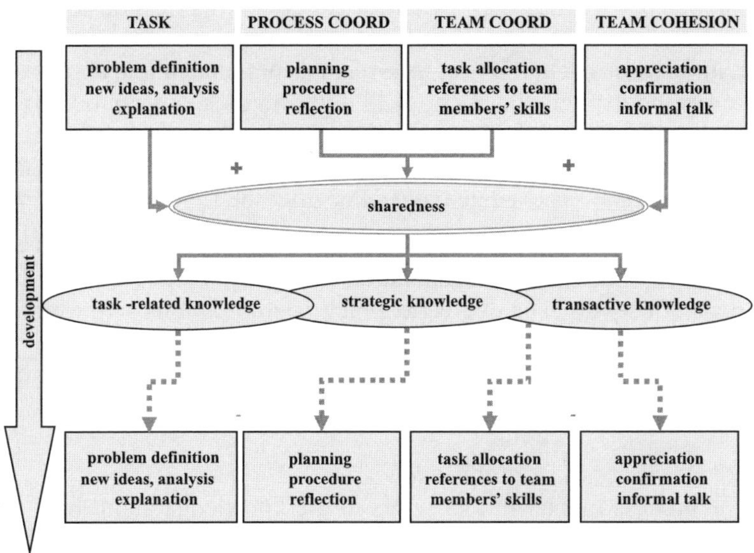

Key: + denotes the more of the activity the more the sharedness during forming, storming and norming
– denotes the more of the activity the less the sharedness during the performing stage

Figure 1. A model of the development of sharedness in teams.

task. Transactive knowledge is knowledge acquired about group members; transactive knowledge contributes to transactive memory as described by Wegner, Giuliano and Hertel[14] and Wegner[15]. Transactive memory is memory about who knows what rather than each person remembering everything.

14. Wegner, D.M., Giuliano, T., and Hertel, P. (1985) Cognitive Interdependence in Close Relationships in Ickes, W.J. (ed) *Compatible and Incompatible Relationships*, Springer, pp. 253–276.

15. Wegner, D.M. (1986) Transactive Memory: A Contemporary Analysis of the Group Mind in Mullen, B. and Goethals, G.R. (eds), *Theories of Group Behavior*, Springer, pp. 185–208.

1.1.1 *Task mental models*

The exchange of information and thus the understanding of the problem and the sharedness of the task mental model are expected to increase with task progression. Information exchange is particularly related to problem definition and evaluation, whereas analysis and evaluation refers to the solution space. Explanations concern questions asked due to missing knowledge. In terms of time, an increase of exchange of information is expected in the first phase and a decrease in the second.

1.1.2 *Team mental models*

In terms of functionality two different team mental models can be distinguished: group coordination and group cohesiveness. Group coordination demands task coordination by allocating roles to team members and referring to individual abilities, knowledge, skills and experience. We assume that both aspects are essential in the first phase and decrease once a shared understanding about these issues has been achieved. In addition to group coordination there seems to be a more affective part of group membership supporting sharedness. Group cohesiveness describes the desire of members to be part of a particular group[16]. There are three communication acts that contribute to cohesiveness: informal talk, appreciation and confirmation. Informal talk is non-task-related communication; appreciation is the explicit statement of liking towards a contribution of a team member; and confirmation is a shorter version of appreciating a contribution by another team member. Group cohesiveness communication is expected to decline after the first phase.

16. Goodman, P.S., Ravlin, E., and Schminke, M. (1987) Understanding Groups in Organizations, in Staw, B.M., Cummings, L.L. (eds) *Research in Organizational Behavior*, JAI Press, Vol. 9, pp.121–175.

Both team coordination and process interventions are means to achieve collective action regulation[17] and help a team to reach its goals effectively. The degree to which coordination is achieved by one individual indicates to what extent the group has a formal or informal leader. If several members contribute, responsibility and authority is more likely to be shared in the group.

17. Weber, W.G. (1997) Analyse von Gruppenarbeit, *Kollektive Handlungsregulation in soziotechnischen Systemen*, Huber.

1.1.3 *Process mental models*

The process mental model refers to the knowledge about how to solve a task. Three aspects can be distinguished: first, planning when to do what; second, the procedural aspects of how to solve a problem and which strategies or methods to use; third, reflection as meta-communication about the process, including intra-team feedback and self-correction mechanisms to adapt the process responding to changing conditions or to unsuccessful results. During a meeting, we expect an increase of planning and procedural aspects in the first phase and a decrease in the second phase. Once a decision is made about how and when to do what, only minor adaptations should be necessary, whereas reflection might be useful throughout the whole process.

2 METHODS

2.1 *Participants and task*

The two transcripts for engineering meetings E1 and E2 were selected for analysis as they involved members of a multidisciplinary design team working together on a larger project. The two meetings followed each other as part of the same project with a three-day interval and were approximately the same length (96 and 100 minutes respectively). Although the thematic focus of the two meetings was different, the two meetings can be regarded as two instances within the same overall design process. As is normal for projects in industry, there is some fluctuation of attendees between E1 and E2 and there are no clear boundaries for what constitutes membership of this design team. Meeting participants need to develop team mental models that can account for variations in focus and member attendance so as to work effectively on the project, thus we treat the two meetings as part of one overall process. Gericke, Schmidt-Kretschmer, and Blessing[18] also showed that the basic activities remained the same across E1 and E2 and that they can therefore be regarded as part of the same process.

18.
Gericke, K., Schmidt-Kretschmer, M., and Blessing, L. (Chapter 11) The Influence of the Design Task Description on the Course and Outcome of Idea Generation Meetings.

The task was to develop a thermally printing 'digital pen', and team members were aware that each meeting was to generate ideas for this. Prior to the first meeting, the participants had been asked to consider analogies or possible solutions for the assignment and were briefed about the major topics to be discussed during each session. Both meetings were attended by seven members. In E1, a business consultant (Alan), an electronic and business developer (Tommy), three mechanical engineers (Jack, Todd, and Chad), an ergonomics and usability expert (Sandra), and an industrial designer who also functioned as project leader (Rodney) took part. In E2, instead of the business consultant and two mechanical engineers, two electronics and software experts (Patrick and Roman) and one electrical engineer (Stuart) participated.

2.2 *Research approach*

Mental models as cognitive entities cannot be observed as such, so those wishing to study them have to try to elicit the content of someone's mental model and make inferences based on the analysis of externalisations. In team settings, verbal communication provides a natural angle into the mental models of the members.

Our approach was therefore to code verbal utterances in terms of our proposed model and to infer the development of team mental models from that.

The development over time was analysed by comparing the two meetings E1 and E2 as well as the first and second half of each meeting. A timeline of the frequency of codes was established for segments of the meeting. In order to compare frequencies, the segments were not based on content but had to be of equal length and not too short allowing a statistical analysis of frequencies between segments. Each meeting was divided into five segments.

The main part of the data analysis concerns the utterances and was carried out at two different levels:

1. Cognitive acts: Analysis of mental representations in terms of cognitive acts referring to the main requirements of working on a design task in a team: task, team and process.
2. Strategic acts: Analysis of strategies and reasoning processes employed by the team members during the two meetings of the engineering group.

2.3 *Coding schemes*

Three categorisation systems were developed for the purpose of this analysis, each being mutually exclusive and exhaustive: CONTENT, COGNITIVE ACTS, and STRATEGIES. Three broad CONTENT categories, TECHNICAL, BUSINESS, and USER identified the topic of discussion (Table 1). For COGNITIVE ACTS, a categorisation scheme was developed using the above-mentioned four types of team mental models: TASK, PROCESS, GROUP COORDINATION, and GROUP COHESIVENESS (Table 2). Segments were also coded for the use of STRATEGIES and REASONING behaviour (Table 3). If more than one topic was addressed within one utterance, the statement was split so that each part could be assigned a single category per categorisation system. COGNITIVE ACTS and CONTENT were coded by three members of the research team, and Cohen's Kappa coefficient was found to be 0.72.

STRATEGIES such as BEING AWARE OF UNDERLYING REASONS, TRADE-OFFS and LIMITATIONS enable team members to keep these considerations in mind if they need to modify the design. The categorisation for REASONING behaviour is a subset of a coding scheme developed for

Table 1. CONTENT categorisation system.

TECHNICAL	Technical aspects of the problem or the solution
BUSINESS	Business aspects such as costs and market considerations
USER	User, target group and context of use

TASK	
PROBLEM DEFINITION & ELABORATION	Defining the problem, elaborating and analysing the constraints, requirements, and the goal of the task.
NEW SOLUTION IDEA OR NEW SOLUTION ASPECT	Stating a new product idea or a new solution for an earlier defined problem or sub-problem or identifying new aspects building on an earlier solution idea.
ANALYSIS	Analysis of the properties and feasibility of a solution idea, analysis of the usage of a product idea and its potential applications, for example by referring to similar products, and evaluation of a solution idea by appraising its feasibility or analysing failure and safety aspects.
EXPLANATION	Clarification questions and explanations about specific (technical) issues, for example by referring to specialised knowledge.
SOLUTION DECISION	Making a solution definitive by the whole team accepting it.
PROCESS	
Utterances about the organisation of when to do what (PLANNING), about how to approach the task, for example how to apply a method (PROCEDURE), and utterances about what and how the team is doing (REFLECTION).	
TEAM COORDINATION	
Role allocation to team members and references to personal abilities, knowledge, skills, or experience.	
TEAM COHESIVENESS	
All utterances about group coherence are included in this category: appreciations about a solution idea or a problem definition, confirmation, and informal communication (for example, joking).	
OTHER	All utterances that are not defined in the categories above.

Table 2. Categorisation system for COGNITIVE ACTS.

understanding how designers acquire and process information[19]. Coding was undertaken by one person. A coder reliability check over 138 utterances (with a second coder not from this research

STRATEGIES	
CONSIDER ISSUES	Awareness of relevant issues, deciding which are the most important.
AWARE OF REASON	Awareness of the reasons behind the use of, for example, a particular component or manufacturing process in a particular design. Knowledge of how the design/technology works.
REFER TO PAST DESIGNS	Referring to past designs with similar issues, designs, or technology. This includes direct previous experience and analogies from other products.
QUESTION DATA	Questioning data for the applicability to current context, including values given for standards, stress models, etc.
KEEP OPTIONS OPEN	Rejecting an option (or delaying a decision) if it then limits later decisions in the design process.
AWARE OF TRADE-OFFS	Awareness of the relationships between any one issue or system and the effect on other issues or systems.
AWARE OF LIMITATIONS	Awareness of the limitations of the current design task and what is expected to be achieved considering the task and information available.
WORTH PURSUING?	Assessment of pursuing a particular solution by predicting what is expected to be achieved.
REASONING	
All aspects indicating reasoning including INTENDED BEHAVIOUR i.e. the desired function of the design, product, assembly, component, material or feature; FORM, defining the geometry and material of the concept, PREDICTED BEHAVIOUR and OBSERVED BEHAVIOUR	

Table 3. STRATEGIES and REASONING categorisation.

19.
Aurisicchio, M., Bracewell, R., and Wallace, K. (2006) Characterising in Detail the Information Requests of Engineering Designers in *Proceedings of ASME 2006 Design Engineering Technical Conferences and Computers and Information in Engineering Conference*, Philadelphia, Pennsylvania, USA.

team) resulted in a Kappa coefficient of 0.645, which was considered reasonable given that the coder was new to the categorisation scheme.

3 RESULTS

In meeting E1 a total of 1513 utterances were identified as compared to 1217 utterances ($\chi^2 = 16.1$, $p < 0.01$) in meeting E2, hence the group was more active in E1 facilitated by a professional moderator. As Figure 2 shows, the most frequent categories in both meetings were SOLUTION ANALYSIS and GROUP COHESIVENESS. In the following sections the results are discussed in relation to the predictions relating to our model of the development of sharedness.

3.1 *Task: Unexpected rise of PROBLEM DEFINITION and EXPLANATION*

According to our prediction the frequency of PROBLEM DEFINITION utterances was expected to decrease as the team achieved more problem clarity over time. Also the frequency of SOLUTION ANALYSIS should decrease over time, as important issues needing further analysis and explanation should have already been discussed earlier. The category EXPLANATION means that team members provide each other with explanations or elicit information, leading to increased sharedness. Assuming that shared understanding is low at the beginning, a high number of EXPLANATIONS is expected. By asking questions and providing information, the team creates a common knowledge base. Therefore, EXPLANATIONS were expected to increase during E1 but decrease during E2.

However, the data present a different picture (see Figure 2). The frequencies of PROBLEM DEFINITION increased significantly from E1 to E2 ($\chi^2 = 24.26$), similarly those for EXPLANATIONS ($\chi^2 = 23.8$). The frequencies of SOLUTION ANALYSIS decreased, albeit not significantly. PROBLEM DEFINITION significantly decreased from the first to the second half of E1 ($\chi^2 = 10.14$), but rose again at the start of E2 (Figure 3). The frequency of PROBLEM DEFINITION was significantly higher in the second half of E2 – an unexpected finding and one also addressed by Gericke, Schmidt-Kretschmer, and Blessing[20]. CLARIFICATIONS also became more frequent towards the end.

20.
Gericke, K., Schmidt-Kretschmer, M., and Blessing, L. (Chapter 11) *op. cit.*

These findings suggest that the group felt the need to return to the problem definition and clarify information, which could indicate that the sharedness they developed earlier was not sufficient. However, a qualitative analysis of the latter half of E2 showed that a more likely explanation is that the group began to tackle more intricate problems and discuss issues at a deeper level. The problem definition was not actually redefined – the issues

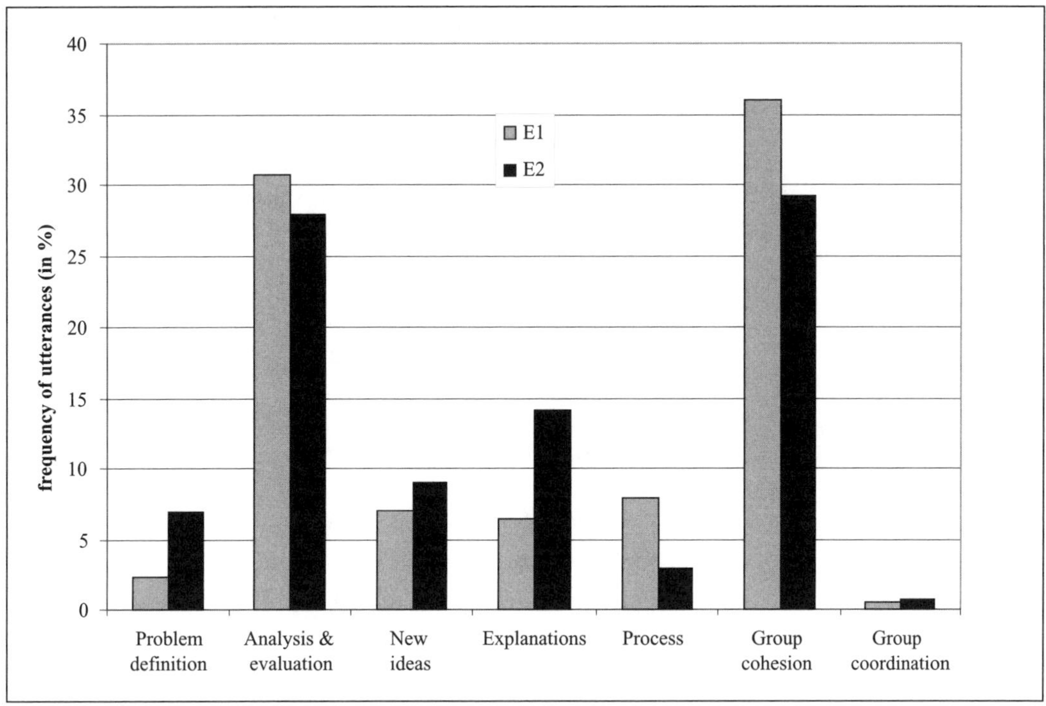

Figure 2. Comparison of the frequencies of COGNITIVE ACTS during meetings E1 and E2.

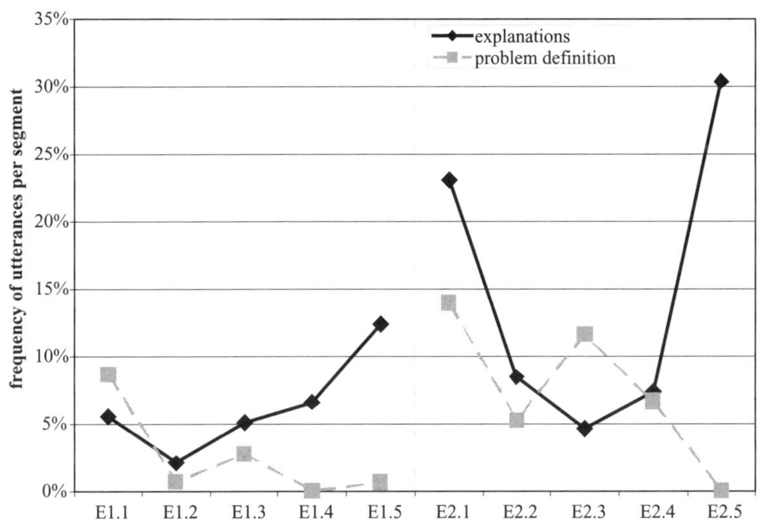

Figure 3. Frequencies of PROBLEM DEFINITION and EXPLANATIONS for E1 and E2, each meeting divided into five equal parts.

were the same, such as power, charging, cost and heat. Some members reiterated these issues to underline their importance, and new solution ideas were judged against the requirements already agreed.

The sudden rise of EXPLANATIONS at the very end of E2 (see Figure 3) is an indication that the team started a new topic, the IT architecture, for which they had to share their mutual knowledge of technology and costs. The change in topics was initiated by Tommy's comment (E2, 1479–1481) that he would like to discuss the architecture, this moment is also identified by Atman et al.[21] as a shift, which in their analysis they mark as the start of a new, final episode in the meeting. At this point the usability expert takes her leave and the rest of the team starts again to build up sharedness on this new aspect.

21.
Atman, C., Borgford-Parnell, J., Deibel, K., Kang, A., Ng, W.H., Kilgore, D., and Turns, J. (Chapter 22) Matters of Context in Design.

3.2 PROCESS *related utterances decrease as expected*

According to our model we expected that after a phase of explicit coordination in terms of planning and procedure the team would develop shared understanding of these issues. More sharedness would lead to more implicit coordination. Thus, the number of PROCESS utterances should decrease. REFLECTIONS on what the team is doing and how they are proceeding should not decrease dramatically because the team should maintain awareness about their process throughout the meeting.

This is exactly what happened. There was a significant decrease in PROCESS utterances from E1 to E2. PROCESS utterances decreased

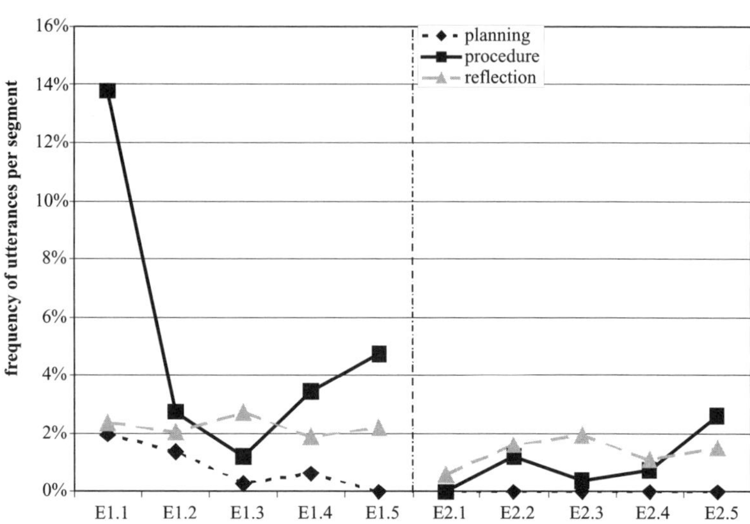

Figure 4. Frequencies of the three process categories PLANNING, PROCEDURE and REFLECTION in E1 and E2.

most dramatically between the first and second part of E1, whereas the frequency of PROCESS utterances stayed the same during E2. As the distribution of the categories PLANNING, PROCEDURE and REFLECTION (Figure 4) shows there are different developmental acts of planning and procedure on the one hand, and reflection on the other.

3.3 *TASK and PROCESS strategic knowledge*

The transcripts were analysed for instances of the use of strategies (see Table 4). The most dominant strategies for both meetings were AWARE OF REASON and CONSIDER ISSUES. This result is in line with previous observations[22], although according to these studies REFER TO PAST DESIGNS was expected to be observed more frequently, which may be a result of the differences in the nature of the design tasks. In this case, there was only one reference in the second meeting to past designs from the first meeting, however many references to products were made. Atman et al also associate CONSIDER ISSUES with more experienced designers in terms of the number of factors considered. STRATEGIES were observed for 39% of the duration of E1 and 56% of the time in E2.

22. Ahmed, Wallace and Blessing (2003), *op. cit.*

The analysis focused on the contribution of strategies to the progression of the design task, defined as the observance of any of the REASONING categories, i.e. utterances related to either the behaviour of the product (including INTENDED BEHAVIOUR, PREDICTED BEHAVIOUR and OBSERVED BEHAVIOUR) or towards defining the FORM. However, this does not necessarily relate to the progression of the final design, as ideas (both rejected and accepted), analysis and evaluation are included. The strategies were analysed against the REASONING categories, and it was found that 57% of the time the strategies corresponded to design progression (55% in meeting 1 and 59% in meeting 2). The strategies were also used when not directly contributing to the progression of the design. The predominant reasons for this were for SOLUTION EVALUATION and CLARIFICATION, ANALYSIS and EXPLANATION. This time was used towards obtaining shared understanding; during E1 this was mainly for understanding issues surrounding USAGE, whereas in E2 this was predominantly CLARIFICATION, and UNDERSTANDING OF TECHNICAL ISSUES. For those utterances identified as TEAM COHESIVENESS (total of 779), there was hardly any use of STRATEGIES (6%), which is not surprising as the category included APPRECIATION and INFORMAL TALK during which strategies would not be expected to be observed. A similar observation was made about the difference between net and gross discourse in design talk by Goldschmidt and Eshel[23] in their study of the architectural design data.

23. Goldschmidt, G., and Eshel, D. (Chapter 18) Behind the Scenes of the Design Theatre: Actors, Roles and the Dynamics of Communication.

Innovation, i.e. utterances containing new ideas and new aspects of ideas, was analysed for the use of strategies employed; 216

Instances of STRATEGIES	E1	E2
CONSIDER ISSUES	230	210
AWARE OF REASON	189	279
KEEP OPTIONS OPEN	18	8
QUESTION DATA	2	8
REFER TO PAST DESIGNS	2	15
AWARE OF LIMITATIONS	7	4
WORTH PURSUING?	8	26
AWARE OF TRADE-OFFS	1	3
Total	457	533
Duration of STRATEGIES		
CONSIDER ISSUES	19%	21%
AWARE OF REASON	17%	28%
All STRATEGIES observed	39%	56%

Table 4. Instances and duration of STRATEGIES employed in E1 and E2.

instances were identified, and 58% of these correlated to the use of a strategy. The predominant strategies were AWARE OF REASON (28%) and CONSIDER ISSUES (25%). The remaining 42% of new ideas or new aspects are not correlated to the strategies employed at that instance. This could be that the analysis did not consider the use of strategies immediately before the new idea or aspect is uttered, or it could also be due to the team environment, where one would expect individuals to verbalise less than in protocol studies of individual design tasks for example.

The meetings were also analysed to understand how the strategies were employed over time (Figure 5). During E1 there is a period in the middle of the meeting where the team spends the majority of time CONSIDERING ISSUES; this period is followed by switching to AWARE OF REASON. The team is identifying relevant issues before trying to understand the technology or how solutions could work. The designers were expected to spend a period of time understanding the problem and technology before proceeding to a synthesis phase during which the design would progress. The strategy AWARE OF REASON, relating to understanding how the technology or the product is being designed, was expected to be associated with problem definition activities. Both initial periods of the meetings, E1 Segment 1 and E2 Segment 1, were observed to be high in the problem definition phase and high AWARE OF REASON confirmed this (Figure 6). However, in both meetings an increase in AWARE OF REASON is observed at the end of the meetings, which is no longer related to problem definition, but more likely to solution generation, analysis and evaluation, something also observed by Atman et al[24].

24.
Atman, C., Borgford-Parnell, J., Deibel, K., Kang, A., Ng, W.H., Kilgore, D., and Turns, J. (Chapter 22) ***op. cit.***

3.4 *Team*

Mental models concerning the team are assumed to serve two functions: providing GROUP COHESIVENESS and ensuring GROUP COORDINATION. Both aspects are discussed separately below.

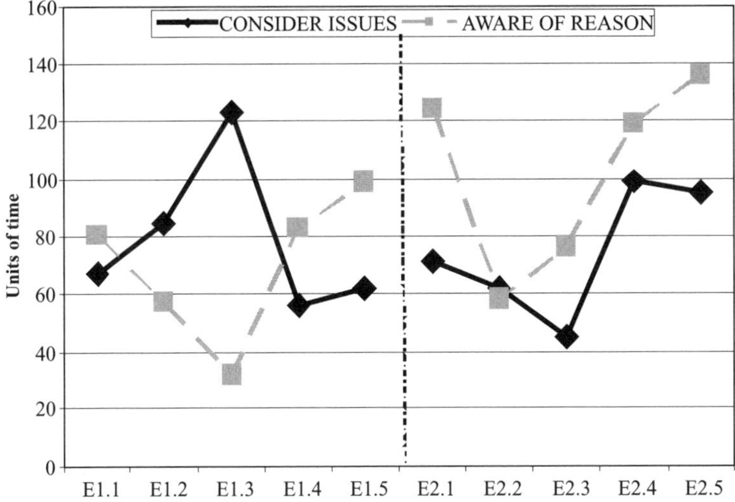

Figure 5. Use of CONSIDER ISSUES and AWARE OF REASON over time.

3.4.1 *Group cohesiveness*

Although mainly a signal of attentive listening, GROUP COHESIVENESS was the category with the highest frequency in both meetings. Similar results of the high prevalence of informal talk and 'yeah' utterances were also found in the architecture meeting[25] Following Owen's[26] finding that members of cohesive groups are more likely to engage in active communication, this is very likely to be a prerequisite for the development of shared team mental models. Based on our model, a decline of GROUP COHESIVENESS utterances over time was expected (see Figure 6). Although there was no significant difference between the two meetings, within each meeting GROUP COHESIVENESS first increased and then declined.

25.
See Goldschmidt, G., and Eshel, D. (Chapter 18) *op. cit.*

26.
Owen, W.F. (1985) Metaphor Analysis of Cohesiveness in Small Discussion Groups, *Small Group Behavior*, 16, pp. 415–24.

3.4.2 *Roles and disciplines in the team*

In order to coordinate their activities, teams allocate tasks to individual members on the basis of abilities and experience. GROUP COORDINATION was expected especially in the first phase and to a lesser extent in the second when the group should have achieved a shared understanding on these issues. The data are very clear; there is hardly any GROUP COORDINATION with only 17 utterances made in both meetings. This surprising result can be attributed to the nature of the meetings, which were set up as brainstorming sessions; both meetings focused on gathering and developing new ideas with no further responsibilities being discussed.

3.4.3 *Analysis of individual contributions*

The existence of team mental models indicates whether views on important task and team issues are common within a team. Individuals might have similar background knowledge, share

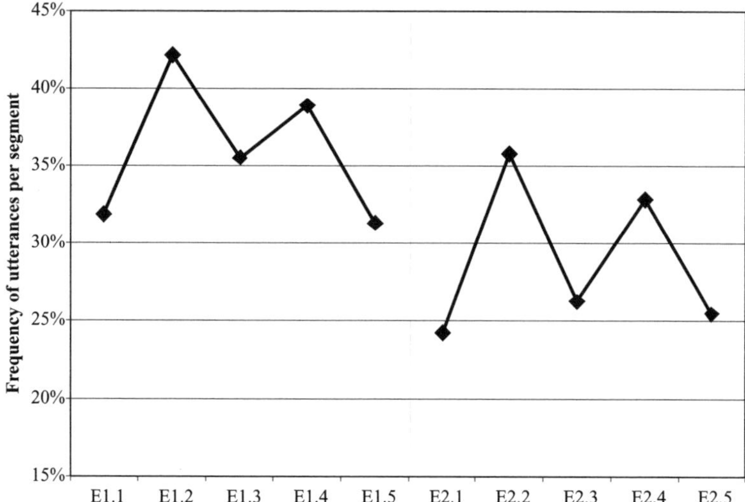

Figure 6. Frequencies of utterances classed as relating to GROUP COHESIVENESS.

working experiences, or they might develop similar solution ideas. However, individual mental models are not necessarily shared equally among all members. Those with the same disciplinary background often also share the same jargon, which in turn makes it even more difficult, with a different background, to acquire the same mental model[27]. Searching for dyads within a team, by looking at interaction patterns between members of a team, can provide a means to look for shared mental models.

27.
See Adams, R., Mann, L., Jordan, S., Daly, S. (Chapter 19) Exploring the Boundaries: Language, Roles and Structures in Cross-Disciplinary Design Teams.

A contingency analysis for all speakers was used to determine who talked to whom and how often. Figure 7 shows a network of the individual contributions. The thickness of the lines indicates the amount of individual talking in a given dyad, indicating how the ideas were shared within the team. The most obvious dyad was Tommy and Todd, who talked to each other more than to any other members. It is also noticeable that the facilitator, Alan, spoke considerably more to Tommy, Todd and Jack than to the other members. They, in turn, talked relatively more to Alan. E2 revealed a clear dyad between Tommy and Patrick, with Tommy replying to Patrick with 34.5% of his utterances and Patrick replying to Tommy with 55.8% of his. All other members talked mainly to Tommy, who could be seen as an informal group leader. However, there is a second moderator visible in the group, as Roman and Sandra also interact with Patrick.

Tables 5 and 6 show the categorised utterances per person and the content for both meetings. In addition to the categories mentioned above, the content of all task related utterances was coded in three ways; BUSINESS, USER, and TECHNOLOGY. This analysis illustrates to what extent team members contributed from their disciplinary perspective.

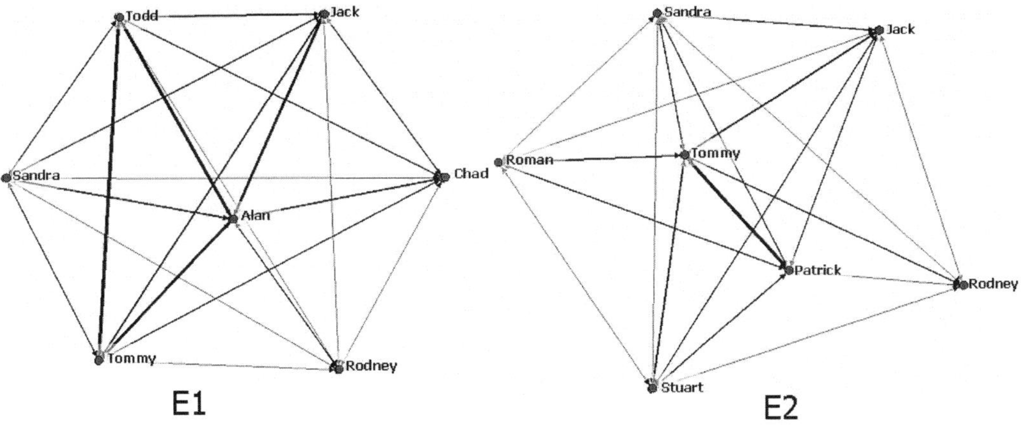

Figure 7. Contingency analyses based on individual contributions.

In the first meeting, E1, Alan led the group by guiding the process (56 utterances). Everyone apart from Sandra and Rodney contributed substantially to new IDEAS, with Tommy and Jack generating the most (24 and 25 respectively). Tommy, Todd, Alan, and Jack contributed most to ANALYSIS, EXPLANATIONS and CLARIFICATIONS. Jack contributed most new solution IDEAS and EXPLANATIONS (11.7% and 10.75%). Sandra did not contribute to EXPLANATIONS and only mentioned the USER twice, which is surprising given that she is the usability expert. PROBLEM DEFINITION and ELABORATION were mainly done by Alan.

During E2, with Alan absent, Tommy led the team with 23 PROCESS utterances and 323 GROUP COHESIVENESS related utterances (Table 6). He also introduced most IDEAS, although new IDEAS were relatively evenly distributed amongst the others. ANALYSIS and EXPLANATIONS were dominated by Tommy and Patrick. Again, Sandra did not participate in EXPLANATION but mainly contributed to ANALYSIS, specifically the sub-category 'how to use' ANALYSIS. PROBLEMS were mainly defined and elaborated by Tommy and Jack. Interestingly, Rodney, who did not contribute to PROBLEM DEFINITION in E1, contributed the most in E2 (14.6% of all his utterances).

The meetings were also analysed to understand how STRATEGIES were employed over time. During meeting E1 there is a period in the middle of the meeting where the team spends the majority of time CONSIDERING ISSUES; this period is followed by switching to AWARE OF REASON.

3.4.4 *Who uses which* STRATEGIES?

In terms of strategies, all participants employed a minimum of three STRATEGIES. However only two participants, Alan and Tommy, were

	IDEA	ANALYSIS	EXPLANATION	PROBLEMS	PROCESS	COHESIVENESS	BUSINESS	USER	TECHNOLOGY
Alan	17	87	17	14	56	260	3	19	131
Tommy	24	104	20	5	21	194	3	13	143
Todd	16	94	23	6	11	216	6	11	132
Jack	25	80	23	3	17	161	13	6	118
Chad	12	44	9	3	6	83	1	8	58
Sandra	4	30	0	4	3	65	0	2	37
Rodney	5	29	3	0	4	43	2	2	34

Table 5. Categorisation of individual utterances in E1.

	IDEA	ANALYSIS	EXPLANATION	PROBLEM	PROCESS	COHESIVENESS	BUSINESS	USER	TECHNOLOGY
Tommy	32	75	68	25	23	323	24	20	176
Patrick	17	78	50	9	7	202	27	26	108
Jack	15	45	11	20	3	113	10	13	70
Stuart	15	43	24	6	0	96	15	10	70
Roman	12	40	14	7	0	86	12	7	56
Sandra	9	42	1	4	1	84	6	9	43
Rodney	10	15	3	7	1	36	7	1	28

Table 6. Categorisation of individual utterances in E2.

observed to use all eight, hence these are expected to be the most experienced. In addition, Jack and Patrick both showed evidence of a high number of STRATEGIES (seven out of eight observed). This is interesting as these four were amongst the highest contributors to ANALYSING and EXPLAINING solutions (Tables 5 and 6). STRATEGIES lead to the ability to ask the right questions and may assist in communication between members.

4 DISCUSSION

4.1 *Summary of main findings*

This chapter has focused on team mental models as a theoretical concept that sheds light on both the cognitive and the social process in a design team. The emphasis in our analysis was on the development of team mental models over time, as indicated by external verbal communication. Based on our proposed model (Figure 1), the frequency of cognitive acts indicating sharedness was predicted to first rise, in order to achieve sharedness, and then decline over the course of a meeting. This prediction was largely confirmed, apart from three unexpected findings.

Contrary to our expectations, the level of PROBLEM DEFINITION did not decrease significantly over the two meetings, instead increasing significantly from E1 to E2. The same is true for the amount of

EXPLANATIONS which, instead of decreasing, increased significantly from E1 and E2. According to our model both results indicate a low sharedness in the group-related to task-specific knowledge. However, a qualitative analysis revealed that these two findings were related to deeper discussion of issues already discussed. The findings therefore do not violate the assumptions of the model, indicating that the process of creating sharedness started again at a more detailed level – as is perfectly normal for the progression of design work.

The third unexpected finding was that there was virtually no TEAM COORDINATION. In our previous field research we have also found only a limited amount of TEAM COORDINATION but typically there were at least some utterances on TASK ALLOCATION and PROJECT PLANNING beyond the meeting. The team in E1 and E2 does neither of these, which is presumably related to the very particular situation; as these meetings were designated as brainstorming sessions decisions would be deferred, only some members having direct responsibilities for the execution of work beyond the meeting.

In terms of STRATEGIES, the two most frequent categories were CONSIDER ISSUES and AWARE OF REASONING. Although no decisions were made in these two meetings, it is likely that in the further progression of the project these strategies will contribute to the team's strategic knowledge about the rationale of the design, for example when making design modifications.

4.2 *Implications of the findings*

The two-stage model of team mental models proved useful for analysing and making sense of the DTRS7 data. It can also be used to develop indicators for the lack of sharedness, and to illustrate to practitioners what kind of processes are required for the development of more effective team mental models. The predictions also held true for the overall group development across both meetings, despite the fact that the purpose of the two meetings was different and only half of all group members attended both meetings. Our rationale of treating the two meetings as part of one overall project therefore seems justified. The unexpected findings largely relate to a change in the level of detail, which the model currently does not explicitly incorporate.

Not all of the predictions from the model could be tested, as the two meetings only form a small subsection of the natural interaction of an existing team. It is rather unusual for a team not to make any DECISIONS or engage in TEAM COORDINATION, which in this case can be attributed to the nature of the idea generation sessions. DECISIONS and TEAM COORDINATION processes should occur in ad hoc teams who only interact for a given time, as well as in those working together over a longer term. A longer field study would enable researchers to know more about the history of this team and the context of the design work than was possible using the DTRS7 data.

10
Variants and Invariants of Design Cognition

Ömer Akın

This chapter is a re-examination of the thesis that, in different fields of design, cognitive processes have significant similarities and differences that help us discriminate as well as bridge them. Some of the notable *variants* of design, that have been reported in the past, such as rich representations, inventive strategies, problem (de)composition, and complexity management[1], have helped draw interdisciplinary comparisons. This chapter is an attempt to revisit these findings in the context of the DTRS7 data and seek new insights about *variants* and *invariants* of design cognition.

1.
Akın, Ö. (2001) Variants in Design Cognition, in Eastman, C., McCracken, M. and Newstetter, W. (eds), *Knowing and Learning to Design*, Elsevier, pp. 105–124.

This chapter is about cognitive models of designers and how these models depict disciplinary differences between engineers and architects. Review of the literature in this area suggests that architects who win awards and reach the highest levels of the profession do so primarily because, and in spite of, the miscalculated risks they have taken; while engineering failures have been condemned for the mistakes made in service of a blind trust in technology. From among innumerable cases that can illustrate these claims, for brevity, let us just note two legendary ones: the unabated praise showered over *Fallingwater* designed by Frank Lloyd Wright[2] validating the former, and the spectacular but educational failure of the Tacoma Narrows Bridge[3], to support the latter.

2.
Toker F. (2003) *Fallingwater Rising*, Alfred E Knopf, New York.

3.
Petroski, H. (1982) *To Engineer is Human,* St Martin's Press, New York.

Are there significant differences in engineering versus architectural cognition that can account for these behavioural characterisations? Are there differences in the representations, solution strategies, problem composition approaches, alternative solution generation or complexity management behaviours observed for engineers and architects? Can we codify these differences? What are the appropriate metrics and measures that should be used?

In this chapter, I will consider some of these questions. My approach is empirical. This type of research is about recognising predictability of measurable events through the detection of *invariants* among data variables. Towards this end, I will cull a handful of salient features of design cognition from the DTRS7 data set compiled by others. The availability of data both for architects and engineers gave me the opportunity to draw intra-disciplinary comparisons and attempt to validate the findings of my earlier work.

1 APPROACH

The specific features I am interested in studying here are based on previous research reported in Akın[4]. While not critical, it is important for readers who are interested in issues of protocol data coding and code validation to review this source as some of the coding categories are taken from it. The primary reason for this selection is due to two things:

4. Akın (2001) *op. cit.*

1. The opportunity to validate some of its conclusions by comparing cognitive behaviours of engineers and architects, both of which are available in the DTRS7 data, and
2. The ongoing and recently heightened recognition of the work reported in Akın[5] by architectural educators as a 'salient' source for design cognition.

5. *ibid.*

Two decades of sustained research I conducted in the area of design cognition that ultimately led to the writing of the 'variants' paper suggests several persistent cognitive patterns across individual differences that characterise architects' design behaviour. First, expert architects use a hybrid search strategy of *breadth-first-depth-next*[6]. They explore many more alternatives than do novices[7]. In addition, they manage complexity by using specific partial solution merging strategies based on *Pairwise Integration*[8]. Finally, there are notable differences in the cognitive approaches used in puzzle-solving, decision-mAking, and designing that can illuminate differences between designers[9].

6. Akın, Ö. (1986) *Psychology of Architectural Design*, Pion.

7. Akın, Ö. (1988) Expertise of the Architect, in Rychener, M. (ed), *Expert Systems for Engineering Design*, Academic Press, pp. 173–196.

8. Akın, Ö. (1994) Psychology of the Early Design Process in Timmermans, E. (ed), *Proceedings of Design Decision Support Systems Conference*, Vaals, The Netherlands.

9. Akın, Ö. and Akın, C. (1997) On the Process of Creativity in Puzzles, Inventions and Designs, *Automation in Construction*, 7, pp. 123–138.

In order to address cogent research questions pertaining to the *variants* as well as the *invariants* of design behaviour, I mapped relevant aspects of the above studies and the findings of the 'variants' paper against the coverage of the DTRS7 data. This yielded several specific questions that can potentially highlight significant behavioural patterns distinguishing architects and engineers:

1. Do architects, as opposed to engineers, navigate their search space differently; and how?
2. Do architects, as opposed to engineers, merge partial solutions through *Pairwise Integration*?
3. Do architects, as opposed to engineers, generate more alternatives and restructure design problems more frequently?

Let us now consider each of these questions in order to define the hypotheses I will address and the codification strategies I will use to analyse the data.

1.1 *Navigation of the design search space*

A prevalent view of the design process is that designers search for solutions in a domain that resembles a multi-dimensional space[10,11]. They begin at a point in this space that is given by a specific formulation of the problem and then they apply methods of 'navigation' that are particularly suitable to this space, in order to move *forward* towards a solution (forward should be assessed with respect to the initial problem position).

10.
Newell, A. and Simon, H.A. (1972) *Human Information Processing*, Prentice Hall.

11.
Akın, Ö. (2006b) The Whittled Design Space *Artificial Intelligence for Engineering Design, Analysis and Manufacturing*, 20 (2), pp. 83–88.

Here, I intend to look for evidence to support the hypothesis that architects conduct their search by first navigating laterally, *breadth-first*, followed by forward movement, *depth-next*. I will explore the proposition that other designers, engineers in this case, navigate the space in a different manner. Towards this end I will codify the DTRS7 data into design entities, acts, and tasks that will be defined in the next sections; and study these, within the design search space.

1.2 *Pairwise Integration*

Due to cognitive limitations designers parse the content of their design problem into ever smaller parts. In this way they make complex, unruly, ill-defined problems smaller and more manageable[12,13]. While this is one of the philosophically cogent proposals towards objectifying design behaviour[14] little is known about the design outcomes of such a strategy. How do designers put these small, partial solutions into a larger, integrated format? Is this a simple act of piecing together jigsaw parts so carefully decomposed that their return to their original 'place' in the puzzle is unambiguous and trivial to find?

12.
Eastman, C. (1969) Cognitive Processes and Ill-defined Problems: A Case Study from Design in Walker, D. and Norton, L. (eds) *Proceedings of the International Joint Conference on Artificial Intelligence*, Washington DC.

13.
Simon, H.A. (1973) The Structure of Ill-structured Problems, *Artificial Intelligence*, 4, pp. 181–201.

My hypothesis in this chapter is that this integration of partial solutions is a laborious, deliberate, systematic integration of pairs of partial solutions until no orphan pieces of the whole are left behind[15]. The key attribute of this process is that, in conformance with the cognitive limitations of the designer, the integration is achieved by merging two solutions at a time. More than two is neither empirically observed nor theoretically supported[16]. Content analysis based on the segmentation of the DTRS7 data will be used to explore this hypothesis.

14.
Akın, Ö. (2006a) *A Cartesian Approach to Design Rationality,* Faculty of Architecture Press, Middle East Technical University, Ankara, Turkey.

15.
Akın (1994) *op. cit.*

16.
Newell and Simon (1972) *op. cit.*

1.3 *Alternatives*

All movement within the design search space is not in the forward direction, i.e., from the initial position towards the goal position. Aside from getting lost, getting stuck, or engaging in wilful digressions, designers move also laterally to consider alternatives.

Alternatives are often seen as the stuff of 'expert' design and the insurance for finding *satisficing* solutions[17]. Estimates vary between 2.5–4 alternatives generated per design issue or *content item*[18,19]. However, the jury is still out on whether this is a sufficient or even necessary condition. It may well be a feature of design that varies

17.
Simon, H.A. (1969) *The Sciences of the Artificial*, MIT Press.

with its domain, expertise of the designer, stage of the problem, or merely the style of design[20]. As Ball and associates observe:

> "taken as a whole, then, the empirical evidence seems to support a view of solution development in design that deviates from the prescriptive position that designers should explore multiple alternatives at all levels of design abstraction"[21].

Based on segmentation of the DTRS7 data, I will explore the hypothesis that architects traverse the search space *laterally* generating more alternatives than their engineering counterparts, as they consider the content of the design problem.

18. Akın (1988) *op. cit.*

19. Ball, L.J., Lambell, N.J., Reed, S.E., and Reid, F.J.M. (2001) The Exploration of Solution Options in Design: A 'Naturalistic Decision Making' Perspective, in Lloyd, P. and Christiaans, H. (eds) *Proceedings of DTRS5: Designing in Context*, Delft University Press, The Netherlands.

20. Chan, C.S. (1995) A Cognitive Theory of Style, *Environment and Planning B: Planning and Design*, 22, pp. 461–474.

21. Ball et al. (2001) *op. cit.*

2 DATA CODING

The coding and analysis described here is applied to the DTRS7 data for architectural design meeting A1 and engineering design meeting E1. Since the goal of this paper is to compare design cognition in different design disciplines, a 2-hour plus transcript relating to each discipline is considered sufficient. Furthermore, the material relating to the meetings A2 and E2, being a continuation of the same projects by the same design teams, was not expected to add substantially to the findings reported here.

Besides the difference in the gap between the two sessions in each data set, there are other important differences between them, such as the participation of the client in the architect's session and the relatively advanced stage of the architectural task as opposed to those of the engineers. The latter is considered in the context of the analysis offered in the later sections of this chapter. The former issue is worth discussing here before we commence with data analysis.

In the architect's protocol, the client has a significant part in the design discourse that is recorded and transcribed. The client primarily provides the user need aspects of this discourse (specification and assessment of design requirements), while the architect is concerned with the physical design proposals side of the same. User needs and design proposals are two inseparable aspects of the design act[22]. The dialectic relationship between them – needs propagating physical design responses and these responses in turn helping refine the need specifications or initiate new ones – constitute the essence of design. This dialectic discourse is also present in the engineers' protocol; but the dual function of requirement and design specification are undertaken exclusively by the participant designers.

22. Özkaya I. and Akın, Ö. (2004) Emerging CAD Processes: The Case of Computer Aided Requirement Management in *Proceedings of the GCADS-04 International Symposium*, Pittsburgh, USA.

In my view, the transcriptions of the DTRS7 protocol contain two design discourse examples, one for architects and the other for engineers, that are entirely equivalent as integral sets of requirement

Design Modality	Task [as applied to design content]	External Representation (ER)	Act [as applied to external representations]	Content [design information]
Criterion (T)	Acquire (A)	Drawing (D)	Initiate (I)	'park'
Feature (F)	Project (P)	Sketching (S)	Modify (M)	'sequence of entry'
	Verify (V)	Literature (L)	Accept/Reject (R)	'garden'...
		Oral (O)		*[open ended list]*

Table 1. Protocol analysis data codification taxonomy.

and design statements, albeit the former by designers and clients and the latter by designers alone. My analysis of the data will not further consider the distinctions that may exist between client provided requirements versus designer provided ones. Since neither adequate hypotheses regarding this topic nor techniques of neatly separating requirement statements from design ones are currently available, this aspect of the transcriptions will remain outside of the scope of the present study. Furthermore, this topic requires deeper analysis than can be provided along side the principal topics of interest in the work presented here. It deserves consideration in a future study.

However, it is important to note that the findings of this work must be taken in the context of the different participants that provide the design requirements in the architects' transcripts versus that of the engineers.

The hypotheses stated in Section 1 require the codification of several categories which have been developed in my earlier research. Below, I will provide a formal notation and brief description of each category. For further exploration, the reader is directed to a primary source[23] which contains most of these categories that are in turn based on yet earlier work[24]. Table 1 is a summary of the coding categories and the nomenclature described below. It includes four major data coding taxonomies: Design Modality, Design Task, Design Act, and of course Design Content, here, the data segmentation.

23. Akın (1986) *op. cit.*

24. Akın, Ö. (1978) How do Architects Design? in Latombe, J.C. (ed), *Artificial Intelligence and Pattern Recognition in Computer Aided Design*, North-Holland, pp. 65–104.

2.1 *Design Modalities*

Design Modalities are based on the most basic design behaviours: generation and evaluation, the two dialectic design mechanisms that have been identified throughout design methods and process literature over the course of the last four decades[25,26]. In correspondence with this, I recognise two coding categories: *design criteria* and *design features*. Design Criteria (encoded as {T}, where '{}' is 'set') are the equivalent of design requirements against which design features are evaluated. Design Features {F} are the equivalent of aspects of design proposed as solutions that satisfy criteria.

25. Jones, J.C. (1973) *Design Methods: Seeds of Human Futures,* Wiley.

26. Özkaya and Akın (2004) *op. cit.*

For example, *T* may state that sufficient space is needed to move a coffin from a hearse to lobby; and the *F* that satisfies this would be an 8' sliding-door.

2.2 Design Tasks

Design Tasks are strategic moves that address high level purposes of design that help the designer manage information. Design information is obtained, modified, and set aside either because its desirable effect is fulfilled, postponed to a later time, or discarded. Based on Akın[27] these tasks are defined in the following way [where '/' is 'or', '::=' is equivalent, '{}' is set, '()' is 'function of' and the super-scripted term refers to the specific Design Content (Section 2.4)]:

27. Akın (1978) *op. cit.*

- *Acquire Information* (A) is all actions that inquire about, obtain, or admit new information into the design space. Here, 'A' is any one of the Design Acts in a Design Modality {T,F} for a given Design Content {c} for the purpose of admitting new information. Formally encoded as: {**A**(T/F)^{C}}
- *Project Information* (P) is all actions that infer new information from the existing through reasoning, design, or speculation; encoded as {**P**(T/F)^{C}}
- *Verify Information* (V) is all actions that consider or finalise the appropriateness of the information at hand; encoded as {**V**(T/F)^{C}}.

2.3 Design Acts

Design Acts are low level moves that directly manipulate External Design Representations (ER). These representations include drawing, sketching, literature sources, and oral feedback, coded as {D/S/L/O}.

Design Acts include *Initiate Action* (I), which is when something first appears in the data or the protocol; *Modify Action* (M), which is the subsequent manipulation to a Design Representation (ER) already initiated; and *Accept/Reject Action* (R), which is the final endorsement of a Design Representation for further consideration, all subject to the design content item (c). These are coded as: {[Design Act (Design Representation)], or {I(ER)^{C}/M(ER)^{C}/R(ER)^{C}}, where, as before, '/' is 'or', '::=' is equivalent, '{}' is set, '()' is 'function of' and the super-scripted term refers to the specific Design Content.

2.4 Design Content

Finally, Design Content (c) is the ever changing informational setting of the design process, creating a focus for Design Modalities, Tasks, and Actions. As the focus of the designer shifts from the 'park' to the 'sequence of entry' to the 'garden', for instance, so does the Design Content. This is the only open ended aspect of the codification which gets to be enumerated as the protocol is coded. Another important item of data coding is data segmentation, which, in the case of this codification, is linked one to one to Design Content.

I recognised after codification commenced that the Design Content needed to be discriminated from Process Management (PM) items. A PM is any self-conscious act of the designer to organise design behaviour. Some instances of this set include *begin, organise, communicate, break,* and *stop*. This is also an open ended list that grew as codification advanced.

2.5 *Sample coding*

The sum total of the coding categories include Segmentation, Design Modality, Design Content, Design Act, Design Task; in addition to five items (Begin Line, End Line, Timestamp, Participant, and sample text) I used to collate this codification with the DTRS7 transcription data.

Extract 1 shows a sample of the transcript from meeting A1 which corresponds to a single segment (#22); the coding for this sample is shown in Table 2.

Extract 1, A1, Sample transcription corresponding to a single coded segment (#22).

Anna yes they will probably want to know how [*points*] how + how far that is from the because they're going to be possibly carry the coffins in and most of the men are sort of in their seventies and eighties [*laughs*] carrying the coffin
Adam that's only eight metres there's not a huge difference
Anna that's alright so long as I yes I mean they're not criticising they're just wondering how the flow of vehicles goes and where I mean they're looking at it from their angle we're looking at it obviously trying to make sure everything's covered so they wanted to know roughly where they went how would they park how would they then offload the coffin normally by carrying on what we call shoulders so they would carry on a shoulder so that then obviously the area the length of that is important to them

In Table 2, each coded line starts with the segment number and the Design Content issue (and ID number) under consideration. In this case, this is the 'coffin carrying distance'. Note that the segment number and the Design Content ID number can be different if the content item occurs multiple times. This is followed by Begin Line, End Line, Time Stamp, Participant and sample text entries. The last three columns indicate the coding of Design Acts

Segment	[#] Content	Begin Line	End Line	Time Stamp	Participant	Sample Text	Act	Task	Modality
22	[15] coffin carrying distance	227	230	18:00	Anna	Yes they will probably want…	I	A	T
		231	231		Adam	That's only eight meters there's…	R	P	F
		232	239		Anna	That's alright so long as I…	R	V	T

Table 2. Sample codification from A1.

as applied to Design Representations, the Design Tasks as applied to design information under the participant's focus of attention, and Design Modality.

In the segment shown in Extract 1 and Table 2, Anna (the client) asks about the potential problem of the distance a coffin must be carried by hand due to the particular configuration she sees. This is an *inquiry* about gaining more information about the issue of 'coffin carrying distance'. This prompts Adam to interpret the design as well as the query, projecting new information about the age of pall bearers, the carry distance, and the configuration of the layout. Finally, this segment is concluded when Anna verifies the need to make the coffin carrying problem clear, which leads to the 'bier' Design Content issue, in the segment following this one. This is an example of a 'well-formed' codification sequence consisting of A -> P -> V not always observed with such clarity in most design protocol transcriptions.

2.6 *Coding validation*

Segmentation turns out to be central to several key issues explored in this chapter, including navigation, alternative generation, and Pairwise Integration of partial solutions. Consequently, validation of codification of this variable is important. A validation experiment was carried out by comparing the coding generated by multiple coders. At the time of this experiment, coders were graduate students and a visiting scholar in the School of Architecture at Carnegie Mellon University. Three randomly selected sample passages from each transcription, each 100-lines long, were given to at least two different coders. Coders were trained on samples from the transcriptions of A2 and E2 that are not included in this study. The sample coding results were compared and feedback was given to coders before they coded the transcription samples in A1 and E1.

The results of the validation experiment were sorted into three categories: *matched*, error of *commission* (partially matched), and error of *omission* (not matched). Any two segments within 10% line-number-overlap were considered to be a match. Overlaps of less than 90% but more than 10% were considered partial matches. Overlaps below

10% were considered errors of omission. Overall mean percentages for these categories are 28%, 60%, and 11%, respectively. I consider the last category, omission, the most serious category of coding deficiency. While this exceeds the goal of 5% I established at the outset, the outcome of 11% omission is still acceptable because the 100-line samples given to the coders were randomly selected and they did not have the benefit of knowing the context of this data. This made the validation experiment particularly conservative.

The last three categories (columns) in Table 2 were coded by the author and were not validated against other coders.

2.7 Coding strategies

Before moving on to the analysis of the DTRS7 transcripts let's try to place this coding scheme in the context of the multitude of coding schemata that exist in the field. This is a research topic

Current coding scheme	Proposal for mapping the two schemata	FBS coding scheme[28]
1. Design Feature (Modality)	Design Feature is a physical description; in FBS most closely resembles the 'description' category; is also related to 'synthesis'	Design description (defined through context)
2. Design Criteria (Modality)	Design Criteria modality is the basic evaluative functionality; in FBS, corresponds to evaluation	Evaluation (defined through context)
3. Design Content (Segmentation)	Design Content is all that is addressed by the functionalities of a design process, it is the content in the design framework, so to speak; 'behavior', expected or derived, in FBS it is the one category that comes closest to it	Behaviour (defined through context)
4. Initiate Act/Acquire Information Task	Initiate and Acquire are similar functionalities operating on different parameters (content issue and information); in FBS, for lack of a better one they correspond to 'formulation'	Formulation ::= 'interpretation of requirements or representations'
5. Modify Act/Project Information Task	Modify and Project are similar functionalities operating on different parameters (content issue and information); in FBS, they correspond to 'analysis' but also subsume aspects of 'synthesis'	Analysis (defined through context)
6. Accept-Reject Act/Verify Information Task	Accept-Reject and Verify are similar functionalities operating on different parameters (content issue and information); they correspond to 'synthesis' in FBS but also subsume aspects of 'analysis'	Synthesis (defined through context)
7. Representation ::= {drawing, sketch, literature, oral}	'Representation' is an external physical entity where as in FBS 'documentation' appears to be a function; closest fit, not a perfect match	Documentation ::= 'transform structure to design description'

Table 3. Comparison of the current coding with the FBS scheme.

28.
Kan, J. and Gero, J. (Chapter 12) Using the FBS Ontology to Capture Semantic Design Information in Design Protocol Studies.

that deserves its own concerted effort; however, in the interest of taking a modest step towards that goal, I offer a modest mapping of my coding approach that has been used consistently over several decades against other reliable coding schemata, particularly those in this volume.

To be specific, let's consider the FBS coding scheme used by Kan and Gero[29]. Table 3 contains my coding categories (left column), the corresponding elements of the FBS scheme (right column) and the rationale of mapping the two schemata (centre column).

29.
ibid.

Two elements in the FBS scheme are defined by Kan and Gero: formulation and documentation[30]. The definitions of the others are surmised contextually from the current and previously published papers[31]. At times the mapping in Table 3 is convincing, in particular elements 1, 2, 5, and 6; and at other times difficulties remain. The latter is due to the overlaps between one category in one scheme and multiple categories in the other one (one-to-many or many-to-one mapping). In general, however, the agreement of scope between the two coding schemata seems to be remarkably similar.

30.
ibid.

31.
Suwa, M., T. Purcell, and J. Gero (1998) Macroscopic Analysis of Design Processes Based on a Scheme for Coding Designers' Cognitive Actions, *Design Studies*, 19, pp. 455–483.

I draw two conclusions from this exercise. Firstly, the current coding scheme has good coverage of the issues explored by researchers in the field, as does the FBS scheme which is widely known and used; and secondly there are important differences between these schemata and the substitution of one for the other is a non-trivial goal. Given these, I consider the use of the codification scheme I devised for this paper appropriate.

3 ANALYSIS

In analysing the results of my codification scheme, several quantitative and qualitative analysis techniques were employed, including descriptive statistics (mean, standard deviation, variance, range, count, confidence levels, and histograms) and variance analysis methods (two-tail t-tests) to compare the protocols. In addition, some segments of data were interpreted using content analysis.

While each method helped develop a comprehensive understanding of the codification results, in the remainder of this chapter, I will refer only to a subset of this analysis that is relevant to the three general questions I posed in Section 1.

3.1 *Content analysis*

Coding of transcriptions of A1 and E1 yielded several hundred segments and nearly as many Design Content items. In the 2343 lines of the transcript of A1, there were 163 distinct segments defined

Rate of Segmentation of the transcript	First 1/6th	Second 1/6th	Third 1/6th	Fourth 1/6th	Fifth 1/6th	Sixth 1/6th
A1 Transcript						
Partitioning line number	390	780	1170	1560	1950	2340
Net number of segments	40	42	22	23	22	14
% of total	24.3	25.8	13.5	14.1	13.5	8.6
E1 Transcript						
Partitioning line number	336	672	1008	1344	1680	2016
Net number of segments	32	17	24	17	31	25
% of total	21.9	11.6	16.4	11.6	21.2	17.1

Table 4. Rate of segmentation in A1 and E1.

by Design Content items ranging from 'new design', 'design concept', and 'precedents' to 'coffin carry distance', 'outdoor seating', and 'mirrored glass'. In the 2019 lines of the transcript from meeting E1, a total of 146 segments were coded including, 'constant pressure', 'angle constraint', 'end over paper', 'overlay', and 'colour marks'. While the designers returned to earlier Design Content items from time to time, these were extremely infrequent. Since they were considered as a continuation of the same Design Content and were considered the beginning of a new segment, they were not given new segment-ID numbers.

The segmentation of each transcript revealed a slightly different pattern of duration for the Design Content items, measured in terms of the number of transcription lines. In A1 the rate was rapid at first and levelled off monotonically; the rate of segmentation in E1 was high at first, lower in the middle and high at the end, like an inverted bell shape (Table 4).

This observation motivated further analysis. Comparison of the segmentation numbers, using a two-tail t-test, indicated that dissimilarity between the two protocols A1 and E1 were not statistically significant ($p \leq 0.01$, for all) for cumulative number of segments, the net number of segments, and the percentage of the net number to the total.

The Design Content items were also tabulated in a hierarchical tree structure for E1 and A1. In the latter case, there were five tiers: root node [Tier-1], inferred categories [Tier-2], major design content ideas [Tier-3], minor design content ideas [Tier-4] and minor design content ideas subsumed under Tier-4 [Tier-5]. Both hierarchies turned out to be belly-heavy, populated mostly in the middle tiers.

3.2 *Process analysis*

To uncover the pattern of behaviour characterising the designers, I considered the tactical (Design Acts) and strategic (Design Tasks

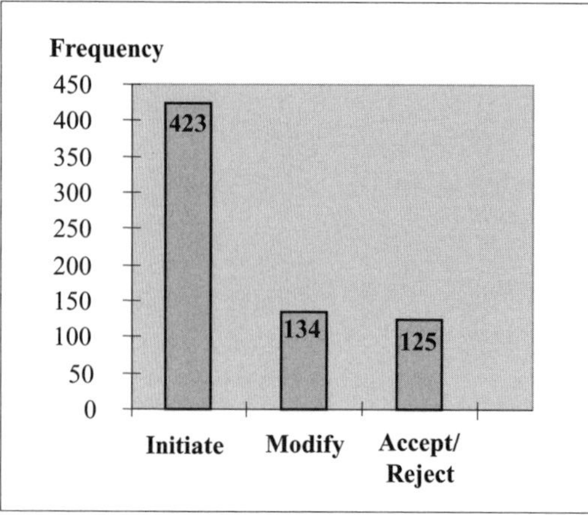

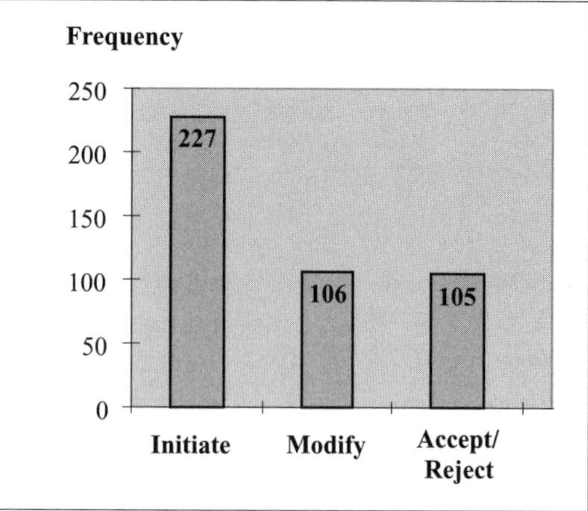

Figure 1. Histograms of Design Acts in A1 (left) and E1 (right).

and Design Modalities) patterns of behaviour discerned in the codification.

First, I examined the tactical behaviours of the two design teams through the Design Acts. These are the lowest level moves of the designers and consist of Initiating (I), Modifying (M), or Accepting/Rejecting (R) the Features or Criteria (Design Modalities) brought to bear on a given Design Content. For each Design Content a combination of Design Acts (I, M, R) would apply. As shown in Figure 1 a distinct pattern is shared by both transcriptions. Initiating is the dominant act. Modifying and Accepting/Rejecting are less frequent. A two-tail t-test confirmed the similarity of these distributions at confidence level of $p \leq 0.027$.

As a matter of fact, this pattern signals experienced designer behaviour, where initiated Design Content meets tacit acceptance after minor modification. It resembles designer behaviour of generating design alternatives reported in Ball et al[32] and others[33,34]. Expertise often enables the designer to start with the most promising options very early on, which suggests that there should be little if any evidence of lateral (alternative) searching behaviour.

Next, I analysed the strategic behaviours of the designers within each segment of the coded transcripts by looking at the Design Tasks applied to the information contained for each Design Content item. For each transcript, the normal sequence of Acquiring, Projecting, and Verifying relevant design information was observed with relatively little distortion but variable frequency of occurrence (Figure 2). Often the sequence was initiated with the Acquire Task and followed by Project and Verify Tasks in different combinations. In longer segments, the Acquire Task appeared to feed the Projection Task (inference making with new information). Mostly, these sequences were concluded with the Verify Task.

32.
Ball et al. (2001) *op. cit.*

33.
Darke, J. (1979) The Primary Generator and the Design Process, *Design Studies*, 1, pp. 36–44.

34.
Cross, N. and Clayburn Cross, A. (1996) Observations of Teamwork and Social Processes in Design, in Cross, N., Christiaans, H. and Dorst, K. (eds.) *Analysing Design Activity*, Wiley, pp. 291–317.

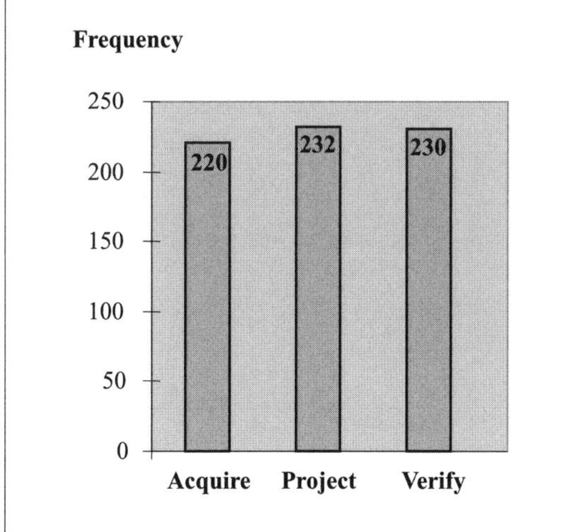

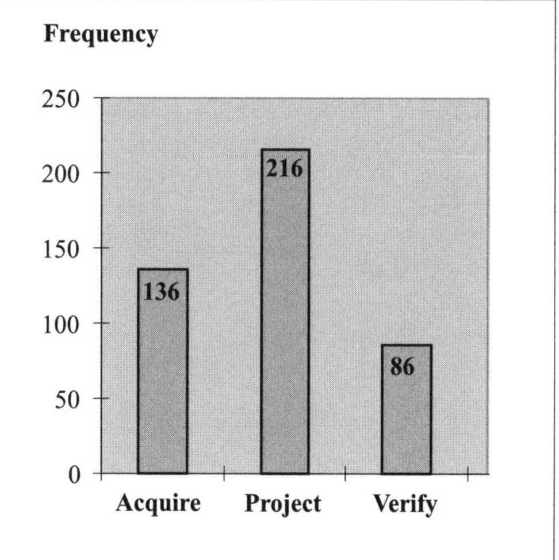

Figure 2. Histograms of Design Tasks in A1 (left) and E1 (right).

A two-tail t-test was used to compare the two distributions, which did not confirm their similarity ($p \leq 0.054$). The major difference between A1 and E1 in this regard was the predominance of the Projection Task in E1 as opposed to the uniformity of frequency of all Design Tasks in A1. This is likely to be due to the fact that the brainstorming strategy used in E1 generates a preponderance of new ideas which leads to a lot of inference activity in the form of the Project Task. Whereas in the A1 transcription, all Tasks are balanced as one expects in 'normal' design.

3.3 *Content and process analysis*

Finally, I analysed the codification for the interactions between the Design Content items and the design process aspects of A1 and E1. This particular approach is intended to provide the material directly useful in discussing the navigation strategies, design alternatives, and Pairwise Integration in design.

3.3.1 *Navigation strategies*

First, I looked for evidence indicating navigational strategies. How did architects 'move around' in the design space? Was this different from engineers?

In A1, there are three clear cases where a partial solution is clearly rejected, causing a redefinition of the design constraints and a turn in the direction of the solution path (Table 5).

In the first example (Segment 48), Anna is trying to adjust the current design to accommodate funeral ceremonies for children in the main chapel. In the second case (Segment 60), Adam considers

Segment	[#] Content	Begin Line	End Line	Time Stamp	Participant	Sample Text	Act	Task	Modality
48	[35] Main chapel	450	451	0:31	Anna	You'd use the main chapel for that...	I	P	T
48	[35] Main chapel (reject)	463	464	0:31	Anna	Would think yes we would have...	M	V	T
48	[35] Main chapel	471	471	0:32	Anna	That's all right then that's ok...	R	V	T
60	[46] Laptop use	569	57	0:38	Adam	But there's no reason why there++	I	A	T
60	[46] Laptop use	581	582	0:38	Anna	The problem with that would be yes	M	V	F
80	[65] Movies	745	751	0:49	Adam	Well I mean the other thing is...	I	P	T
80	[65] Movies	761	761	0:5	Adam	No I...	R	P	F
80	[65] Movies	762	762	0:5	Anna	No	I	P	F

Table 5. Design problem restructuring in A1.

the use of a laptop to support programming in a space that may not be otherwise useable for that purpose. And in the final example (Segment 80), he is considering video and movie viewings during ceremonies and the alteration of space to accommodate the appropriate space and screening surface for this. In all three cases, the problem space is restructured by the rejection of an existing solution configuration.

In contrast, the E1 coding was full of new Design Content items being introduced into the problem space. As implied by the spike of Projection Activity in Figure 2, they were quickly introduced (Acquire), elaborated at length (Project), before being accepted (Verify) or in some cases left to languish for a while longer. In these episodes of Design Tasks, the normal restructuring of the problem space was not observed. For this reason, I examined the role of analogical reasoning in E1 since a vast majority of the ideas introduced (approximately 80%) came by way of analogical reasoning.

3.3.2 *Analogical reasoning*

Analogical reasoning[35,36] was predominant in E1. In a particular span of 24 segments (Table 6), eight contain analogical reasoning. Various attributes of the digital pen lead to it being first viewed as a soldering iron, then a sledge, then a pen stabiliser (as in a bicycle

35.
Ball, L. and Christensen, B. (Chapter 8) Analogical Reasoning and Mental Simulation in Design: Two Strategies Linked to Uncertainty Resolution.

36.
Ball, L.J., Ormerod, T.C., and Moreley, N.J. (2004) Spontaneous Analogising in Engineering Design: A Comparative Analysis of Experts and Novices, *Design Studies*, 25, pp. 495–508.

Segment	Last Two/Each Segment Design Acts	Tasks	Elements	Analogy Description	Result
8 (L: 85–107)	M → M	V → P	T → T	Soldering iron	To solve pressure and angle
12 (L: 136–150)	M → M	P → P	F|T → F|T	Sledge	Guiders down the side of pen tip
13 (L: 153–162)	M → R	V → V	F → F	Pen stabiliser (bicycle, wind surf)	Universal joint as stabiliser
23 (L: 224–235)	I → I	V → P	F → F	Hot ballpoint	Eliminate pressure problem repeat action + tracker
24 (L: 236–253)	M → R	P → V	T → F	Switch	Control continuity through rapid turn on and off of heat
26 (L:272–322)	R → M	V → P	T → F	Man's shaver (wind-surf)	Use spring action to control pivot and multiple pivot point
30 (L: 335–343)	M → M	P → P	F → F	Gimble	Three-dimensionality
32 (L: 354–364)	M → R	P → V	F → F	Toys	Drag on
MODE primary secondary	M → M M → R	V → P P → V/P	F → F T → F	Effective brainstorming analogies with specific, relevant design features	

Table 6. Design analogies developed in E1.

or a wind-surf device), hot ballpoint, a switch, a man's shaver, a gimble, and a toy.

This is largely due to the relatively early stage of the design process and the use of brainstorming. However, this may also be due to the designers being engineers, drawing their inspiration from technology rather than 'conceptual reasoning' of architects, as in A1, usually driven by guiding metaphors.

On the other hand, there was very little evidence in E1 to support the prolific generation of design alternatives. A rare example occurs between lines 1679 and 1701 of the transcript where the engineers develop two rival ideas for switching the pen on-and-off, one based on a mechanical and the other on a control feature.

In A1, however, a number of design alternatives were developed; two in the first third of the transcript, three in the second one-third, and seven in the final third (Table 7). This adds up to twelve segments out of a total of 163. Typically, each alternative generating segment ended with a combination of Initiating and Accepting/Rejecting Acts, plus the Projecting and Verifying Tasks, and a balanced Modality (Criteria-Feature) consideration. This is typical expert designer behaviour[37], signifying the search for a 'better' or *satisficing* design in the absence of the best[38].

37.
Akın (1988) *op. cit.*

38.
Simon (1973) *op. cit.*

Segment	Last Two/Each Segment Acts	Tasks	Modes	Description of Alternative	Result
15 (L:169)	R → I	V → A	T → T	Outdoor seating area	Ok
25 (L:277)	R → R	V → V	F → T	Move up hearse, create two car space	Ok
50 (L:474)	I → I	P → P	F → F	Little slit window replace colored glass	Maybe
66 (L:628)	I → I	V → P	T → T	Door or window for visibility	Maybe
78 (L:720)	M → R	V → P	F → F	Bold architectural shape to create area	No
primary pattern	I → I	V → P	F → F	Note: All small moves except Segment #78	
secondary pattern	R → I/R	P → P	T → T	Key: → indicates temporal transition	

Table 7. Partial coding of design alternatives developed in A1.

3.3.3 *Pairwise Integration*

In previous work, I reported on the use of Pairwise Integration (PI) as a primary strategy for combining partial solutions developed in the context of decomposed design problems into a single, comprehensive solution proposal. In a step by step manner, designers take two partial solutions responding to two distinct Design Content items and merge them into one solution[39]. This is repeated until all partial solutions are merged into a single comprehensive solution. In the present codification, for both architects and engineers, there is evidence of this strategy but with an interesting twist.

39. Akın (1994) *op. cit.*

The evidence supporting PI is the concurrent application of Design Acts and Tasks to two Design Content items, i.e., oscillating between them persistently for over at least 20 lines of transcription without shifting the focus of attention onto another Design Content. The rectangular boxes in Figures 3 and 4, and labelled A, B, C, D, mark just such episodes of PI. The Design Content items are indicated as column headings.

The evidence for PI is prevalent throughout A1. The coding of lines 1184 and 1999 in Figure 3 illustrate both the clarity of this strategy in A1 and its effectiveness in dealing with the complexity of integrating multiple design features. The arrows in the figure show the exact sequence in which the Design Content items have been considered.

For instance, in box A, which is a PI episode between FORM and CONSTRUCTION DRAWING, the form of the *roof* is considered. This leads to a discussion about the *appearance* of the roof. Next, *materials* to be used to construct the roof are considered. A *cubist* or *modern* appearance is raised by the material considerations. After a sideline dealing with *costs*, the architect goes back and forth between *materials* and *roof appearance* several times before arriving at a solution that accommodates both requirements at once.

The sideline between the *roof* form and *cost*, labelled *donations* [effect on](*cost*), is a typical example of how integration of one pair is in turn merged with that of another pair of Design Content

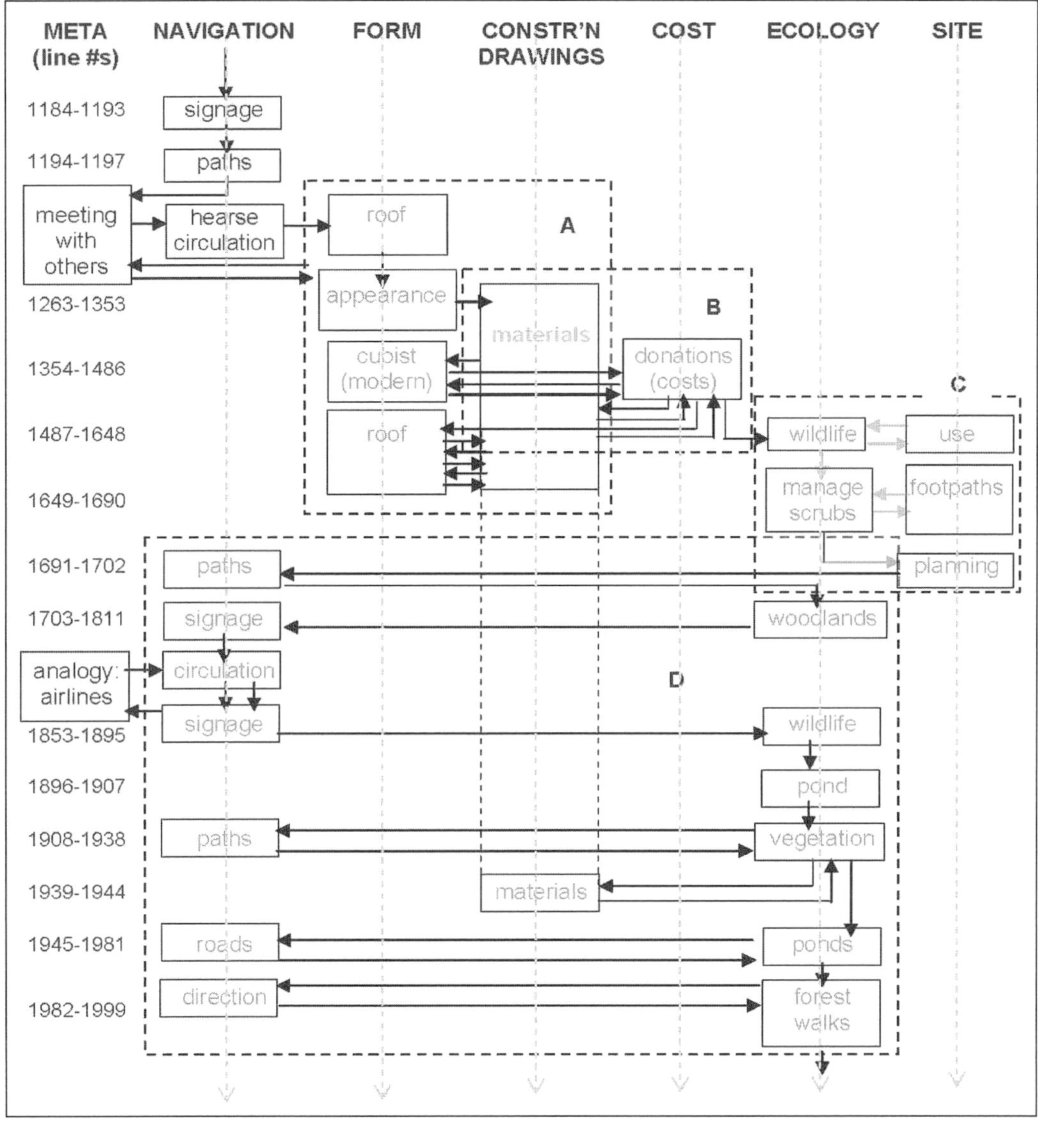

Figure 3. Pairwise Integration of design features in A1.

issues, in this case, between *materials* and *cost*. In this fashion, the coded segments shown in Figure 3, defined by Design Content issues under the general categories of ECOLOGY, SITE, NAVIGATION, FORM, CONSTRUCTION DRAWINGS and COST are integrated pairwise.

In the case of E1, the entire design process is driven by analogical reasoning. Consequently, when in evidence, the PI strategy is based on the features evoked by the same process of analogical reasoning. During the course of 312 transcribed lines, starting with 1479 and ending with 1791, the engineers introduced

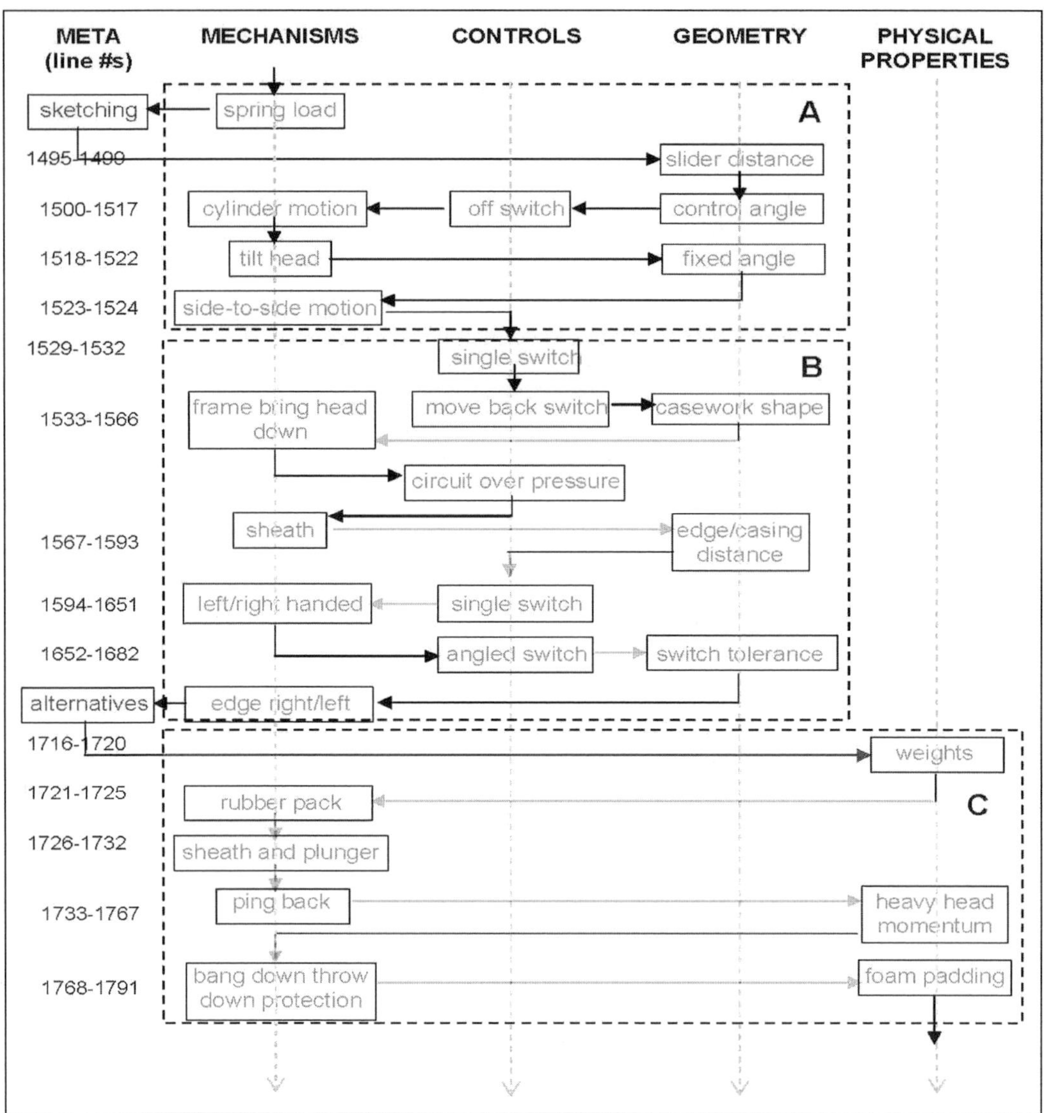

Figure 4. Pairwise Integration of design features in E1.

27 design features (Figure 4): *spring loading, slider, angle control, off switch, cylinder motion, tilt head, fixed angle, side to side, single switch, move back switch, shape of casework, frame-bring print head down, circuit over pressure, sheath, distance between edge and casing, single switch, left/right handed, angle switched, switch tolerance, edge right/left, weight, rubbery pack, sheath and plunger, ping back, heavy head momentum, bang down-throw down strength,* and *foam pad.*

As in the case of A1, the engineers oscillate between two *or more* items of Design Content, at a time; first between MECHANISMS and GEOMETRY; then between MECHANISMS, CONTROLS and GEOMETRY; and finally between MECHANISMS and PHYSICAL PROPERTIES. The notable anomaly here is the *threesome* shown in boxes A and B in Figure 4, which violate the only-two-at-a-time 'rule' of PI (Section 3.3.3).

4 DISCUSSION

In the above sections I introduced three research questions (Section 1), a coding scheme (Section 2), and an analysis of the coding in terms of these research questions (Section 3). Now let us discuss the implications of the analysis for each of the questions posed.

4.1 *Navigation*

The two design teams, architect-clients and engineers, exhibited different navigation strategies. While the architects apply a strategy of 'feeding forward' from the design features and criteria they have developed, the engineers use a strategy of 'feeding backward' from the features found through analogical reasoning. In other words, architects take design features at hand, such as ECOLOGY, SITE, NAVIGATION, FORM, CONSTRUCTION DRAWINGS and COST (Figure 3), and refine their design specifications (features) until the underlying requirements (criteria) are satisfied. This is 'forward' movement in a design search space[40]. After some adjustment of the design specifications to the requirements of the Design Content items, an acceptable *satisficing* solution emerges.

40. Newell and Simon (1972) *op. cit.*

The engineers, on the other hand, developed dozens of analogical solutions. These represent 'end points', or goals, in the design search space (Table 6). Framed in this way, the design task can be seen as that of finding the path from these end points, 'backwards' to the problem at hand. We can also argue that each of these analogies constitute the root of a potential depth-first branch. In this way, the breadth-first-depth-next strategy[41] is supported by the E1 data.

41. Akın (1994) *op. cit.*

In A1, there is no evidence of a similar breadth-first-depth-next strategy. This is in contradistinction to the hypothesis I posed at the beginning.

4.2 *Pairwise Integration*

The PI strategy helps tame complex problems to conform to the cognitive limitations of the human designer. These include the span of the short term memory, limitations of visualising multi-dimensional problems, exhaustive enumerations of precisely defined information[42]. In spite of these limitations, designers harness the complexity of design problems by first parsing them into ever smaller sub-problems, solve them, and then combine these partial solutions into comprehensive ones[43]. Through the divide-and-conquer strategy, they adapt the unruly problem to fit within the bounds of the human cognitive instrument.

42. Newell and Simon (1972) *op. cit.*

43. Simon (1973) *op. cit.*

Earlier evidence on this subject[44] points to an integration mechanisms that manipulates only two partial solutions at a time. In the codification of the A1 transcription (Figure 3), there is evidence that supports Pairwise Integration. In the E1 protocol, however, while there is evidence for PI there is also evidence for the integration of three Design Content items at once (Figure 4, boxes A and B).

44. Akın (1994) *op. cit.*

One plausible explanation for this is the team design nature of the E1 data. The original evidence for PI comes from protocols by solo designers. Is it possible that team design, aside from the known benefits of co-design, may help overcome cognitive limitations of the individual; or in effect function as a collective, super-cognition tool? Would it be plausible that two or more designers engage in a cognitive division-of-labour that effectively integrates the design features found in three, or maybe even more, partial-solutions into one comprehensive solution during one persistent episode? While I cannot answer this question with the current data and its coding, this poses a potentially important venue for sub-solution integration research and indeed research on complexity management strategies in design.

4.3 *Alternative generation and problem restructuring*

The hypothesis I set out to test under this heading is that experienced designers explore parallel alternatives before developing one of them in detail.

In A1, the participants undertook a task of refining a design that was relatively well developed where some of the early structuring of alternatives was no longer under consideration. In spite of this, twelve alternative ideas were developed and discussed with the client, leading to major modifications and improvements in the design (Table 7).

Is this a lot? How does this compare to the engineering design data? Typically, novices hardly generate any alternatives[45]. However, neither of these groups of designers can be considered novices. Instead they are experts with different approaches to design. Furthermore, the coding of A1 shows that only two alternatives are considered at any time alternatives are considered, which puts the average number of alternative generated well below the average

45. Akın (1988) *op. cit.*

observed in other research, namely 2.5 to 4 alternatives per Design Content issue.

I conjecture that there are other tacit instances of problem restructuring in this data, which do not appear in the current coding. As illustrated in Figure 1, there are 125 Accept/Reject segments in A1 and 105 in E1. These are the points when a new problem structure may spring from the assessment of the Design Content at hand. In particular, when this matches the Design Criterion (T), i.e., the primary reason for problem structuring[46], conditions are ripe for altering the problem constraints.

46. Akın (1988) *op. cit.*

Furthermore, there are roughly 230 Verify Tasks in A1 and 86 in E1 (Figure 2), all of which present a similar potential for problem structuring instances, which may be tacit.

5 CONCLUSIONS

In addressing the design cognition of architects versus engineers, some of the hypotheses based on previous work reported above were confirmed, while others if not rejected are candidates for reconsideration.

In summary, the state of my discoveries on the variants and invariants of design cognition indicates that:

1. *Navigation*: within their own search spaces, architects (A1) used a feed-forward strategy as opposed to engineers' (E1) feed-back strategy; breadth-first-depth-next strategy was in evidence with engineers but not supported by the architectural design data;
2. *Pairwise Integration*: both architects (A1) and engineers (E1) used a clearly observed PI strategy; furthermore, engineers demonstrated a triple-integration (TI) strategy which can be possibly be explained as a consequence of team design, however, this is certainly subject to more investigation through specially designed experiments to validate such hypotheses; and
3. *Alternative generation and problem re-structuring*: in spite of the relatively advanced state of the design in A1, there was moderate evidence of alternative generation; yet the rate of alternative generation was well below those observed in earlier work; on the other hand, there are many points in the coding that can be markers for tacit (undiscovered) alternative generation and problem re-structuring points; this requires a more detailed coding analysis of the data than the one that was conducted.

The coding presented in this chapter has emphasised quantifiable aspects of the data accompanied by a focus on statistical analysis. This provided useful insight about the overall patterns of designer behaviour while yielding relatively small attention to the properties exhibited locally and in detail. This has been an intentional strategy, yet in order to further examine the results of this work, analogue representations such as graphs, flow diagrams, and temporal simulations would be necessary.

Finally, some of our analysis and conclusions have been vulnerable to unmeasured influence by a few uncontrolled, independent variables, such as the scope, phase, and knowledge context of the two problems tackled in A1 and E1. While this was an unavoidable result of the available data and the choice of research topic, it is something that should be avoided in any future version of this work.

Part 4

Design Process Models

11
The Influence of the Design Task Description on the Course and Outcome of Idea Generation Meetings

Kilian Gericke, Michael Schmidt-Kretschmer & Luciënne Blessing

This chapter presents the results of an explorative, comparative analysis of two design meetings. It focuses on identifying how the style of the Design Task Description (DTD) influences the course and the outcome of idea generation meetings. A framework is used to classify potential influencing factors including the DTD, and to discuss the various relationships and influences between the factors in order to understand the differences in the course and the outcome of the meetings. We show that the style of the DTD was found to influence the length of the analysis phase in each meeting and led to differences in the course of the two meetings and to the rate of idea generation.

Our research goals are to understand and improve the process of designing with the aim of developing better products. In the particular study presented here we focus on how the design task is formulated. In the context of engineering design where products have to satisfy the needs of all stakeholders Design Task Descriptions, which include specifications of the set of design requirements, play a crucial role throughout the design process[1,2]. Here, we define a requirement as a characteristic which a designer is expected to address in the design solution[3]. During the early phases of the design process the DTD is developed in detail as further requirements are formulated. The DTD is a prerequisite for a focused idea generation meeting.

Design literature suggests several generally applicable methods for supporting requirement formulation, for example checklists[4], matrices[5] and Quality Function Deployment/House of Quality approaches for transforming customer needs into engineering characteristics[6]. The recognition that requirements need to be managed addresses the fact that views of the various stakeholders on needs and requirements can differ, and that consequently there can be different perceptions of the value of a particular solution[7].

Requirements and solutions are strongly interrelated, as has been observed in laboratory environments[8,9,10] and in industry[11,12]. The idea that requirements and solutions co-evolve during the design process[13] invites a reconsideration of the emphasis, in some approaches to design, on the formulation of a requirements list

1.
Schmidt-Kretschmer, M. and Blessing, L.T.M. (2005) Longitudinal Analysis of the Impact of Requirements Management on the Product Development Process in a Medium Sized Enterprise in *Proceedings of International Conference on Engineering Design (ICED'05)*, Melbourne.

2.
Pahl, G., Beitz, W., Feldhusen, J., and Grote, K. (2007) *Engineering Design – A Systematic Approach,* Springer-Verlag.

3.
Chakrabarti, A., Morgenstern S., Knaab, H. (2004) Identification and Application of Requirements and their Impact on the Design Process: A Protocol Study, *Research in Engineering Design* 15, pp. 22–39.

4.
Pahl, Beitz, Feldhusen and Grote (2007), *op. cit.*

5.
Ward, J., Shefelbine, S., and Clarkson P.J. (2003) Requirements Capture for Medical Device Design in *Proceedings of International Conference on Engineering Design (ICED'03)*, Stockholm.

6.
Akao, Y. (2004) *Quality Function Deployment: Integrating Customer Requirements into Product Design*, Productivity Press.

7.
Schmidt-Kretschmer, M. and Gries, B. (2006) Bug or Feature? Möglichkeiten und Grenzen des Fehlermanagements in der Produktentwicklung, in *Proceedings of 17th Symposium Design for X*, Friedrich-Alexander-Universität Erlangen-Nürnberg.

8.
Fricke, G. (1993) *Konstruieren als flexibler Problemlöseprozeß – Empirische Untersuchung über erfolgreiche Strategien und methodische Vorgehensweisen beim Konstruieren*, Technische Hochschule Darmstadt, Darmstadt.

9.
Nidamarthi S., Chakrabarti, A., and Bligh, T.P. (1997) The Significance of Co-evolving Requirements and Solutions in the Design Process in *Proceedings of International Conference on Engineering Design (ICED'97)*, Tampere.

10.
Sipilä, P.P. and Perttula, M.K. (2006) Influence of Task Information on Design Idea Generation Performance in *Proceedings of International Design Conference – Design 2006*, Dubrovnik.

11.
Almefelt, L.F., Andersson, P., Nilsson, and Malmqvist, J. (2003) Exploring Requirements Management in the Automotive Industry, in *Proceedings of International Conference on Engineering Design (ICED'03)*, Stockholm.

12.
Schmidt-Kretschmer and Blessing (2005), *op. cit.*

13.
Reymen, I., Dorst, K. and Smulders, F. (Chapter 4) Co-evolution in Design Practice.

prior to the generation of solutions. Almfelt et al.'s work[14] found that solutions result which do not meet the requirements when reference to requirements is not made *throughout* the design process. Fricke's investigations[15,16] indicate that the depth and breadth of the DTD can have an impact on the solution finding process. He showed that a detailed DTD, with clearly specified requirements, given at the start of the process reduced the number of solutions resulting from the solution finding process. Similarly, Hansen and Andreasen observed that in the early conceptual design activities: "a product specification, which contains a lot of specification statements about product properties is not productive to support the synthesis of a product idea"[17]. Published material on DTD and on solution finding methods does not support a detailed understanding of the relationship between DTD and solution finding. It remains unclear as to which aspects of the DTD influence the process in which way, what the proper level of detail or style of DTD is, and how requirements should be dealt with during the development process.

1 RESEARCH QUESTION AND HYPOTHESES

This chapter discusses the results of an explorative, comparative study based on the analysis of transcripts from the two engineering design meetings, E1 and E2, aimed at contributing to our understanding of the following research question: *How does the style of the Design Task Description influence the course of the process and the outcome of idea generation meetings?*

Our analysis is based on pursuing the following hypotheses:

1. The course of the design process depends on the style of the DTD; different behaviour in the task clarification stage of a process will affect the course of the overall process.
2. The number of ideas resulting from the process depends on the amount of time used for task clarification. An extensive analysis leads to more ideas for solving the task.
3. A longer analysis of the task supports the exploration of issues not mentioned in the DTD.

The framework we use to structure the elements of our research is based on the framework proposed by Frankenberger, Badke-Schaub and Birkhofer[18] to structure the factors influencing teamwork in engineering design practice. We have simplified the original framework to highlight the basic relations between the elements (groups of influencing factors) we address, as shown in Figure 1.

The framework sets out the relation between the elements relevant to our hypotheses, namely the Task, the ideation Process and the

Result. The framework also supports recognition of the influence of Boundary Conditions. These include individual prerequisites, group prerequisites and external conditions.

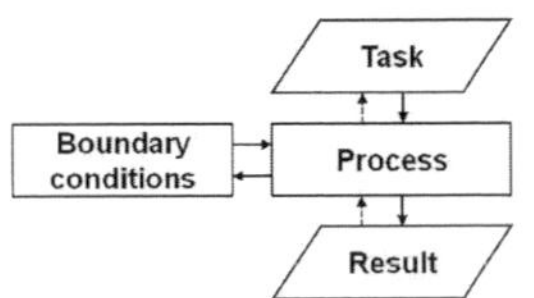

Figure 1. Basic framework for elements of the ideation process, after Frankenberger, Badke-Schaub and Birkhofer[19].

Our hypotheses are concerned with relationships between the elements Task, represented by the DTD, the Process and its outcome, the Result. Factors which may provide alternative explanations of influence on the Process beyond the relationships between Process, Task and Result are contained in the element Boundary Conditions. In contrast to the original framework, we have modified the relationships between the elements to correspond with the literature on co-evolution of requirements and solutions to show possible feedback from Results to the Process and from the Process to the Task.

2 OVERVIEW OF THE DATA

Our analysis focused on the two engineering design meetings. Each of the two meetings addressed predominantly different issues for the same product (E1 addressed mechanical engineering problems, and E2 addressed electronic engineering problems) and the groups meeting on the two occasions differed slightly. The goal of both meetings was to use brainstorming to find solutions for the given design task. Brainstorming is an intuitive solution finding method which is well established in industry. The goal of a brainstorming process is to generate a large number of ideas through group creativity, without criticising or evaluating the ideas generated[20]. Most ideas will subsequently be discarded, but with perhaps a few novel ideas being identified as worth following-up[21].

The brainstorming method was introduced by the moderator at the beginning of meeting E1. Matthews' analysis[22] found that in these two meetings: "designers' adherence to the rules of brainstorming are noticeably tempered by their orientation to social order". However, this finding does not alter the fact that the main objective of brainstorming – namely the generation of a large number of ideas without criticising or evaluating them – was also the objective of these two meetings.

The data available for our analysis comprised:

- videos, transcripts of the meetings, room layouts and seating plans;
- copies of sketches and photographs of flipcharts generated during the meetings;
- the DTD (in the form of a design brief and slides of the presentations given to the design teams at the start of the meetings);
- a technical drawing of an existing prototype given to the design teams;
- some artefacts (e.g. the prototype, a shaver).

14.
Almefelt, Andersson, Nilsson and Malmqvist (2003), *op. cit.*

15.
Fricke (1993), *op. cit.*

16.
Fricke, G. (1999) Successful Approaches in Dealing with Differently Precise Design Problems *Design Studies*, Volume 20, pp. 417–429.

17.
Hansen, C.T. and Andreasen, M.M. (2007) Specifications in Early Conceptual Design Work in *Proceedings of International Conference on Engineering Design (ICED'07)*, Paris.

18.
Frankenberger, E., Badke-Schaub, P., and Birkhofer, H. (1998) *Designers The Key to Successful Product Development*, Springer-Verlag.

19.
ibid.

20.
Paulus, B.P. and Nijstad, B.A. (2003) *Group Creativity*, Oxford University Press.

21.
Cross, N. (2000). *Engineering Design Methods*, Wiley.

22.
Matthews, B. (Chapter 2) Intersections of Brainstorming Rules and Social Order.

The data contained little about the background of the participants and almost no information about the context of the meetings. These and other limitations are discussed in Section 4.5. The analysis focused on the transcripts, the videos were used to verify our interpretation of the transcripts. The use of the artefacts and the sketches in the meetings did not form part of our analysis.

3 RESEARCH APPROACH

The comparison of the two engineering design meetings consisted of the following steps: formulation of hypotheses, development of coding scheme, coding and inter coder reliability checks, analysis of events in each meeting and finally comparison between the meetings. These steps are briefly outlined in Section 3.1 below and the coding scheme is described in more detail in Section 3.2.

3.1 *Method summarised*

Initially the data was studied by reading the transcripts and viewing the videos to obtain a general insight into the nature of the meetings, with a particular focus on the DTD. This allowed us to identify factors that might have an influence on the course and the outcome of the meetings, viewed as idea generation meetings. Based on the impressions from this initial review, and supported by findings from published work to which we have already referred, Frankenberger, Badke-Schaub and Birkhofer's framework[23] was selected and adapted as described above (Section 1) and hypotheses were formulated.

23. Frankenberger, Badke-Schaub and Birkhofer (1998), *op. cit.*

Coding schemes were developed (see Section 3.2 below) based on the hypotheses and on the Boundary Conditions, as far as these were known. The coding was executed by two of the authors. To test inter-coder reliability Kappa-coefficients were calculated. These were found to lie between 0.79 and 0.96, with a mean value of 0.93. This is accepted as satisfactory. Once meeting transcripts had been coded into categories of events time-stamps were added for each coded event so that the duration of different types of events could be determined.

Based on this preparatory work, the number of events for each of the defined classes was counted. This allowed us to examine distribution of the events for each participant of the meetings. In addition to the number of events, the time spent on each event category was also calculated. Finally, the meetings were compared using the distribution of events in terms of frequency and time. Additionally, the number and allocation of ideas generated during the meetings was analysed. Because only two meetings were analysed and compared our findings are presented descriptively and do not contain statements about statistical significance.

3.2 *Coding scheme*

The conversational exchanges of the meeting participants recorded in the meeting transcripts were the main data source. The data was coded by identifying types of EVENTS to which a contribution (an individual's turn in the conversation) or part of a contribution related. EVENTS were categorised as follows:

EVENT:

APPLICATION EVENT: (part of a) contribution referring to a 'use case', which might lead to new requirements, for example a left-handed user:

"if you're /right handed\ you drag if you're left handed you push" (E1, 1050)

REQUIREMENTS EVENT: (part of a) contribution referring to a requirement, for example the age of the user, the length of the print head:

"yes we could just up the age limit it's not a problem" (E1, 1392)

SOLUTION EVENT: (part of a) contribution referring to an idea that solves an ISSUE:

"you could have thumb holes or something like that" (E1, 732)

OTHER EVENT: EVENTS that do not belong to one of the other categories, such as moderation remarks, jokes, etc.:

"syringe I can spell that [*laugh*]" (E1, 1201 – 1202)

ISSUE:

particular aspect of the product for which a SOLUTION EVENT is sought by the designers, which – in the observed meetings – came mainly from the DTD (see Table 1 for details)

REQUIREMENTS:

characteristic, which a designer is expected to fulfil in the design SOLUTION[24]

IDEA:

creative resolution of an ISSUE.

24. Chakrabarti, Morgenstern and Knaab (2004), *op. cit.*

4 FINDINGS

We present our findings organised in terms of the elements of the framework given in Figure 1, Task (as DTD), Process, Result and Boundary Conditions. In Section 4.1, the differences in the DTDs of the two meetings are described. In Section 4.2, the differences in the courses of the two Processes are discussed and linked to the differences in the DTD. The same is done in Section 4.3 with respect to the process outcome (Result). To identify potential alternative explanations for the findings, Section 4.4 discusses the possible influence of the Boundary Conditions on the Process as depicted in Figure 1. Finally, in Section 4.5 we briefly review limitations.

4.1 *Design Task Description (DTD)*

The DTD explicitly provided for the designers in the two meetings included ISSUES to be addressed and the REQUIREMENTS to be met by the design SOLUTION. The DTDs also contained a description of the goals for the meetings, corporate intent and brainstorming rules. The REQUIREMENTS were not listed explicitly but contained within the text. The DTD for meeting E1 differed in style and content from the DTD for meeting E2, as can been seen from the extracts in Figures 2 and 3. The DTD from meeting E1 contains a list of ISSUES to be addressed; the DTD from meeting E2 does not.

The main feature of the DTD of meeting E1 was a table with the ISSUES to be addressed during the brainstorming session and the problems related to each ISSUE, as the extract in Figure 2 shows. Hansen and Andreasen[25] state that a detailed product design specification does not support the synthesis of a product idea and we observed here that a more formal style of the design task description constrained the possible SOLUTION space.

25.
Hansen and Andreasen (2007), *op. cit.*

The DTD of meeting E2 did not include an explicit description of the ISSUES. As can be seen from the extract in Figure 3, the style of the DTD of meeting E2 was more informal with a more enquiring character. This should support the expansion of the solution space to correspond with Hansen and Andreasen's finding[26] that a productive product design specification consists of statements

26.
ibid.

Focus of generating ideas to overcome specific problems:

Problem	Issue in the concept
Wobbly arm movement of the user (5-11 year old)	Print head needs to stay in contact with thermal paper to print
Keeping the print head within an optimum angle range	The print head needs to activate the paper at the right angle to ensure good quality printing

Figure 2. Part of the Design Task Description for meeting E1.

Product features
■ What can we do with it
▪ A thermal print head looks like a linear line of dots (about 7mm long)
▪ Each of which can be individually addressed
▪ Normally moved relative to the media with the dot line held perpendicular to the line of motion.
▪ Here, user provides transport
■ Control and selection of user features
▪ How to provide a suitable UI for the intended user group
▪ Select features and control the device in applications envisaged.

Figure 3. Part of the Design Task Description for meeting E2.

which express value, contain context information, and articulate key functions.

The DTDs also differed in granularity. The ISSUES to be addressed in meeting E1 are expressed in more detail than in the task description for meeting E2. For instance, an ISSUE identified for attention in meeting E1 was to consider ways to detect when the print head should or should not fire. This provided a clear focus on one sub-function (print head activation), allowing the participants to find solutions without a reformulation of the problem. Given such a prompt, the idea generation process can start immediately. The DTD of meeting E2 is less specific. The ISSUES are listed as open questions, for example, by asking how to provide a suitable UI for the intended user group. To address this ISSUE it is necessary for the designer participants to determine what suitability means and to define functions which have to be fulfilled by the user interface (UI).

4.2 *Process*

Firstly, we analyse which ISSUES and REQUIREMENTS were addressed during the Process. Secondly, the differences in the courses of the Processes are analysed by looking at the sequence and frequencies of analysis and synthesis activities. Finally, differences in design time are considered.

Meeting E1 focused on mechanical engineering ISSUES and meeting E2 on electronic engineering ISSUES. Five ISSUES were listed in the DTDs for each meeting. The differences in the ISSUES itemised for each meeting reflect these different emphases but they also differ in their specificity, as shown in Table 1 which summarises the five ISSUES for each meeting.

Our analysis of the contributions showed that in meeting E1 only the ISSUES on the agenda were addressed whereas in meeting E2 one derived ISSUE was also addressed. This might be seen as an indication that the informal and more enquiring character of the DTD of meeting E2 enhances the creativity of the participants by widening the problem space.

Table 2 summarises our comparison relating to REQUIREMENTS. We categorised as *given* REQUIREMENTS any that were mentioned in the DTDs or during the introductory presentations by the meeting moderators. Those that were introduced during the meetings we categorised as *derived* REQUIREMENTS.

There is little difference in the two meetings between the numbers of *given* REQUIREMENTS. The number of REQUIREMENTS which was derived during the meetings showed slightly more difference, 12 in meeting E1 and 16 in meeting E2. Again, it may be the case that the less formal and less precise formulation of the DTD of meeting E2 may have encouraged the consideration of more requirements in E2.

The conclusions regarding Process only relate to the EVENTS of the meetings analysed in this study, and not to the larger design process of which they were a part. According to the categorisation of Dörner[27] and Fricke[28,29] problems can be classified into dialectic problems (search and application problems) and synthesis problems. To solve problems, either steps of *analysis* (dialectic problems) or *synthesis* (synthesis problems) have to be undertaken. *Analysis*: "is the resolution of anything complex into its elements and the study of these elements and their relationships"[30], whereas *synthesis:* "is the fitting together of parts or elements to produce new effects [involving] search and discovery, and also composition and combination"[31].

27.
Dörner, D. (1976) *Problemlösen als Informationsverarbeitung*, Kohlhammer, Stuttgart.

28.
Fricke (1993), *op. cit.*

29.
Fricke (1999), *op. cit.*

30.
Pahl, Beitz, Feldhusen and Grote (2007), *op. cit.*

31.
ibid.

To distinguish *analysis* and *synthesis* steps as elemental components of the design process, the APPLICATION and REQUIREMENTS EVENTS of the meetings were grouped into the category *analysis,* whilst the SOLUTION EVENTS were assigned to the category *synthesis*. The other EVENTS classed as OTHER remained unassigned to either *analysis* or *synthesis*. APPLICATION EVENTS were classified as *analysis* because these EVENTS do not describe solutions for the product under development but deliver information by analysing the potential use and other aspects of the product.

Meeting	ISSUES
E1	Keeping the print head within an optimum angle range
	Detect when the print head should not fire (overheating)
	Find SOLUTIONS to save the print against over pressuring
	Ensure contact to paper (print head level, wobbly arm movement)
	Ensure smooth running over contours
E2	Define a suitable user interface
	Select features
	Enable control
	Find options for electronics architecture
	Search other applications, e.g. B2B
	Energy [derived ISSUE]

Table 1. ISSUES to be addressed in each meeting.

REQUIREMENTS	E1	E2
given in design brief (DTD)	9	9
given in presentation (DTD)	6	8
derived	12	16

Table 2. Numbers of *given* and *derived* REQUIREMENTS in meetings E1 and E2.

The sequence of *analysis* and *synthesis* steps in the two meetings is different. In meeting E1, the participants continuously generated ideas (*synthesis*) and only occasionally dealt with the REQUIREMENTS and APPLICATIONS (*analysis*) (see Figure 4). Two distinct periods can be observed in meeting E2 (see Figure 5), a long initial *analysis* phase (APPLICATION and REQUIREMENTS) changing into a very clear *synthesis* phase (SOLUTION).

Figures 4 and 5 refer to the occurrence of the EVENTS, not to their duration. Each OTHER EVENT turned out to be rather short compared with the duration of the *analysis* and *synthesis* EVENTS, as can be seen in Figure 6.

The difference between the two Processes is illustrated in Figure 7. In meeting E1 ISSUE after ISSUE is addressed, focusing in particular on *synthesis*. In meeting E2 all ISSUES are addressed together, first by means of *analysis* and later by means of *synthesis*. A similar difference was found in the processes studied by Fricke: some designers tend to focus on the functions to be fulfilled, i.e. they tend to go through all steps of the process for each function before addressing another function, others tend to focus on the process, and to address all functions in each step before starting the next step.

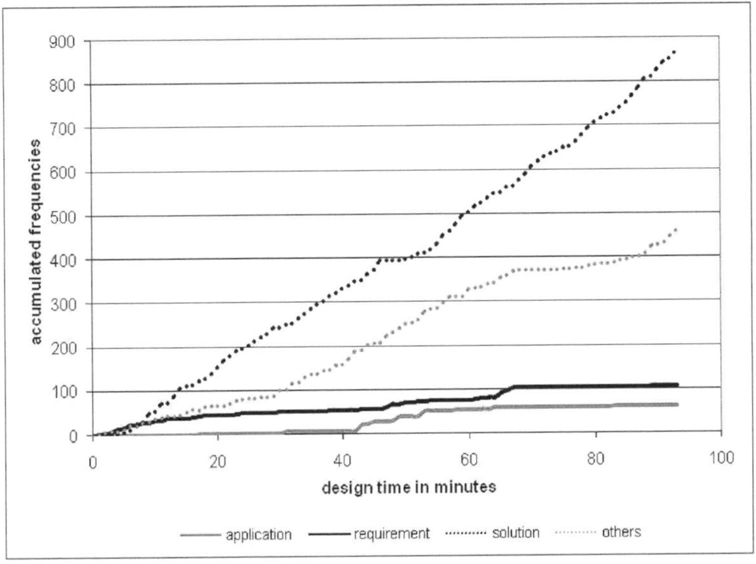

Figure 4. Accumulated EVENTS for meeting E1.

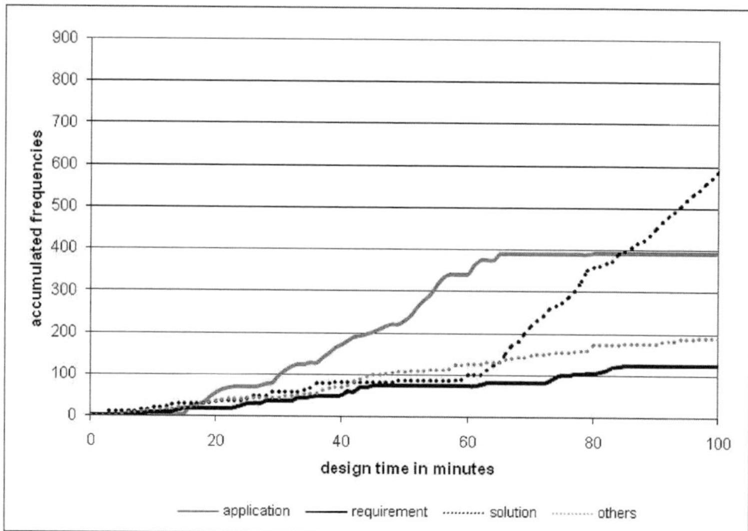

Figure 5. Accumulated EVENTS for meeting E2.

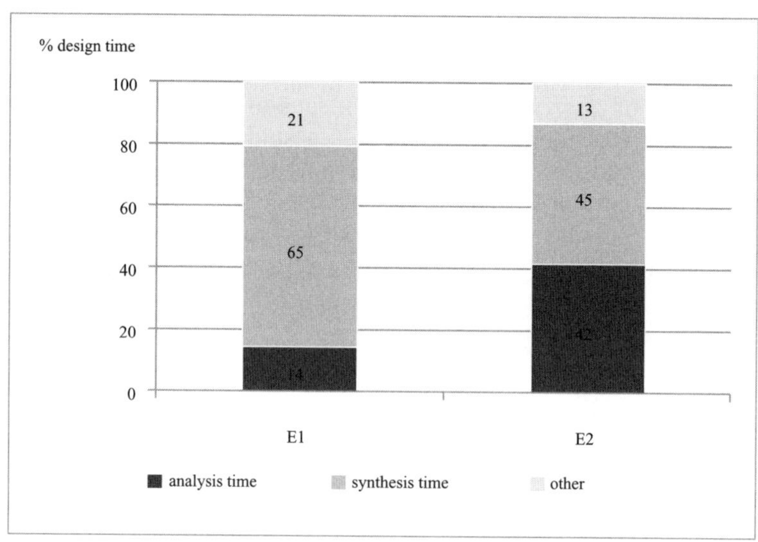

Figure 6. Comparison between E1 and E2 of allocation of design time.

The long initial *analysis* in meeting E2 could be the consequence of the style of the DTD. The DTD suggests applications and according to Jansson[32] designers tend to stick to principles that have been suggested to them. The participants first analysed the applications to get a better understanding of the problems related to the issues to be addressed. This is the input for the subsequent *synthesis* phase.

32. Jansson, D.G. and Maulin, R.G. (1988) Conceptual Design: Model, Methodology, and Experiment in *Proceedings of International Conference on Engineering Design (ICED'88)*, Budapest.

In meeting E1, which started with a more formal DTD with specified issues, such an initial *analysis* phase cannot be observed. The *analysis* was separate for each ISSUE.

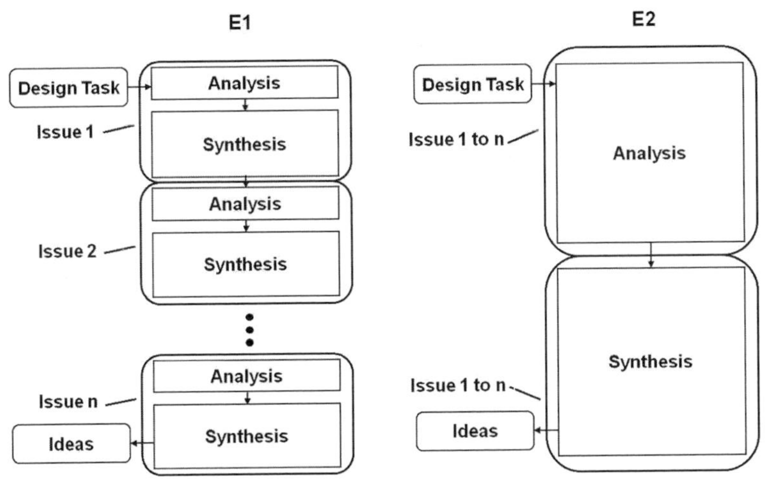

Figure 7. Comparison between E1 and E2 of the course of the process.

The *analysis* focus in E2 continued until it changed abruptly due to an intervention of the E2-moderator after 58 minutes. (Tommy sums up some requirements and directs the group towards talking about implementation.) The course of any process can obviously be strongly influenced by its moderator.

4.2.1 *Allocation of design time*

The total duration of both meetings was similar (E1: 98 minutes, E2: 100 minutes). The total time less the initial moderation is what we refer to in the following as design time. The design time was 93 minutes for meeting E1 and 100 minutes for meeting E2. As Figures 4 and 5 suggest, the time spent on *analysis*, *synthesis* and OTHER EVENTS in the two meetings differs considerably (see Figure 6).

The time spent on *analysis* during meeting E2 (42%) was much more than during meeting E1 (14%). Accordingly, in meeting E2 less of the time (45%) was spent on *synthesis* than in meeting E1 (65%). Interestingly, in E2 equal amounts of time were spent on *synthesis* and *analysis*.

4.3 *Outcome*

Because only two meetings from a larger design process could be analysed, it was not possible to relate the findings to the quality of the solution or other such success criteria. The focus of the analysis, as reflected in the research question, was chosen to be the number of IDEAS generated during the meetings, as this was the (brainstorming) aim of the two meetings and thus represents a measure of success of the meetings. The underlying assumption is that the chance to generate a good SOLUTION increases with the number of ideas generated[33].

33. van der Lugt, R. (2007) Relating the Quality of the Idea Generation Process to the Quality of the Resulting Design Ideas in *Proceedings of International Conference on Engineering Design (ICED'07)*, Paris.

To obtain an overview of the outcome of both meetings the IDEAS were identified and categorised according to the ISSUE each one addressed (see Table 3). In meeting E1, 47 IDEAS were generated compared with 46 IDEAS in meeting E2.

Although the total number of IDEAS overall in both meetings is nearly identical, the IDEA generation rate during the *synthesis* EVENTS of meeting E2 is approximately 20% higher than the rate of meeting E1. (See Table 4, which also shows the design time and the *analysis* time.) A possible explanation is that the long initial *analysis* phase in meeting E2 increased the IDEA generation rate. A similar phenomenon was observed by Klaubert and Blessing[34]. Both brainstorming meetings lasted longer than the 45 minutes recommended in the literature[35,36], without a diminishing flow of ideas.

34.
Klaubert H.L. and Blessing, L.T.M. (1997) An Analysis of Tiny Design: Design Process for MicroElectroMechanical Systems (MEMS) in *Proceedings of International Conference on Engineering Design (ICED'97)*, Tampere.

35.
Cross (2000), *op. cit.*

36.
Pahl, Beitz, Feldhusen and Grote (2007), *op. cit.*

4.4 *Boundary Conditions*

Due to the limitations of this study, which we refer to in Section 4.5, comparatively few factors associated with the Boundary Conditions could be analysed. Two influencing factors that we are able to consider are 'design time' and 'subjects involved'. The influence of the available design time as an explanation for the observed differences between meeting E1 and E2 can be rejected given that the design time for the meetings was 93 minutes and 100 minutes respectively. Both meetings involved seven participants, four of whom were present in both meetings. The participants had different backgrounds and contributions; these are itemised in Table 5. Contributions such as

E1		**E2**	
ISSUE	No. of IDEAS	ISSUE	No. of IDEAS
print head angle	16	user interface	9
overheating	3	features	10
over pressuring	22	control	11
smooth running over contours	6	electronics architecture	9
contact to paper (wobbly arm)	2	energy	7
other	1	other	2
Total	47	Total	46

Table 3. Distribution of ideas across ISSUES in E1 and E2.

general laughter are classified as group contributions. The table shows some relationship between the emphases of the meetings (E1 predominantly mechanical, E2 predominantly electronics) and the backgrounds of the participants. In both meetings the role of the moderator was allocated to the participant with a business consultant background.

	No. of IDEAS	Design time (min)	*Analysis* time (min)	*Synthesis* time (min)	Synthesis time IDEA generation rate (IDEAS/min)
E1	47	93	13.3	60.5	0.78
E2	46	100	41.4	45.3	1.02

Table 4. IDEA generation rates E1 and E2.

Table 5 clearly shows that in both meetings the majority of contributions came from only a few participants. The highest percentage of contributions came from the moderators of the meetings. The contributions of the usability specialist and the industrial design student who attended both meetings were relatively small, with the contribution of the student being the smallest of all participants.

We could not observe a clear effect on the number of contributions of the fact that someone was a specialist in a meeting. For example, Jack, as a specialist for mechanical engineering, made nearly the same percentage of contributions in both meetings. The specialists Todd and Patrick attended only the meeting for which their background was relevant. Their contribution in that meeting was large. However, the other specialists who attended a particular meeting because of their background contributed far less.

Looking at the content of the contributions, it was found that in meeting E1 the distribution over the four EVENT categories of the EVENTS for each participant showed a very similar pattern irrespective of their total contribution or their background (see Figure 8). For example, Tommy as a specialist for electronics produced a similar amount of SOLUTION EVENTS as each of the specialists for mechanical engineering.

In meeting E2, however, clear differences in the contribution of specific categories of EVENTS depending on background can be seen. The majority of the EVENTS which Sandra and Rodney contribute are APPLICATION EVENTS. The difference for Sandra in meeting E2 com-

Name	Background	E1	E2
Group		5%	2%
Alan	business consultant	26% (moderator)	
Tommy	electronics, business consultant	18%	32% (moderator)
Jack	mechanical engineer	14% (specialist)	12%
Sandra	ergonomics, usability	6%	9%
Rodney	industrial design student	4%	4%
Chad	mechanical engineer	7% (specialist)	
Todd	mechanical engineer	20% (specialist)	
Roman	electronics, software		9% (specialist)
Stuart	electronics		10% (specialist)
Patrick	electronics, software		22% (specialist)

Table 5. Distribution of contributions to the meeting by participants as percentages of the total number of contributions in each meeting.

pared to the group average can be explained by the fact that Sandra left the meeting after Tommy introduced the solution finding phase that focused on the ISSUE of electronics architecture. Rodney stayed but contributed hardly anything after this point in the process. Interestingly, Sandra contributed with more EVENTS in this second meeting than in the first one, although she attended only 60% of the time. Rodney contributed in this same period only slightly fewer EVENTS than in the first meeting. A possible explanation is the type of process that followed from the DTD. The more open formulation of the DTD of E2 and the issues that were suggested led to discussions about applications rather than technical details. For a usability expert and industrial design student, considerations of application are central to their field of expertise. This does not take away their ability to contribute with SOLUTION EVENTS as shown for

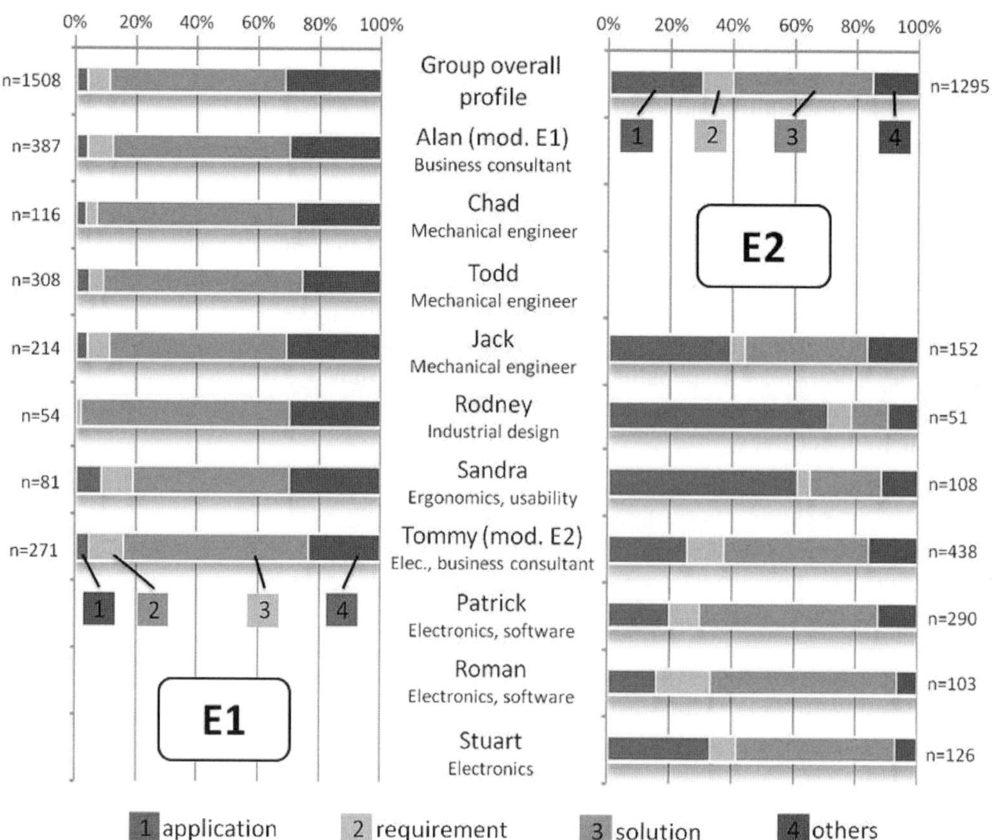

Figure 8. Relative frequencies of EVENTS in the two meetings.

E1, in which their relative contributions were the same as the other participants.

4.5 *Limitations*

The analysis of meeting transcripts as a method of empirical design research can support a deeper understanding of design practice. However, this type of data is only really suitable for the analysis of relatively short periods of some particular kinds of design activity and therefore seldom covers a complete design process. Analysis may be richly detailed but the extent of the design process to which it gives access is limited.

The results of the comparison described in this paper have to be viewed in the light of the limitations inherent to the setup of the study and the analysis method chosen. Several factors other than differences in the DTD that could have influenced the process and the outcome could not be observed, e.g. the experience of the participants related to the topic, their motivation, previous work and other work on the design of the product outside the meetings, the appropriateness of the available time for solving the issues, and obviously what the participants were thinking while others were speaking. Furthermore, the fact that only two meetings could be observed limits the strength of the statements that we can make. Nevertheless, the study indicates an effect of the DTD on the process and its outcome, which is both plausible and for which corroborating evidence is found in literature.

5 DISCUSSION

Our intention was to explore how the style of the DTD influences the course of the process and the outcome of these two ideas generation meetings. Here we briefly review our findings organised around the three hypotheses we set out in Section 1.

Our first hypothesis was: *The course of the process depends on the style of the DTD; different behaviour in the task clarification stage of a process will affect the course of the overall process.*

The DTDs from the two meetings differed in style and granularity. The style of the DTD in meeting E1 is formal. It provides a list of short statements about the issues which should be addressed during the meeting. The informal style of the DTD in meeting E2 can be described as more questioning (compare Figures 2 and 3). The courses of the processes in the two meetings show different patterns when we compare the accumulated frequencies of EVENTS (compare Figures 4 and 5). During meeting E1 *synthesis* (SOLUTION

EVENTS) and *analysis* (APPLICATION and REQUIREMENTS EVENTS) show a steady increase. Meeting E2 started with a long initial *analysis* phase which led into a *synthesis* phase initiated by the moderator. In meeting E1 there was no such initial analysis for all issues. The issues were analysed consecutively with a following synthesis phase (as depicted in Figure 6). We trace the long initial *analysis* in meeting E2 back to the more questioning style of its DTD.

Our second hypothesis was: *The number of ideas resulting from the process depends on the amount of time used for task clarification. An extensive analysis leads to more ideas for solving the task.*

In meeting E1 relatively less time was spent on *analysis* than during meeting E2 (14% compared with 42%). Using idea generation rate as an indicator of the result of the meetings, we found that a preceding analysis phase in E2 lead to a higher idea generation rate (approximately 20% higher), but did not result in an increase in the total number of ideas generated over E1. The observed higher idea generation rate in meeting E2 indicates a potential to enhance the outcome of idea generation meetings by optimising the amount of time spent on *analysis* so as to provide an appropriate balance between *analysis* and *synthesis*.

Our third hypothesis was: *A longer analysis of the task supports the exploration of issues not mentioned in the DTD.*

The *analysis* time in meeting E2 was longer than in meeting E1. During meeting E2 one additional ISSUE and 4 additional REQUIREMENTS were derived compared to meeting E1. This is consistent with the findings of Fricke[37], who states that an informal DTD avoids the impression of comprehensiveness of the DTD and promotes identification of additional REQUIREMENTS. Based on the observed differences in our study and the findings in other reported work it can be said that the combination of an informal questioning DTD combined with an adequate duration of the *analysis* phase supports the exploration of new issues. The observations we have made about Boundary Conditions (above in Section 4.4) do not suggest we should qualify our discussion of the hypotheses regarding the course of the process and the outcome.

37. Fricke (1993), *op. cit.*

6 CONCLUSIONS

Our main research question for this study was: *How does the style of the Design Task Description influence the course of the process and the outcome of idea generation meetings?* Our hypotheses addressed the relations between the DTD, the course of the process and its outcome. As observed, an informal style of the DTD leads to a longer analysis phase. This enhances the idea generation rate during the synthesis phase that follows. Another observation is that a longer analysis phase supports the identification of additional issues and requirements that are not mentioned in the original DTD. The exploration of further issues may influence the outcome of idea generation meetings. Interestingly, this finding from our observation of a partial, but real engineering project in an industrial context, confirms the findings of Fricke[38,39], which were generated in a laboratory environment.

38. Fricke (1999), *op. cit.*

39. Fricke (1993), *op. cit.*

During the observed meetings the outcome measured in terms of the total number of ideas generated did not differ. We did not see a diminishing flow of new ideas during either of the meetings. Further research might explore the relationship between the DTD and the outcome of ideation meetings, such as brainstorming sessions, more extensively. For example, it would be interesting to compare outcomes in terms of the number of ideas for different types of DTD in settings where the amount of time spent on different phases might be controlled.

ACKNOWLEDGEMENTS

We are grateful for the opportunity to participate in the DTRS7 project that has made this study possible. We also thank Johannes Fischer, Philipp Kotsch and Pit Schwanitz of the Technische Universität Berlin for their support.

12
Using the FBS Ontology to Capture Semantic Design Information in Design Protocol Studies

Jeff Kan & John Gero

This chapter presents a method for capturing semantic information from design protocols. We report on a preliminary study that analyses the transcript of a design meeting using the FBS ontology and derives processes within this ontological framework by employing linkography. The usefulness of the method is examined by applying it to the first engineering meeting, E1, as a case study. We show that the original 1990 FBS ontology captures 66% of all the meaningful derived processes, while the situated FBS ontology captures 92%. According to the ontology, the session is characterised by a high percentage of structure reformulation, followed by behaviour reformulation, and analysis.

The Delft workshop in 1994[1] documented a variety of coding schemes that were applied to study the same design protocol; these help to gain a rich understanding of different aspect of designing. However, the diversity and uniqueness of each scheme makes it hard to reuse and compare results either qualitatively or quantitatively. The motivation behind the work presented here is to develop a general coding scheme that yields high quality, uniform results, that maps well to the behaviour of designers, produces a deeper understanding of design thinking and activities and that can be applied across datasets independently of the domain and the number of participants. The general coding scheme is based on an ontology of the domain of designing and as a consequence is not an ad hoc development specific to a particular dataset but one that can be used uniformly across design protocols independently of the specific design activity being studied and unrelated to the number of participants in a design team.

1. Cross, N., Christiaans, H., and Dorst, K. (1996) Introduction: The Delft protocols workshop in Cross, N., Christiaans, H., and Dorst, K. (eds) *Analysing Design Activity*, Wiley, pp. 1–14.

This chapter explores the use of the FBS ontology as a general coding scheme to study designing. Its aim is to capture semantic information that can then be used to explore different aspects of designing according to the focus of interest and to locate different types of design transformation processes.

1 FBS ONTOLOGY

In this section a brief summary of the FBS framework with its relation to design and design creativity is presented. The FBS ontology framework[2] models designing in terms of three basic classes of

2. Gero, J.S. (1990) Design Prototypes: A Knowledge Representation Schema for Design *AI Magazine*, 11, pp. 26–36.

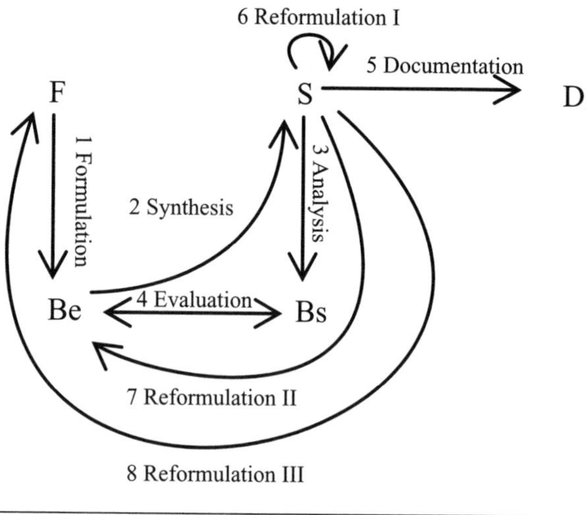

Figure 1. The FBS ontology of designing[3].

3. After Gero (1990), *ibid.*

variables: function, behaviour, and structure. In this view the goal of designing is to transform a set of functions into a set of design descriptions (D). The function (F) of a designed object is defined as its purposes or teleology; the behaviour (B) of that object is how it achieves its functions and is behaviour which is either derived (Bs) or expected (Be) from the structure, where structure (S) comprises the elements of an object and their relationships. A design description is never transformed directly from the function but undergoes a series of processes among the FBS variables. These processes include: a formulation which transforms functions into a set of expected behaviours; a synthesis, wherein a structure is proposed that is likely to exhibit the expected behaviour; an analysis of the structure which produces its derived behaviour; an evaluation process which acts between the expected behaviour and the behaviour derived from structure; and documentation, which produces the design description. Based on the structure there are three types of reformulation where new variables are introduced: reformulation of structure, reformulation of expected behaviour, and reformulation of function. Reformulation of function is relatively rare, as it changes or redefines the design problem. Figure 1 shows the relationships among the eight transformation processes and the three basic classes of variables.

These eight processes are claimed to be the fundamental processes for designing. For example, analogical reasoning – reported to play an important role in creative designing[4,5] – is considered as

4. Gero, J.S., and Maher M.L. (1991) Mutation and Analogy to Support Creativity in Computer-Aided Design, in Schmitt, G.N. (ed.) *CAAD Futures '91*, ETH, Zurich, pp. 241–249.

5. Goldschmidt, G. (2001) Visual Analogy, in Eastman, C., McCracken, M. and Newstetter, W. (eds) *Design Knowing and Learning: Cognition in Design Education*, Elsevier, pp. 199–219.

part of the reformulation processes. Analogies can be of function, behaviour or structure.

2 DATA SUMMARY

The data selected for this study was the transcript of the first engineering meeting, E1. This particular engineering session was selected because of its content. From viewing the video recordings of all four meetings comprising the DTRS7 data we judged the architectural meetings to be predominantly concerned with presenting and communicating with clients; relative to the engineering meetings they contained fewer design activities. Although both engineering meetings concerned creating a new thermal printing pen, the session studied, E1, involved generating novel ideas based on analogies. The other session (E2) related more to generating ideas for usage and control of the device.

Engineering meeting 1 can be divided into two episodes; the first one concerned the problem of keeping the print head in contact with the media and the optimum angle to the media, despite wobbly arm movement. The second episode dealt with protecting the print head from abusive use and overheating. In the first episode participants were asked to generate ideas from available products that follow a contour. There were seven participants involved in this session. Gericke, Schmidt-Kretschmer and Blessing[6] describe the participants disciplinary backgrounds and Adams et al.[7] present a detailed analysis of each of their roles in the meeting. During the meeting several products were mentioned, for example, a sledge, snowboard, wind surfboard, shaver, snow mobile, train, and slicer. Other concepts such as wheels, spirit level, and laser leveler were also discussed[8]. Loosely related to these analogical references, a few shapes, such as mouse-type pen, were proposed. User behaviour as well as product behaviour was also considered during the meeting. Sketches and drawings were produced throughout the meeting. Most of the analogies were observed to occur in the first episode, thus the transcript associated with this part of the first engineering meeting was selected for analysis in this study.

6.
Gericke, K., Schmidt-Kretschmer, M., and Blessing, L. (Chapter 11) The Influence of the Design Task Description on the Course and Outcome of Idea Generation Meetings.

7.
Adams, R., Mann, L., Jordan, S., and Daly, S. (Chapter 19) Exploring the Boundaries: Language, Roles and Structures in Cross-Disciplinary Design Teams.

8.
Stacey, Eckert and Earl (Chapter 20) describe the roles these and other types of object references play during this and the other meetings in some detail.

3 CODING SCHEME

The coding scheme applied to the data consists of the FBS classes; the five codes F, Be, Bs, S and D described in Section 1 above and shown in Figure 1 with the addition of two additional codes denoting requirement (R) and others (O) to accommodate data that did not fit within the first five codes. In Gero's FBS computational model[9], designing was assumed to start with function; later in

9.
Gero (1990) *op. cit.*

Table 1. Examples of coding E1 using the classes from the FBS ontology.

FBS class	Examples from the transcript of E1
R	"quite important is its about the thermal-incli- inclis ( ) pen" (E1, 43) "design a-a prototype" (E1, 56–57)
F	"that's the standard plain thermal paper err and then it can draw" (E1, 54)
Be	"either atoms or line types" (E1, 55) "we can print thermo reactive dyes onto media substrates" (E1, 68)
Bs	"it'll be about fifty percent more expensive" (E1, 199) "if you lift an optical mouse slightly off the page you'll see the pattern it creates" (E1, 672–674)
S	"...sledge" (E1, 137) "show the relative size of the pen if you've got an example" (E1, 171)
D	
O	"yeah we'll come to that in a minute" (E1, 737)

10. Gero, J.S., and Kannengiesser, U. (2004) The Situated Function-Behaviour-Structure Framework, *Design Studies*, 25, pp. 373–391.

Gero and Kannengiesser's situated FBS framework[10] (a cognitive model), designing was viewed as starting with requirements. The R code is included for this study since the designing activities during meeting E1 start with requirements instead of function. Table 1 shows examples of each code drawn from the meeting transcript.

3.1 *Data segmentation*

The transcript was segmented such that each segment contains only one class of FBS code. Segmenting and coding are done simultaneously by discerning whether an action or utterance expresses the FBS aspect of designing (F, Be, Bs, S, D) or concerns requirements (R) or other matters (O). If an utterance, a speaker's turn, contains more than one class it is further divided. This also applies to the 'O' and 'R' codes. Drawing and writing actions are also considered as segments of structure. As a consequence, segments are of varied lengths, usually very short, typically being a contribution lasting just a few seconds.

The two distinct episodes of the meeting are about the same length, approximately 57 minutes each. For this study of the first episode we disregard the first 5 minutes as it involves the management of the meeting rather than the design process and mainly concerns rehearsing the rules for brainstorming. The part of the transcript analysed was coded twice by the same coder with a ten day separation and then self-arbitrated using the Delphi method proposed by Gero and McNeill[11]. The agreement over the two codings was over 86%. The 52 minutes in the first episode we coded amounted to 475 segments. The average segment length was 6.5 seconds. Of the segments coded, 27 were identified as O (other), these

11. Gero, J.S., and McNeill, T. (1998) An Approach to the Analysis of Design Protocols, *Design Studies*, 19, pp. 21–61.

consist mostly of jokes or communications that are not related to the design process or the resulting artefact. Setting these aside, Table 2 shows the percentages for each of the FBS categories of the remaining 448 segments. From this we see that requirement (R) and function (F) account for only about 5% of the transcript segments. The highest percentages are in the structure and behaviour classes. These high percentages are due to the frequent use of analogies with other products and situations that characterise the participants' contributions during this first half of the meeting.

FBS class	% of segments
R	1.6
F	3.8
Be	15.4
Bs	28.1
S	40.2
D	10.9

Table 2. Percentages of each FBS class represented in the segmented data for 52 minute episode in the first half of meeting E1.

3.2 *Linking the segments*

The coding of the segments alone does not give us the distributions of the design processes (the arcs between the FBS classes shown in Figure 1). To discern the connections between segments we used Goldschmidt's linkography technique[12]. Figure 2 presents an extract from the segmented and coded transcript (E1, 137–157) together with the constructed linkograph.

12. Goldschmidt, G. (1990) Linkography: Assessing Design Productivity in Trappl, R. (ed) *Cyberbetics and Systems '90*, World Scientific, Singapore, pp. 291–298.

Two participants, the moderator Alan (A) and a mechanical engineer, Jack (J), were involved in this extract which shows a cluster of links. The focus of the discussion at this point in the meeting was examples of products or situations where something needs to follow a contour. Jack suggested an object (structure) – 'sledge' (segment 38) – and continued to explain the behaviour of the sledge, namely how it maintains contact and being level on the snow (segments 40 and 48). The sledge was compared with a set of skis (segment 43) in terms of the structure (segment 44) and behaviour (segments 45, 47 and 48). The coding of segment 50 is worthy of some explanation; it was coded as expected behaviour (Be) as we interpreted Jack to be borrowing the behaviour of the analogised objects and targeting these to be the expected behaviour of the designed object. Finally, the structure of stabilisers (segment 53) was suggested. Segment 39 was linked to segment 38 because the *writes sledge* action was a response to the initiation and suggestion of the 'sledge' in segment 38. Jack started explaining in segment 40 why a sledge was a proposed candidate solution so segments 38 and 40 were linked. A linkograph for the whole episode was constructed by examining the potential relationship of each segment with each of the preceding segments as we have illustrated here for segments 38 to 55.

Figure 3 shows a larger part of the linkograph of this session that includes the cluster of links (labelled sledge) we have shown in more detail in Figure 2. Other clusters were also labelled. These clusters were distinguished by the visual inspection of link density. Linkographs allow us to trace the structure of reasoning[13]. Although by using linkography the strength of an idea can be compared

13. *ibid.*

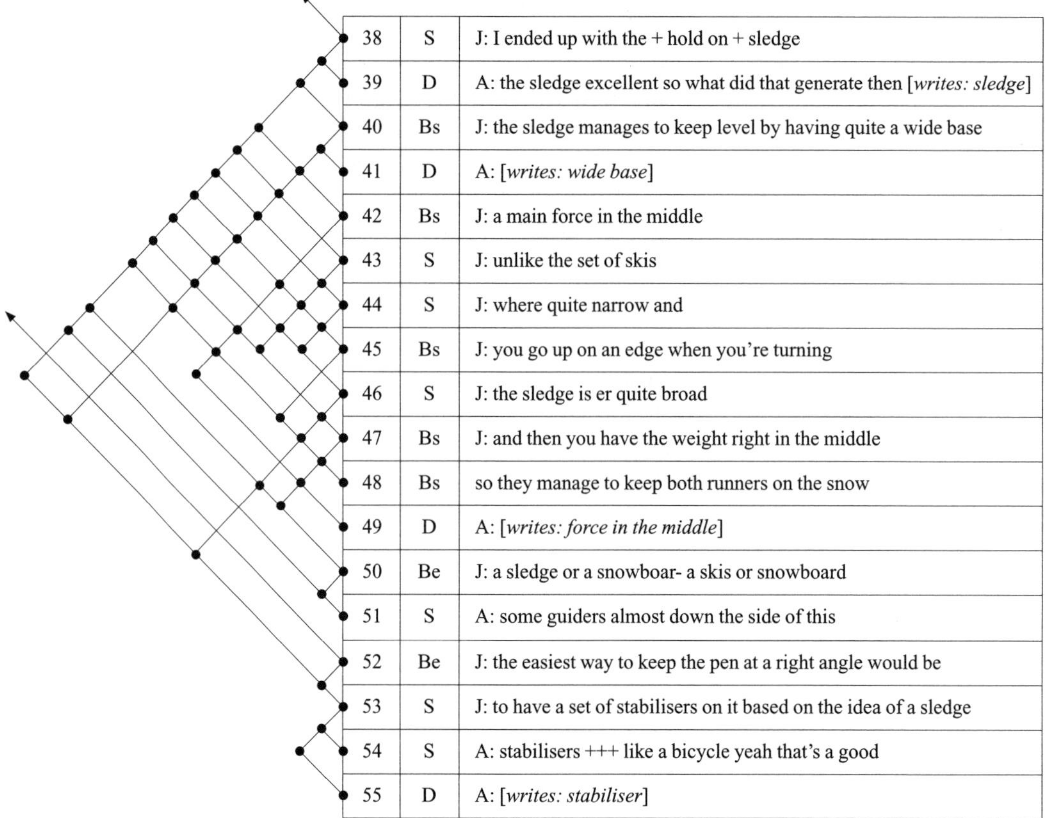

38	S	J: I ended up with the + hold on + sledge
39	D	A: the sledge excellent so what did that generate then [*writes: sledge*]
40	Bs	J: the sledge manages to keep level by having quite a wide base
41	D	A: [*writes: wide base*]
42	Bs	J: a main force in the middle
43	S	J: unlike the set of skis
44	S	J: where quite narrow and
45	Bs	J: you go up on an edge when you're turning
46	S	J: the sledge is er quite broad
47	Bs	J: and then you have the weight right in the middle
48	Bs	so they manage to keep both runners on the snow
49	D	A: [*writes: force in the middle*]
50	Be	J: a sledge or a snowboar- a skis or snowboard
51	S	A: some guiders almost down the side of this
52	Be	J: the easiest way to keep the pen at a right angle would be
53	S	J: to have a set of stabilisers on it based on the idea of a sledge
54	S	A: stabilisers +++ like a bicycle yeah that's a good
55	D	A: [*writes: stabiliser*]

Figure 2. Rotated linkograph showing relationships between coded transcript segments (segment number, FBS class, transcript segment).

quantitatively either using the critical move measurement or using entropic measurement[14], we did not pursue this here because the focus of this study concerns combining ontological coding with linkography to capture semantic information from the transcript of a design meeting.

In the whole of the episode coded there are 1,071 links among the 475 segments, so on average each segment has about 2.1 links. However, some segments have many more links than others. Table 3 compares the distribution of the codes of the segments with the occurrences of codes in the links. Compared to the coded segments it can be observed that the codes in the links decrease for documentation, have a moderate decrease in behaviour derived from structure, and a slight decrease in function. The requirement has increased, and there is an increase in expected behaviour and structure as well. This implies the expected requirement, behaviour, and structure segments in this session are more influential. The statistics of 1990 FBS coding, as for Gero and McNeill[15] and McNeill, Gero and Warren[16], do not reflect this.

14.
Kan, W.T., and Gero, J.S. (2007) Comparing Entropy Measures of Idea Links in Design Protocols, *AIEDAM*, 21, pp. 367–377.

15.
Gero and McNeill (1998), *op. cit.*

16.
McNeill, T., Gero, J.S., and Warren, J. (1998) Understanding Conceptual Electronic Design using Protocol Analysis, *Research in Engineering Design*, 10, pp. 129–140.

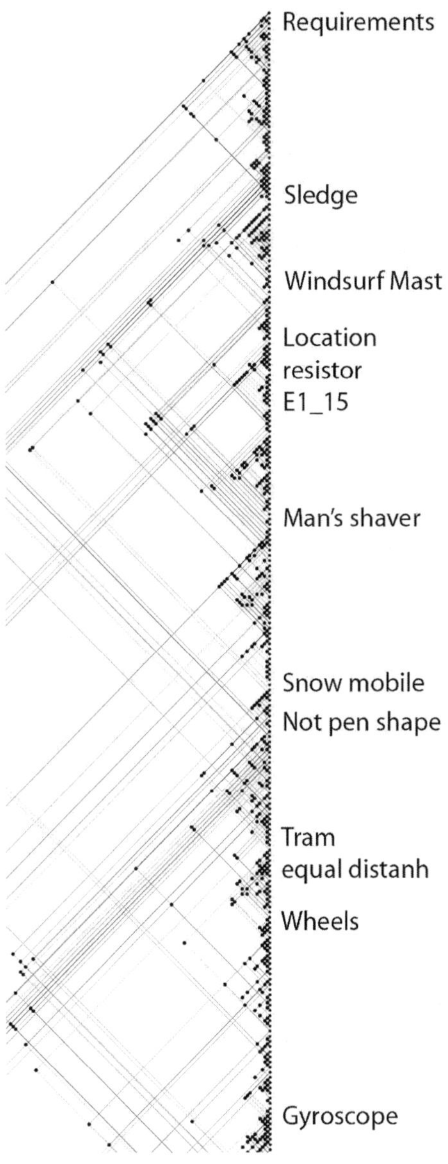

Figure 3. Part of the linkograph of the segmented protocol.

Code	Segments		Links	
R	7	1.6%	36	1.7%
F	17	3.8%	56	2.6%
Bs	126	28.1%	504	23.8%
Be	69	15.4%	396	18.7%
S	180	40.2%	936	44.3%
D	49	10.9%	187	8.8%
Total	448	100%	2115	100%

Table 3. Comparing the distribution of codes in segments and links.

4 DERIVING FBS PROCESSES FROM CODED SEGMENTS AND LINKS

In the following analysis, we use the symbol '>' to denote the link or the transformation between the nth and the (n + i)th segments. For example, looking at Figure 2, segment 38's links to the two subsequent segments (39 and 40) were coded S > D and S > Bs respectively. According to the ontology, the first of these links, S > D can be seen as the documentation process (transformation from structure to design description) and the second link S > Bs as the analysis process (transformation of structure to behaviour). Figure 4 illustrates how these links from the linkograph relate to elements of the FBS ontology (cf. Figures 1 and 2).

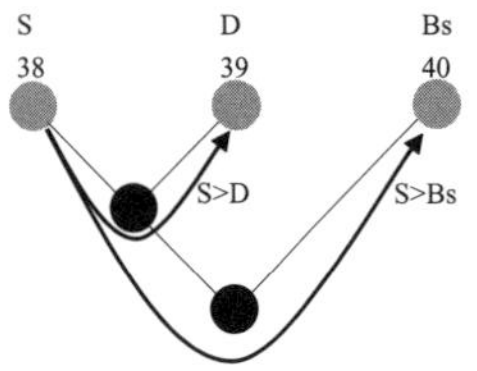

Figure 4. Deriving transformation processes from a linkograph.

If we consider the 1,071 links as design processes, the linkograph represents a network of these as transformation processes. As there are six FBS codes, 36 types of possible transformations are possible. According to the FBS ontology many of these processes are meaningless, however there were instance of processes (e.g. the F > S process) which the framework does not accommodate (cf. Figure 1). Table 4 shows the 30 types of links present in the coded segments; those which represent FBS framework processes are shown in Table 5. Our results are broadly comparable with those of Badke-Schaub et al.[17] in terms of the frequency of cognitive acts (although their analysis relates to the whole of meeting E1). There is a rough correspondence between the distribution of activities. Badke-Schaub et al.[18] identify about 2.5% of design activity as problem definition, this can be compared with our 1.4% for Formulation; they identify 30% as analysis and evaluation, compared with our 22.1% (12.2% Analysis + 9.9% Evaluation); their 6.5% for new ideas is higher than our 3.4% for Reformulation II and III (structural reformulations do not usually contribute to new ideas but in this session there are various structural type of analogies).

17.
Badke-Schaub, P., Lauche, K., Neumann, A. and Ahmed, S. (Chapter 9) Task, Team, Process: The Development of Shared Representations in an Engineering Design Team.

18.
ibid.

In the episode analysed, the reformulations were mostly of structure and behaviour. The fragment of the transcript concerning the sledge shown in Figure 2 contains the reformulation of structure (S > S),

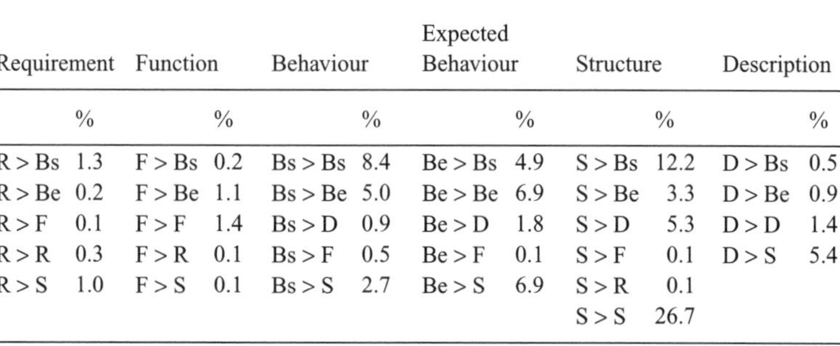

Requirement		Function		Behaviour		Expected Behaviour		Structure		Description	
	%		%		%		%		%		%
R > Bs	1.3	F > Bs	0.2	Bs > Bs	8.4	Be > Bs	4.9	S > Bs	12.2	D > Bs	0.5
R > Be	0.2	F > Be	1.1	Bs > Be	5.0	Be > Be	6.9	S > Be	3.3	D > Be	0.9
R > F	0.1	F > F	1.4	Bs > D	0.9	Be > D	1.8	S > D	5.3	D > D	1.4
R > R	0.3	F > R	0.1	Bs > F	0.5	Be > F	0.1	S > F	0.1	D > S	5.4
R > S	1.0	F > S	0.1	Bs > S	2.7	Be > S	6.9	S > R	0.1		
								S > S	26.7		

Table 4. Percentages of all the processes derived from codes and links.

	Process	Occurrences as links	% of total links
Formulation	R > F, F > Be	14	1.2
Synthesis	Be > S	68	6.9
Analysis	S > Bs	120	12.2
Documentation	S > D	52	5.3
Evaluation	Be <> Bs	48(Be > Bs), 49(Bs > Be) 97	9.9
Reformulation I	S > S	262	26.7
Reformulation II	S > Be	32	3.3
Reformulation II	S > F	1	0.1
Total		646	65.9

Table 5. Frequencies and percentages of the 8 FBS processes.

from the structure of a sledge to the structure of a set of stabilizers like those in a bicycle (segment 38 to 53 and segment 53 to 54).

Other examples of structure reformulations in the larger episode included making analogies with other products, for example a wind surfboard mast and a man's shaver; and considering the thermal pen in the shape of things other than a pen. Examples of behaviour reformulations included using a universal joint to keep the angle; using springs to keep it level; and a suggestion about the locations of resistors (E1, 190) which prompted responses concerning the cost (E1, 195, 199).

The reformulation of function was rare, reflecting the nature of this session, namely brainstorming mechanical engineering ideas for keeping the thermal pen in contact with the media at a correct angle. Some of the functional aspects were deliberately not dealt with in this meeting, being regarded as off the agenda as indicated by the conversation. For example the enquiry: "could we sorry could we actually see what they're doing I mean are they drawing pictures or making invitations or Christmas cards" (E1, 622–623) met with the response: "erm we're going to try to deal with that a fair bit on Monday" (E1, 624).

The 1990 FBS ontology accounts for more than half of the processes derived from the links of the coded segments (the 65.9% shown in Table 5). What are the other processes (34.1%)? Does this result reflect a deficiency of this ontology? Some of the most frequent processes appearing in Table 4 but not accounted for in Table 5 are: Bs > Bs (8.4%), Be > Be (6.9%), D > S (5.4%), S > Be (3.3%), Bs > S (2.7%), and Be > D (1.8%).

Re-examining the coded transcript, in the case of Bs > S, we noted that the large scale of the granularity of our segmentation fails to pick up a Be in the Bs > S processes. If the granularity of segmentation were finer, there would be an expected behaviour (Be) before the structure code (S). Returning to the example in Figure 2, segment 40: "the sledge manages to keep level by having quite a wide base" (E1, 141) was coded as Bs because it analyzes an existing product to get the 'keep level' behaviour. This segment was linked to segment 53: "the easiest way to keep the pen at a right angle would be to have a set of stabilizers on it based on the idea of a sledge" (E1, 153–156) which was coded as structure because it proposed a structure, namely a set of stabilizers. The idea of sledge with the behaviour of 'keep level' is transformed to expected behaviour of 'at a right angle' which leads in turn to the structure of 'a set of stabilizers'.

The F > F and Be > Be can be viewed as reflections of function and behaviour in many cases. An instance of Be > Be during the discussion which noted that the shape of the designed object did not need to resemble a pen occurred when the moderator suggested: "something else that gets pulled behind it for example" (E1, 358) and one of the engineers responded to this with: "[what] they'll do is move the lump around" (E1, 370). These segments were linked and both were coded as expected behaviour (Be).

The D > S is the interpretation of depicted structure. A Bs > Bs is usually a result of further analysis, for example in Figure 2 the link between segments 40 and 42 and links between segments 42 and 48 were concerned with further analysis of the action and reaction of force (weight). Sometimes a Be > D transformation was the depiction of behaviour but the FBS ontology does not distinguish depiction of behaviour from depiction of structure. Using E1, 370 as an example again: "[what] they'll do is move the lump around" (E1, 370) was linked to the subsequent segment where the moderator was writing down 'move lump', which is a depiction of behaviour. These transformations are meaningful processes resulting from the interactions among those present and artefacts. In order to capture these subtleties, we turn to the situated FBS framework[19] and recode this episode.

19.
Gero and Kannengiesser (2004), *op. cit.*

5 SITUATED FBS CODING

The situated FBS framework makes use of new concepts: the notion of situated cognition developed by Clancey[20]; the idea of constructive memory based on Dewey's[21] and Bartlett's[22] work; and the observation by Schon and Wiggins that designing is an 'interaction of making and seeing'[23]. Gero and Kannengisser[24] used these ideas, integrating them into the FBS ontology to form the situated FBS framework by introducing interactions among three worlds – external, interpreted, and expected. An agent interacts and understands the external world through his interpretation of it to form memories of his interpreted world. In order to change the external world (the act of designing) he 'focuses' to transform experiences to produce the expected world before taking action in the

20.
Clancey, W.J. (1997) Situated Cognition: On Human Knowledge and Computer Representations, Cambridge University Press.

21.
Dewey, J. (1896) The Reflex Arc Concept in Psychology, *Psychology Review*, 3, pp. 357–370.

22.
Bartlett, F.C. (1932) *Remembering: A Study in Experimental and Social Psychology*, Cambridge University Press.

23.
Schon, D.A., and Wiggins, G. (1992) Kinds of Seeing and their Functions in Designing, *Design Studies*, 13, pp. 135–156.

24.
Gero and Kannengiesser (2004), *op. cit.*

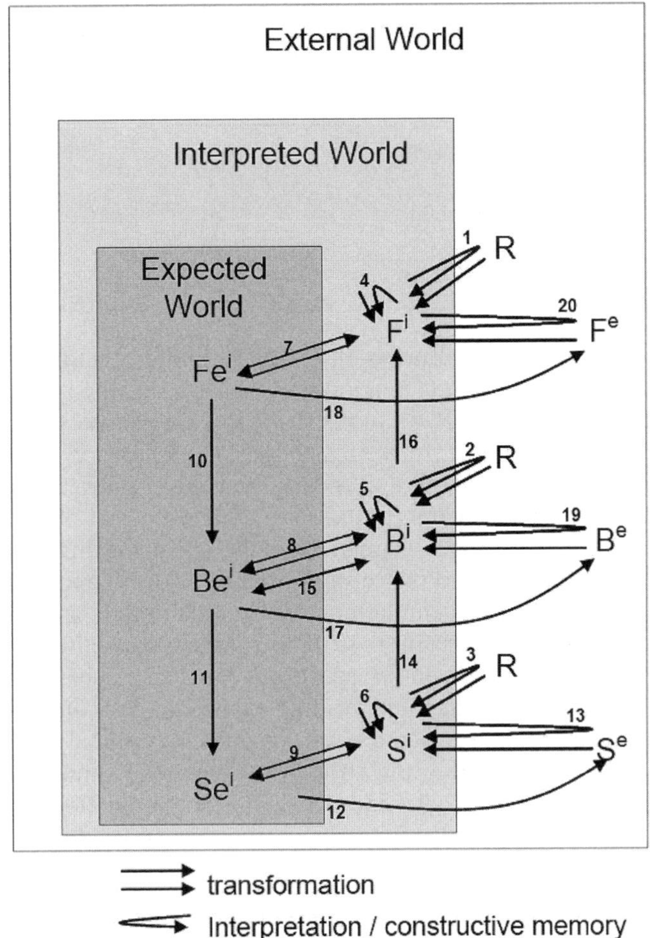

Figure 5. The situated FBS ontology[25].

25. *ibid.*

external world. The situated FBS framework class variables reside in these three worlds. The superscript 'e' indicates the variables are of external world, the superscript 'i' signifies the variables are of internal world and the lower case suffix 'e' represents variables are of expected world. In this more developed framework the original eight processes from the FBS ontology shown in Figure 1 are increased to twenty to account for the activities taking place during designing viewed from this perspective. The resulting situated FBS ontology is depicted in Figure 5.

Table 6 summarises the relationship between the twenty situated FBS processes and the original eight processes[26]. Of particular

26. *ibid.*

FBS process	Situated FBS processes
Formulation	1, 2, 3, 4, 5, 6, 7, 8, 9, 10
Synthesis	11
Analysis	13, 14
Evaluation	15
Documentation	12, 17, 18
Reformulation I	6, 9, 13
Reformulation II	5, 8, 14, 19
Reformulation III	4, 7, 16, 20

Table 6. Relationship between situated FBS processes and the original 8 FBS processes[27].

27. *ibid.*

interest in the situated framework are the formulation and reformulation processes. The formulation process involves: the interpretation of requirements (R) in terms of F^i, B^i, and S^i representations (processes 1, 2 and 3); reflecting, based on experience, on those representations (processes 4, 5 and 6); focusing on subsets of these internalized requirements (processes 7, 8 and 9); and process 10 that corresponds to the original formulation in the FBS framework. The focusing and reflecting (processes 4 to 9) appear in all the three types of reformulations. Reformulations II and III are not limited to being driven by structure alone but also by external representation of function (F^e, process 20) and also behaviour (B^e, process 19). Process 11, transforming expected behaviour (Be^i) to expected structure (Se^i) is the synthesis process; the analysis process involves interpreting the structure (process 13) and deriving the behaviour from structure (process 14). The evaluation process (process 15) is concerned with comparing the expected behaviour (Be^i) with the interpreted behaviour (B^i).

With the introduction of the new FBS classes as shown in Figure 5, a revised set of codes for segments becomes available; these are shown in Table 7. To simplify comparison between our coding using the two FBS ontologies the segments themselves were not revised, so the total numbers of segments and total number of links for the episode studied remained unchanged. As will be seen this has some unexpected consequences.

The segmented data was coded twice by the same coder and then self-arbitrated using the Delphi method as before. The agreement between the two occasions of coding was about 67%. Figure 6 shows the recoded data for segments 38 to 43. Segment 38 was coded as S^i because Jack was showing a picture of a sledge and was about to draw an analogy with the structure of the sledge. The main activity in segment 39 was writing down the word ‘sledge’. It was treated as the documentation of structure as the word ‘sledge’ denotes the object, the segment was coded as S^e. Segment 40 was coded as B^i because it interprets the ‘keep level’ behaviour of the sledge. Segment 41 was coded as S^e as the depiction concerns the structural aspect, the ‘wide base’, of the object.

Situated FBS code	Description
R	requirements derived from the given brief
F^i	interpreted function either derived from requirements or ascribing meaning to depicted structure
F^e	external representation of function, usually in terms of written words
Fe^i	expected function resulted from focusing on F^i
B^i	interpreted behaviour from depicted structure or requirements
B^e	external representation of behaviour, usually in terms of written words
Be^i	expected behaviour derived from expected function or interpreted behaviour which results from requirement or interpreted structure
S^i	interpreted structure either from external structure or from requirement
S^e	depiction that indicates structure
Se^i	expected structure sometime without depiction

Table 7. Code descriptions for situated FBS ontology classes.

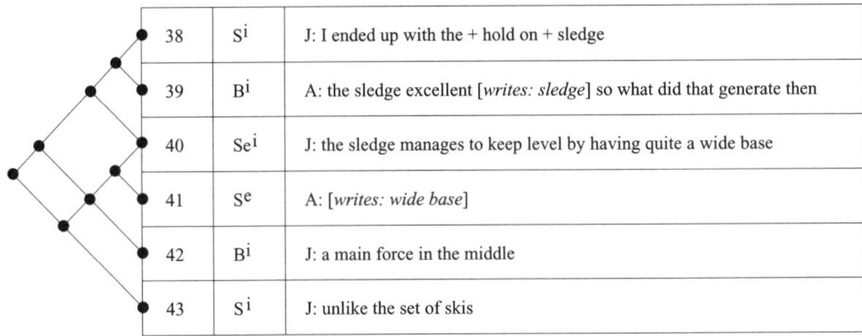

38	S^i	J: I ended up with the + hold on + sledge
39	B^i	A: the sledge excellent [*writes: sledge*] so what did that generate then
40	Se^i	J: the sledge manages to keep level by having quite a wide base
41	S^e	A: [*writes: wide base*]
42	B^i	J: a main force in the middle
43	S^i	J: unlike the set of skis

Figure 6. Examples of situated FBS coding, column one is the segment number and column two is the code.

Segment 42, similarly to segment 40, involves the interpretation of a behavioural aspect, ‘a main force in the middle’ of the object, thus it was assigned the B^i code. Segment 43 concerns a further object, ‘skis’, and was coded as S^i.

Table 8 shows the distribution of codes in the segments and links. Table 8 shows that there is no documentation of function and indicates that expected behaviour, expected structure, and interpreted structure are more influential (through linking) than they appear from a simple segment count. Conversely, looking at the degree to which they are linked, the interpreted function and behaviour, the expected function, and the depiction of behaviour are of less importance than they appear from the segment count.

Using the same approach as described in Section 4 and depicted in Figure 4, the situated FBS processes were derived. For clarity and ease of analysis the reflection categories of processes (4, 5, and 6 in Figure 5) were separated from the formulation processes so that there is no overlapping of processes in any of the categories (compare Table 6 with Table 9). Aggregating the meaningful processes

Table 8. Distribution of situated FBS codes in segments and links.

Situated FBS code	Segments		Links	
R	7	1.6%	36	1.7%
F^i	8	1.8%	18	0.8%
Fe	0	0.0%	0	0.0%
Fe^i	9	2.0%	38	1.8%
B^i	125	27.9%	493	23.3%
B^e	13	2.9%	42	2.0%
Be^i	69	15.4%	396	18.7%
S^i	98	21.9%	485	22.9%
S^e	36	8.0%	145	6.9%
Se^i	83	18.5%	462	21.8%
Total	448	100.0%	2115	100.0%

of situated FBS into the basic eight design processes (shown in Figure 1) gives us a 92% coverage of all the derived processes as shown in Table 9. Compared with the original FBS there is an increase in the capture of the reformulations. The increase is most noticeable for Reformulation II (behaviour) in this dataset.

There are 10 codes from situated FBS (column 1 Table 8) giving potentially 100 possible process types, only 20 of which are meaningful types of links according to the situated FBS framework (those shown in Figure 5). However, 50 types of processes were recorded in the coding of the data. In most cases, those not supported by the framework are thought to have arisen because the granularity of segmentation was not close grained enough to identify transformations occurring at a finer level. For example, the process derived from the link from 39 to 40, $S^i > S^e$ is not one of the 20 situated FBS processes. Closer inspection of segment 39 shows that it should be divided into three segments with 'the sledge excellent' coded as Se^i, 'so what did that generate then' coded as either Be^i or Se^i and '[*writes sledge*]' coded as S^e.

We might also ask what are the other processes (8.1%) not accounted for in Table 9. Among these are processes like $B^i > S^e$ and $B^i > S^i$. Figure 6 contains both examples, the derived process

Table 9. Distribution of FBS processes following recoding using situated FBS codes.

FBS process	Situated FBS processes	Percentages
Formulation	1, 2, 3, 10	3.4%
Synthesis	11	7.5%
Analysis	14	13.4%
Evaluation	15	4.6%
Documentation	12, 17, 18	8.1%
Reformulation I	6, 9, 13	31.7%
Reformulation II	5, 8, 19	21.6%
Reformulation III	4, 7, 16, 20	1.5%
Total		91.9%

from segment 40 to 41 is an example of $B^i > S^e$. The processes from links between segments 42 and 43, and between 40 and 43 are examples of $B^i > S^i$. In the first round of coding, segment 40 'the sledge manages to keep level by having quite a wide base' was code as B^i; in the second round it was coded as S^i. The final arbitrated code was B^i. It should contain two parts – the behaviour part of 'keep level' and the structural part 'wide base'. Segment 43, 'unlike the set of skis', was also a questionable coding (B^i or S^i). It was finally coded as S^i but by carefully examining the context, the analogy of 'unlike' can be seen to be both structural and behavioural. The structural analogy was 'wide base' against 'narrow'; the behaviour analogy was 'force in the middle' versus on one leg. Figure 7 illustrates a proposed refinement of the segments from segment 40 to 43 together with their codes and linkograph fragment. The first column in the table contains the segment reference using an alphabetic suffix to show subdivided (former) segments. The links are revised, showing that they correspond to the processes from the situated FBS ontology.

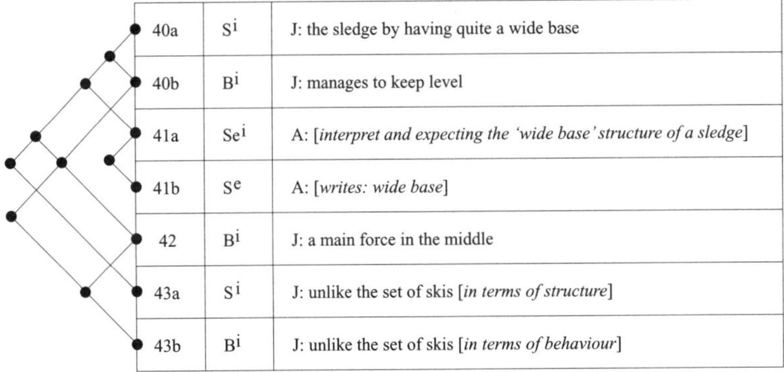

40a	S^i	J: the sledge by having quite a wide base
40b	B^i	J: manages to keep level
41a	Se^i	A: [*interpret and expecting the 'wide base' structure of a sledge*]
41b	S^e	A: [*writes: wide base*]
42	B^i	J: a main force in the middle
43a	S^i	J: unlike the set of skis [*in terms of structure*]
43b	B^i	J: unlike the set of skis [*in terms of behaviour*]

Figure 7. Finer grained segmentation of the transcript.

We conclude that the 'missing' processes (8.1%) were caused by lack of experience in using this coding method; this includes making judgements about the appropriate granularity of segmentation and the identification of appropriate links. In this data some of the segments required a finer grained coding than was used. Further analysis and refinement seems likely to resolve these 'missing' processes as exemplified above.

6 DISCUSSION

The FBS ontology denotes fundamental processes of designing; its ontological variables are disjoint. Unlike most coding schemes, supported by available protocol analysis software, which allow overlapping of codes the ontological approach requires precise association of a single code with each data segment. This clear distinction converts a dataset into unambiguous segments supporting quantification of the amount of effort spent in relation to function, behaviour, or structure. Attention to links not only provides a structural view of the processes but also allows us to locate the dominant codes and the frequency of each design transformation process. The nested representation of links, the linkograph, together with FBS coded segments provides an opportunity to look at a design meeting transcript not in a linear manner but as a network of processes. The study of the interaction among the FBS classes and processes may help to deepen our understanding of designing.

The use of the FBS ontology allows us to see the design process semantics of this transcript. Of particular interest is that formulation and reformulation are the largest activities in terms of events and that the vast majority of reformulation is concerned with behaviour and structure. This maps well to our qualitative understanding of this session which is concerned with generating novel ideas by analogy (for a detailed treatment of the same data from this perspective see Ball and Christensen)[28]. Our analysis provides a quantitative measure of this. Despite immaturity with applying this FBS-based coding scheme, we are still able to account for 92% of all designing activities in the part of the meeting analysed.

28.
Ball, L. and Christensen, B. (Chapter 8) Analogical Reasoning and Mental Simulation in Design: Two Strategies Linked to Uncertainty Resolution.

The FBS-based coding scheme, subsumed in a more comprehensive scheme, has previously been used in protocol studies of individual designers[29]. The method presented in this study has been used to compare face-to-face design collaboration between two architects vs. designing using internet communication via a 3D world[30]. The distribution of the eight FBS processes of the face-to-face session in that work and the data analysed in this study exhibit similar patterns, while the 3D world session looks very different. The results presented in this chapter demonstrate that the method presented can be used to analyse a team of designers working together. The approach is not limited to analysis of individual designers at work or to particular means of communication.

29.
Gero and McNeill (1998), *op. cit.*

30.
Kan, J.W.T., and Gero, J.S. (2008) Acquiring Information from Linkography in Protocol Studies of Designers, *Design Studies*, 29(4), pp. 315–337.

Many coding schemes have been developed for use with design protocols and meeting transcripts. All schemes are based on particular views of the activity of designing. Many schemes are uniquely devised for the data to which they are applied. This limits the applicability of the results obtained. Where more general coding schemes have been attempted, they still tend to lack sufficient generality to allow them to be applied in widely varying circumstances. It is claimed that the use of the FBS ontology and the situated FBS ontology provides a basis for coding which is generally applicable that does not depend on any special circumstance associated with a particular set of data.

ACKNOWLEDGEMENT

This research is supported by an International Postgraduate Research Scholarship, University of Sydney.

Part 5

Language, Discourse and Gesture

13
'Does this compromise your design?' Socially Producing a Design Concept in Talk-in-Interaction*

Rachael Luck

This chapter explores how a design concept was interactionally produced in the talk-in-interaction between an architect and two client representatives. Its intention is to develop our understanding of some actions and practices used to negotiate the properties of a design and interactions in design team meeting settings more generally. In the study described some differences were observed between the properties of the design concept in comparison with the design ideas discussed in conversation. A provisional and qualified analogy is made between the design concept and properties of material objects in studies of science, technology and society.

When the activity of design is viewed as a social process in design interactions, the notion that attributes of a 'design' are modified and negotiated is readily accepted. To set the scene for this research several established concepts associated with the activity of design are first clarified.

Co-evolution[1] is understood as a process in creative design where design moves are made in both problem and solution spaces. A 'design space' is considered to exist in which participants move and modify the properties of a design[2]. The design space is the phrase I use to refer to these phenomena. To differentiate the design space from the status of a design as it is being modified by participants, the notion of a boundary object[3] is meaningful. It is understood that a boundary object serves to mediate communication between people with different knowledge concerns (e.g. different expertise, expert-layperson interaction) and that over time, in the process of design, a boundary object will be modified. A sub-set or distinct category of boundary objects are the intermediary objects of design, IODs[4], which represent the intermediate states of a final deliverable for a project. Intermediary objects have two dimensions: they are related to the action itself and are a means for coordinating design activity[5]. In Vinck and Jeantet's[6] study of the sociotechnical dynamics of design they acknowledge the non-human agency of mediating objects and define 'open objects' as an object category. Aspects of their research will be re-visited when the materiality of an 'object' is considered.

This research is attentive to the actions of the participants in talk which interactionally produce and accomplish changes to the

1.
Dorst, K. and Cross, N. (2001) Creativity in the Design Process: Co-evolution of Problem-Solution, *Design Studies*, 22, pp. 425–437.

2.
This concept is operationalised by Reymen, Dorst and Smulders (Chapter 4).

3.
Star, S. and Griesmer, J. (1989) Institutional Ecology, 'Translations' and Coherence: Amateurs and Professionals in Berkley's Museum of Vertebrate Zoology 1907–1939, *Social Studies of Science,* 19, pp. 387–420.

4.
Arikoglu, E.S., Blanco, E. and Pourroy, F. (Chapter 21) Keeping Traces of Design Meetings through Intermediary Objects.

5.
Boujut, J.F. and Laureillard, P. (2002) A Co-operation Framework for Product-Process Integration in Engineering Design, *Design Studies*, pp. 497–513.

6.
Vinck, D. and Jeantet, A. (1995) Mediating and Commissioning Objects in the Socio-Technical Process of Product Design: A Conceptual Approach, in MacLean, D. Saviotti, P. and Vinck, D. (eds.) *Design Network Strategies*, pp. 111–129.

*Reprinted from: CoDesign Vol. 5 No 1 (March 2009), pp. 21–34,
DOI: 10.1080/15710880802492896

design as it co-evolves, that is as moves in a design space modify a boundary object.

A concern of the work presented here is to study how a range of properties of a design are produced and enacted in and through design interactions that to differing degrees moderate which moves are more acceptable in a design space. The architectural 'design concept' is a phenomenon of interest that is studied in relation to the ideas and concerns that are part of the conversations between an architect and the client representatives.

Darke's notion of the primary generator[7], as observed by Lawson in his studies of architects in practice[8], acknowledges that designers can have an underpinning design concept which acts as a point of reference, while other actions that modify the design are ongoing. But how is a design concept produced in interaction? I attend to how the design concept was produced in and through the interactions at two architectural meetings and consider the interplay between the design concept and design ideas amongst the participants. How is the design concept positioned as having different properties from other design ideas, and known to the participants as a constraint that to some degree moderates what are acceptable design moves and actions?

7.
Darke, J. (1984) The Primary Generator and the Design Process, in Cross, N. (ed.) *Developments in Design Methodology*, Wiley, pp. 175–188.

8.
Lawson, B. (1994) *Design in Mind*, Butterworth-Heinemann.

1 THEORETICAL POSITION AND ITS ASSOCIATED METHODOLOGY

The theoretical and methodological position adopted is from the established and associated positions of ethnomethodology and conversation analysis (EM/CA) and from the social sciences more generally. A central tenet of this position is that talk *is* action. Spoken actions are not only linguistically real (as a referent is semiotically real) but also bring about actions in a situation, for example, moves in a design space, which with minimal ambiguity can be considered to be real. This tenet is acknowledged in design research by Dong[9] theoretically and by Medway[10] and Fleming[11] empirically who observe acts of reification in design interactions that 'bring things into being'.

The design actions performed in speech are analysed in relation to their sequential and structural production in talk, with reference to a repertoire of CA literature. In contrast with analyses of the activities of design from other language-based perspectives, where coding frameworks are developed either deductively or inductively[12], this research investigates the space between what is 'socially produced' and 'socially accomplished' in interaction, to empirically ground some actions of designing social scientifically.

9.
Dong, A. (2007) The Enactment of Design through Language, *Design Studies*, 28, pp. 5–21.

10.
Medway, P. (1996) Virtual and Material Buildings: Construction and Constructivism in Architecture and Writing, *Written Communication*, 13, pp. 473–514.

11.
Fleming, D. (1998) Design Talk: Constructing the Object in Studio Conversations, *Design Issues*, 13, pp. 41–62.

12.
Mazijoglou, M., Scrivener, S. and Valkenburg, R. (2000) Matching Descriptions of Team Design, in Scrivener, S., Ball, L. and Woodcock, A. (eds.) *Collaborative Design*, Springer, pp. 279–287.

An advantage of working with data in this manner is that in part it attends to 'meaning' and 'interpretation', which are pervasively problematic when using language as data. From the CA perspective, instead of making interpretations of what is 'meant' by what is said, or how a recipient interpretes this, what are analysed is the next action, that is, the speech in response to the previous turn-at-talk. The CA position is in contrast with cognitive perspectives and this is consequential for what can be studied and revealed about the activity of design. Only what is spoken is analysed as an action, also there is no speculation on what is happening in the architect's and other participants' minds.

Natural conversations are treated as data (in this setting the data are the talk at the two architectural meetings, A1 and A2). A consequence of this condition methodologically is that only the talk between the architect and clients are treated as data and not the supplementary material; here the conversation between the architect and the person collecting the data (this point will be revisited in the discussion of the inferences from the data analysis and interpretation).

An objective of CA is to describe: "the normative structures of reasoning which are involved in understanding and producing courses of intelligible interaction"[13]. This approach to research is descriptive but not wholly interpretative. I use the terms 'explicate' and 'inference' to differentiate between the claims based on analyses from any broader sociological inferences which are extrapolations from the analyses. The structure of this chapter adopts a pattern used in EM/CA by Heritage[14], where a proposition is presented and the remainder of the argument refers to extracts from the data to explicate phenomena.

13.
Heritage, J. (1988) Explanations as Accounts: A Conversation Analytic Perspective, in Antaki, C. (ed.) *Analysing Everyday Explanation: A Casebook of Methods*, Sage, pp. 127–144.

14.
ibid.

2 WORKING WITH DATA

Working with the data, viewing the videos of the two architectural meetings and parsing the data transcriptions, I noted a difference in the degree to which the design concept was socially negotiable, in comparison with other design ideas that were discussed. This observation hinged on an episode which I later refer to as the 'design impasse', when a tension between the design ideas and the design concept was apparent. The proposition I explore is that a design concept has different properties from the design ideas that were part of the architectural meetings. In particular, that the design concept was socially constructed as being less negotiable, that is exhibiting more intransigent properties in the process of design, than design ideas which can be modified, and were positioned as more negotiable.

To investigate these phenomena, the argument adopts the following logical sequence to explicate

- actions of designing in talk
- the production and acknowledgement of the design concept
- local 'ownership' of the design concept
- a design impasse.

2.1 *Designing in talk*

In the stepwise production of this argument I first consider how the activity of designing is empirically, social scientifically ground in talk, with reference to extracts from the data. In contrast with those who have defined and labelled some actions and activities of design, for example 'indexing' by Fleming[15] and the domains of design[16], I identify actions associated with 'ordinary' conversation that bring about moves in a design space.

15.
op. cit.

16.
Schön, D.A. (1983) *The Reflective Practitioner: How Professionals Think in Action*, Temple Smith, London, pp. 95–98.

In Extract 1 (A1, 682–690), the client representative, Anna, considers the implications of the suggestion to add a door to the design.

Extract 1, A1, Visualising changes to the design as a boundary object.

682	Anna	you don't want a door because that would just mean the funeral
683		directors in and out in and out all the time and they should be in there
684		anyway
685	Charles	well [*points*] no I mean there
686	Anna	from this side I don't want particularly a door [*points*] because it's also
687		going to
688	Adam	a short cut yeah
689	Anna	it's also going to [*begins to point*] interrupt this flow of space here as well
690		isn't it a door people sort of popping in and out there it looks like that

What was socially accomplished in this interaction is of interest to better understand some actions that are part of the activity of design. Anna, a non-architect, has appreciated that including another door will have consequences for the use of the space, as at certain times this means of access will disrupt a ceremony. Without any adjustment to the design representations, Anna has visualised the changes in state of the design as a boundary object and also the use implications from this adjustment. In talk the boundary object has been verbally modified, the use consequences were visualised. This is an accomplished design skill and arguably it is more significant that a non-architect socially accomplishes these actions that are associated with designing. This inference is supported by Glock's[17] observation that during these interactions: "designing

17.
Glock, F. (Chapter 16) Aspects of Language Use in Design Conversation.

does not seem to be solely in the realm of the designer" as the client participants produce moves in a design space.

Extract 2, A1, A display of mutual understanding about design properties.

Adam	I saw the pond initially as being quite still but there's no reason why it
	couldn't have a fountain in it or something to give it
Anna	yes but even if it was still I would think it would be something
	reflective wouldn't it even through the light from the you know yes
	I'm quite keen on sort of developing that that erm that's one of our
	unique sort of things not many I don't know of any other crematorium
	chapel that's got something as unique as that so you know I'm very keen
	to sort of keep that in there and use that
Adam	good so in principle you really like the idea of its spirituality being
	amplified to make it very calm
Anna	yes
Adam	and spiritual
Anna	yes
Adam	relaxing
Anna	yes
Adam	meditative sort of space
Anna	yes not even necessarily particularly for funerals but for sort of

In Extract 2 (A1, 511–527) Adam, the architect explains that he sees the pond as being still, however he makes it known that the pond could incorporate a fountain. This turn accomplishes making it known and shared amongst the participants that the pond is an aspect of the design he is amenable to modify. Anna's next turn is an interruption that gains the conversational platform. In this turn she produces an agreement with his assessment, a preferred turn-shape[18], and an upgrade response: "I'm quite keen on sort of developing that" (A1, 515). As the conversation continues, Anna accomplishes other actions of agreement, which acknowledge the properties 'to make it very calm', 'spiritual' and 'relaxing' that Adam attributes to the space. In the sequential production of talk these back-channel acknowledgements are actions that display mutual understanding, which make some properties known and shared amongst the participants.

18. Pomerantz, A. (1984) Agreeing and Disagreeing with Assessments: Some Features of Preferred/Dispreferred Turn Shapes, in Atkinson, J.M. and Heritage, J. (eds.) *Structures of Social Action: Studies in Conversation Analysis*, Cambridge University Press, pp. 57–101.

The explications of these two extracts from the data begin to empirically ground the activity of design in talk. Meaningfully the actions accomplished in talk have modified the design space and the properties of the design as a boundary object. These actions of designing were accomplished in the structural patterns associated with 'ordinary' conversation. Extracts 1 and 2 illustrate situated examples of designing, the social activity of design in interaction. I next attend to the production of the 'design concept' in interaction.

2.2 *Socially producing the 'design concept'*

The actions of the architect that socially produce the design concept are revealed in several extracts from the data. I first explicate the talk that preceded the introduction of the design concept, to set the context for its production.

Extract 3, A1, The context for the production of the design concept.

Anna	possibly do away with the thought of having cameras down the drive to see the next funeral arriving and that's why er the thought of having the door here might be better for us but I don't know how that upsets the feeling of what you've got here and where the roof space is in a sense there the roof carries over from here doesn't it
Adam	yes the roof flies over I would be resistant to having a
Anna	yes
Adam	door there because I think it would spoil the you know the chapel
Anna	it's not a door but a small room for the music room the music room is sort of swinging round a bit here instead but I don't know how that would spoil the feeling for that not wanting to disrupt too much

Anna's concern is to: "see the next funeral arriving" (A1, 649–650) and her design move is to include a door in the design, however in this turn she also displays uncertainty, whether: "that upsets the feeling of what you've got there" (A1, 651–652). The architect's next turn is an agreement: "I would be resistant to having a door there" (A1, 654–656). Anna's overlapping utterance, 'yes' is an acknowledgement that her design idea does upset the design idea. Adam's latched utterance warrants that: "I think it would spoil the you know the chapel". In Anna's next turn: "I don't know how that would spoil the feeling for that not wanting to disrupt too much" (A1, 658–659) is a second response upgrade[19] that emphasises her previous concern that her move in the design space would not meet with the architect's approval. An upshot from this explication is

19. *ibid.*

Extract 4, A1, Socially producing the design concept.

Adam	yes if I could go back to the architectural concept [*pulls out drawing*] on that show you where I'm coming from I've done these concept diagrams to try and explain how what holds the architecture together because the building as I mentioned before is a combination of four strips of what we call servant space which are low spaces and three strips of served space which are the barrel vaulted spaces and you put those together and you get this combination of ser- servant served servant served servant served and servant so I've been trying my hardest to keep all the important spaces like the chapel and things ( ) entrance so on and so forth vestry and the major staff rooms that more important spaces under the barrel vaults and keep all the supporting accommodation under the served if we then went back to the plan

that, to some extent, the architect monitors moves in the design space against as yet unspecified criteria.

The actions of the architect that socially produce the design concept are next considered , in particular when the concept diagrams are brought into the interaction.

In Extract 4 (A1, 600–671) the architect uses a physical representation of the design concept, the 'design concept drawing' as an aid to explain the concept to the client representatives. The design concept drawing acts as an intermediary object that the participants use to negotiate the design space. The drawings are a graphic representation and are not the design concept. Adam states:

> "I've done these concept diagrams to try and explain how what holds the architecture together because the building as I mentioned before is a combination of four strips of what we call servant space which are low spaces and three strips of served space which are the barrel vaulted spaces" (A1, 661–665).

This explicates the ordering system that underpins the design, namely, to organise the spaces in the building according to function, as either 'servant' or 'served' spaces. Another property, the difference in ceiling height between the two types of space, is also made known in this turn-at-talk.

These analyses have explicated what was accomplished in a brief exchange (six turns at talk). Some properties of the 'architectural concept' were made known and part of a shared knowledge, as well as beginning to show tentativeness concerning what are acceptable moves in the design space.

Although the architect has warranted some properties of the design concept, from an EM/CA perspective, I am attentive to the recipients' acknowledgement of these properties. Anna's actions acknowledge that the property of servant and served space has previously been warranted to the design, when Adam states: "if you remember I described this as servant and served space" Anna's response is an acknowledgement: "that's it" (A2, 751–753).

2.3 *Negotiated ownership of the design concept*

From a cursory glance at the data, the use of the possessive pronoun 'your' is a preliminary indicator that the design concept is to some degree 'owned' by the architect. I now attend to some instances in the data when references to 'your design' are made.

Extract 5, A1, Negotiating ownership of the design.

799	Anna	yes that's what we're looking for but the concern that we have at the
800		moment is whoever is operating and perhaps working some of this the
801		video they need to be able to see sort of in a sense rather than just at an
802		angle so what what in an ideal world and I don't want to compromise
803		your design the door the viewing room at the existing site which is very
804		similar to this we are putting on here so that we can then see down that
805		way so I'm trying to think of a way that we can get them to look
806		through maybe at this angle through here that's what I'm looking at so
807		they're able to view near enough both sides of the chapel
808	Adam	yeah [*begins to sketch*] I wonder if it's possible to do something like that
809		where they'd have a sort of vision spot through there
810	Anna	yes
811	Adam	I mean if I made this feel like a room with its own lid on it that was
812		inside the chapel that was just token toke tucking its nose into the
813		chapel we might be able to get it to work it does go slightly against the
814		grain for me to do that but it does satisfy what you wanted and it means
815		that we could link this up to it actually so- ++++++
816	Anna	OK is that too heartbreaking for you [*all laugh*]
817	Adam	well it's not as pure a summation as I was looking for but I mean
818		maybe there's another way of doing it maybe if I keep my thinking cap
819		on because you can see I'm trying to keep the spaces pure the
820		purer the space the more spiritual I think it will be the more you mess
821		around with it
822	Charles	yes

Anna raises a design concern and in the same turn states: "I don't want to compromise your design" (A1, 802–803) then continues to state what she attempts to achieve by modifying the design. Adam's response is produced while sketching: "I wonder if it's possible to do something like that where they'd have a sort of vision spot" (A1, 808–809). However his next turn: "we might be able to get it to work it does go slightly against the grain for me to do that but it does satisfy what you wanted" (A1, 813–814) makes it known that there are aspects about this solution he is less satisfied with. Anna's next turn: "is that too heartbreaking for you" (A1, 816) acknowledges that there are some design moves which are less acceptable for the architect. I infer from this that Anna is aware that some design moves are more acceptable than others. An upshot from this interaction is that the architect acknowledges Anna's concern to be able to view the chapel, but he keeps the design solution open for further deliberation.

Charles suggests that a bridge be included in the design as an alternative to the architect's previous idea to include stepping-stones. Although the architect acknowledges that a bridge could be included, Anna registers Charles' suggestion as an action that: "compromising your design all the time" (A1, 1152). The architect

Extract 6, A1, Acknowledging the possibility of modifying the design.

1148	Charles	I think if we could put a bridge or something that looks like a bridge
1149	Adam	you can have that as a bridge if you wanted not necessarily stepping
1150		stones
1151	Charles	yeah I think a bridge would be
1152	Anna	compromising your design all the time [*laughs*]
1153	Adam	it's not necessarily a compromise but if there's a good reason for
1154		changing it let's change it I've always wanted to do a stepping stone
1155		type bridge across a pond maybe this isn't

response: "it's not necessarily a compromise but if there's a good reason for changing it let's change it" acknowledges Charles' idea without committing to adopting this. I infer that the architect is making it known that modifications to the design are acceptable actions for the client representatives in this situation, however the architect has local agency (at that moment in time) to decide which design ideas will be taken forward within the scheme.

At the second architectural meeting the design concept is a subject for discussion.

Extract 7, A2, Attentiveness and awareness concerning design moves.

246	Adam	we could we could put a cover over that I think we could extend this
247		line and cover that
248	Anna	would that cause any problems at all for your design at all
249	Tony	we can sketch that out
250	Adam	it completely compromises it
251	Anna	[*laughs*] right we won't do it then
252	Adam	/[*laughs*]\
253	Anna	/[*laughs*]\ OK yeah
254	Adam	I'm sure we could do something there

The architect suggests a solution (Extract 7), (A2, 246–247) to a concern that the wreath court is exposed. Although a solution was provided by the architect Anna questions: "would that cause any problems at all for your design at all" (A1, 248). Adam's response: "it completely compromises it" (A1, 250) is acknowledged by Anna as a joke and Anna develops the humorous tone of the interaction in the next turn: "[*laughs*] right we won't do it then" (A1, 251). This interaction accomplishes many things. It registers that Anna is attentive to not 'compromising' the design, that is she is aware that some design moves are better than others, and that there are underpinning design criteria.

Extracts 5, 6 and 7 have illustrated that Anna in particular has been attentive to not compromise the design when raising concerns and modifications to the boundary object. Anna's repeated reference to 'the design' I infer is a linguistic indicator that she is not differentiating between moves that modify the boundary object from design moves that modify or challenge the attributes of the design concept. It is beyond the boundaries of EM/CA to speculate on the reasons for this, whether Anna has appreciated that there is a difference. However, Anna's action of asking a question accomplishes several things: she actively seeks the architect's assessment[20] and the conversational platform is transferred to the architect. The architect's next turn response is an assessment, an opportunity to accept, modify and challenge (amongst other actions) Anna's design idea. Assessments, evaluations of the properties of the design, were pervasive in these interactions and even when not directly sought assessments were given, as Matthews[21] has previously observed. In this setting the architect was locally in a privileged position to moderate moves in the design space in relation to his conception of the design concept. I infer from this that the architect had locally negotiated ownership of the design concept and used the concept to moderate moves in the design space. This observation will be re-visited.

20. Pomerantz, A. (1984) Pursuing a Response, in Atkinson, J.M. and Heritage, J. (eds.) *Structures of Social Action: Studies in Conversation Analysis*, Cambridge University Press, pp. 152–163.

21. Matthews, B. (2007) Locating Design Phenomena: A Methodological Excursion, *Design Studies*, 28, pp. 369–385.

2.4 *A design impasse*

The extract from the data I refer to as the 'design impasse', Extract 8, is from a longer stretch of talk (A2, 582–693). The architect has previously warranted that 'servant-served' spaces are attributes of the architectural concept. In this interaction another attribute of the concept is revealed.

At the heart of the design impasse is uncertainty surrounding the number of cremators to include in the scheme. The design impasse starts with Anna's question: "the only other thing I've just slightly was the where the cremators are just whether + there's room there for another machine" (A2, 582–583). The architect questions: "so you're considering a third cremator" (A2, 590) and Anna answers: "yes". The client's idea to change the number of cremators as part of the scheme is responded to by the architect in a different manner from other design ideas. In previous extracts from the data e.g. Extract 5, even when an idea challenged the design concept the architect 'worked at' finding a solution. In the interaction shown in Extract 8 the architect's response is to challenge the idea.

The architect warrants: "well this is fairly fundamental ... deciding the number of cremators ... because originally there were going to be no cremators" (A2, 618–622). Anna's acknowledgements agree with the architect's claims. That is there is agreement amongst those present that this idea is a fundamental change to the design. The architect explains that: "this might have a fundamental change on the whole width of this bay" (A2, 626–627) and that: "what we could do is ask CREMCORP to revisit this design to see if three cremators could go in" (A2, 648). Anna's response rescinds the need for a third cremator and she modifies her request to: "just enough space to + put all the add on bits" (A2, 674–675). The architect's next turns are of particular interest as they are dispreferred turn shapes[18] that produce a disagreement.

Extract 8, A2, The design 'impasse'.

674	Anna	yep so you know but it's not so much a third cremator but just enough
675		space to + put all the add on bits that probably might come with it
676		such as a a stairwell that sort of type o- it's just the th- just the two doors
677		on either side just making me think whether they might want to be-
678	Adam	yeah I mean I think this was put in for architectural reasons
679	Anna	right
680	Adam	because it's such a symme/trical\
681	Anna	/metrical\
682	Adam	building
683	Anna	yeah OK
684	Adam	you know I mean if you wanted us to look at seriously at putting in a
685		third cremator I think we'd have to review this whole area the
686		chances are we couldn't + retain the symmetry erm in that way if you
687		would like us to look at a third cremator we can do that I think we
688		need to just clear ++ clear direction from you wha-
689	Charles	yeah
690	Adam	what this was until now
691	Charles	yeah it was
692	Adam	was a two cremator building
693	Charles	it still is

What was negotiated was that the design suggestion proposed by Anna was seen to contravene the 'design concept'. The architect invoked another attribute of the design concept 'symmetry' as backing to support his argument that her idea to modify the design was unreasonable: "I mean I think this was put in for architectural reasons" (A2, 678): "because it's such a symmetrical building" (A2, 680–682), "we'd have to review this whole area the chances are we couldn't + retain the symmetry" (A2, 685–686). This was socially consequential as the client representatives then clarified that it is still a: "two cremator building". In effect, the client representatives acknowledged the architect's concern that adjusting the number of cremators would compromise the architectural design too much.

It was interactionally accomplished that 'symmetry' was a property associated with the design concept. When the client representatives made moves in the design space that breached or contravened this previously unstated design parameter or 'rule' of the concept the architect made these properties of the concept known and part of a shared knowledge. Significantly, in these actions it was also made known that the architect could invoke other properties of the concept during these conversations. These actions illustrate that the architect could moderate which moves in the design space were more acceptable adjustments to the boundary object. Based on these actions and the earlier observation of locally negotiated ownership (Section 2.3) an inference made is that the architect had some degree of ownership of the design concept.

While the actions of warranting the properties of the design concept were indicative of an asymmetrical relationship between the architect and the client representatives, I emphasise that this observation is of a locally negotiated asymmetry, and not the more general association of asymmetry with the participant roles of architect and client. The association of actions with participant roles is examined elsewhere in this volume by McDonnell[22] and by Oak[23] who examines specifically where roles were talked into being.

22.
McDonnell, J. (Chapter 14) Collaborative Negotiation in Design: A Study of Design Conversations between Architect and Building Users.

23.
Oak, A. (Chapter 17) Performing Architecture: Talking 'Architect' and 'Client' into Being.

3 LINGUISTIC MARKER FOR THE DESIGN CONCEPT

I next briefly consider how the design concept was signified. While several participants in these interactions performed the activity of designing, the design moves that modified the design concept were more sensitive to the actions of the architect. I observe a linguistic pattern; this architect used 'architectural' or 'architecturally' as linguistic markers to reference the 'design concept' and to indicate which aspects of the design were less negotiable. This inference is based on observations of each occurrence of 'architectural' or 'architecturally' in the data. The following data extracts are three examples of this pattern of talk.

> "I felt that architecturally it needed to be a great deal more bold to make a statement about it being a very important" (A1, 337–338)
>
> "we wanted to make sure the architectural concept worked through so that flues didn't come from ++ the nicest part of the building" (A2, 771–774)
>
> "that would certainly enable you to get a wider catafalque inside there that might be suitable for two people the architectural idea here is to have like a cylinder which will be top lit" (A1, 379–381).

4 INFERENCES: ACTIONS PRODUCING OBJECTS

Following the logic of the analyses, these inferences were made. I observed a range of objects that were instrumental in the interactions, for instance the design concept as well as the current status of the design in the design space. It was the interplay between these objects that was particularly interesting, especially the difference in status that was given to design ideas in comparison with the design concept. The design concept was observed to have greater intransigence in interaction than design ideas.

I infer that embedded within the design as a boundary object was a design concept that was less negotiable. The design concept was interactionally produced in the actions of the participants, and the client representatives acknowledged its presence. While some properties of the concept were made known to the participants, for example symmetry, the architect was the sole actor in these interactions who warranted what were properties of the design concept. In this respect the architect displayed a local (moment-by-moment) ownership of the design concept.

In collaborative design situations, such as the two architectural meetings, these actions and properties are significant, as other participants, although they produce actions of designing that change the design space, are unaware whether or not their moves comply with the underpinning concept. This pattern of interaction was observed when another attribute of the concept was revealed by the architect at a later stage to argue against a design move. These actions are indicative of a local interactional asymmetrical relationship[24], that is, a moment-to-moment shift in the balance of the interactions between the architect and the client representatives. In the interactions of the two architectural meetings social relations and local identities were incrementally developed, which were transformable at any moment[25]. I infer that in situations when the architect mobilised the design concept this action produced a local asymmetrical relationship (which is distinct from asymmetry associated with participant role).

24.
ten Have, P. (1991) Talk and Institution: A Reconsideration of the 'Asymmetry' of Doctor-Patient Interaction, in Boden, D. and Zimmerman, D. (eds.) *Talk and Social Structure: Studies in Ethnomethodology and Conversation Analysis*, Polity Press, pp. 138–163.

25.
Drew, P. and Heritage, J. (1992) *Talk at Work: Interaction in Institutional Settings*, Cambridge University Press; p. 21.

5 INFERENCES: AGENCY OF THE DESIGN CONCEPT

The next step is to propose that the architectural design concept has agency, that is, to have properties as a non-human agent to influence what were acceptable moves in the design space.

The agency of the design concept is considered to be evident in the data as the design concept imposed constraints within which the architect designed the building, for example following the design parameter of symmetry. Although it was interactionally the actions of the participants that mobilised the design concept, the parameters the participants designed to (symmetry, servant-served spatial configuration) were properties of the design concept.

To substantiate this inference I draw attention to the data conditions of conversation analysis again, where naturally occurring conversations between the participants are the data (the design meeting interactions) and not the supplementary material (the interview with the architect). The actions produced in and through talk were interpreted without assuming privileged researcher' knowledge of the architect's motivations. It is outside the parameters of EM/CA to speculate on the intent of the architect, for example, to infer what influence the Kimbell building designed by Kahn had on the concept and on the design of this crematorium (a matter to which the architect refers in the conversation with the researcher referred to in Section 1).

To question the origin of the design concept, reference to the supplementary material is necessary, as well as making assumptions about what's going on in the architect's mind, but can a researcher know how the architect was influenced by Kahn's work? Influenced in what sense – function, aesthetics, metaphorically, to what extent, etc? These questions are not part of EM/CA research but are problematic in design more generally. Eckhart and Stacey[26] have considered the creative behaviours of fashion designers, including the position that the memory of garments play in the conceptualisation and perceptual evaluation of acceptable future designs.

26.
Eckhart, C. and Stacey, M. (2001) Designing the Context of Fashion: Designing the Fashion Context, in Lloyd, P. and Christiaans, H. (eds.) *DTRS 5: Designing in Context*, Delft University Press.

The architect's reluctance to break the condition of symmetry illustrates that the properties of the concept were beyond the architect's volition. Other actions indicated that breaking a condition of the design concept would compromise the design. However it was the actions of the architect, not other participants, which mobilised the design concept; this in turn moderated which design moves were more acceptable with respect to the concept. The quote used in the title of the paper illustrates this, as in the act of asking the architect whether a design move 'compromised the design' the architect's agency to approve (reject or other actions) a design move was shown. For these reasons and because the concept had properties and parameters that constrained moves in the design space the design concept is considered to display agency.

6 EXTRAPOLATIONS: 'OBJECT' PROPERTIES

The final sections of this chapter break from inferences based on EM/CA analyses to make more provisional associations between properties of the design concept and objects in studies of science, technology and society. I extrapolate from the previous explications and inferences to reflect on some properties of the design concept as a phenomenon, or an object type.

At the outset of this argument actions that bring about moves in a design space were associated with co-evolution where the problem space and solution space co-evolve together. Significantly, Dorst and Cross[27] have observed that: "a chunk, a seed, of coherent information was formed in the assignment information, and helped to crystallise a core solution idea". The notion of a 'core idea' can also be associated with Darke's primary design generator[28].

27.
op. cit., p. 434.

28.
op. cit.

From my analyses I observe that the design concept, which could be viewed as a chunk or seed of information, exhibited several properties including some that were antithetical: exhibiting the property of being fixed yet at the same time flexible, as the design concept was never completely described. In common with the observations of Dorst and Cross[29], 'coherent information' aspects of the design concept were fixed and less negotiable, e.g. symmetry, while other properties of the concept were revealed over time and were attributed in interaction. This observation highlights mobility as another property of the design concept. It was mobilised in interaction (by the participants) yet had properties that were fixed (the agency of the concept). I next refer to studies in science, technology and society to attempt to describe and categorise this phenomenon, that has fixed yet mobile and flexible properties.

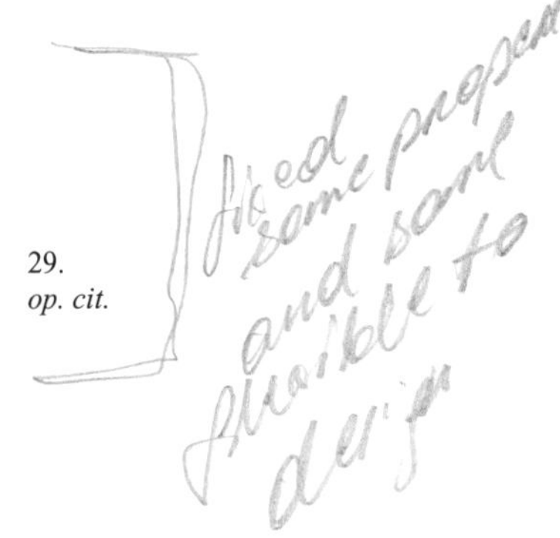

29. *op. cit.*

The descriptions of objects and agency from various approaches to science, technology and society (STS) have particular epistemic and ontological grounds. While Suchman[30] discusses non-human agency and has approached human-machine interaction from a theoretical ground common to the study presented here (namely, EM/CA), her work has focused on interaction with an artefact. In this chapter, the design concept which is considered to have agency has no material reality (the design concept drawing is a representation and not the design concept). The constructionist, social construction of technology view[31] similarly is associated with objects with a material reality. However, actor-network theory acknowledges that 'actants' in a network which have agency can be material-semiotic objects. Law and Singleton's work on object lessons[32] is a development from their earlier difficulties describing the non-object of an alcoholic liver. Several versions of objects are described, including 'fluids'. "[W]e cannot understand objects unless we also think of them as sets of present dynamics generated in, and generative of, realities that are necessarily absent"[33]. This description of an object type has similarities with the properties of the design concept.

30. Suchman, L. (2007) *Human/Machine Reconfigurations: Plans and Situated Actions (2nd Edition)*, Cambridge University Press.

31. Bijker, W.E., Hughes, T.P. and Pinch, T. (1986) *The Social Construction of Technological Systems: New Directions in the Sociology and History of Technology*, MIT Press.

32. Law, J. and Singleton, V. (2005) Object Lessons, *Organization*, 12, pp. 331–355.

33. *ibid.*

Observing the sociotechnical dynamics of design Vinck and Jeantet state: "both things and humans have some opacity, an indispensable presence, a specificity and an active role"[34]. One of their object classes, open objects: "are also elements of language and communication between humans. Things are items in language". However, STS are normally concerned with material and not verbal constructs. Although particles of talk are semiotically real, as they produce semiotic objects, referents, which can be regarded as 'objects', I acknowledge that the proposition that the design concept is an object is provocative.

34. *op. cit.*

To make the analogy between the design concept with objects in studies of science, technology and society I knowingly merge and contravene the epistemic boundaries in STS of constructivism and actor network theory. Yet this line of inquiry has raised the question of how to categorise 'virtual' constructs, the immaterial 'objects' of design. Stacey and Eckhart[35] have begun to engage with this question by acknowledging that boundary concepts are a shared abstraction from a different conceptual structure and in recognising that there are various dimensions to design ambiguity. I highlight the description of the design concept as a 'type' of object as an area for further investigation.

35.
Stacey, M. and Eckert, C. (2003) Against Ambiguity, *Computer Supported Cooperative Work*, 12, pp. 153–183.

7 CONCLUSIONS

The concerns of this paper were to explore how an 'architectural design concept' was produced in the interactions between an architect and client representatives and to consider the consequences of its mobilisation. Interactionally the design concept was mobilised and with reference to some structures of talk from conversation analysis it was inferred that in and through the actions of designing in talk the participants acknowledged that there was an underpinning design concept.

The architect attributed properties to the design concept and although some properties were revealed in interaction, other participants did not comprehensively know the properties of the concept. The design concept was used to moderate preferred moves in the design space by actions of the architect, who was the sole person in these interactions that made assessments whether a design move complied with the design concept.

These actions were consequential for the participants' design moves and for collaborative design more generally. The architect had socially negotiated agency to moderate whether the client representatives' moves in the design space were in keeping with the design concept. This meant that the architect was in a locally asymmetrical position to accept, modify or reject design ideas.

Beware the linguistic markers 'architecture' and 'architectural'! In these interactions I infer that 'architecture' and 'architectural' were used as linguistic markers to reference the design concept. The architect had greater 'ownership' of aspects of the design that were marked by the prefix 'architecture' or 'architectural'.

The design concept was observed to have agency, acting as a non-human agent with properties that constrained and moderated the architect's moves in the design space. In contrast with inferences from the empirical study of the data, the final part of this chapter was more speculative and a qualified analogy between some properties of the design concept and the study of objects in social studies of science were made. In the progression of the argument explications and inferences from analyses were made, which led to more provisional observations that questioned the 'object' status of the design concept. The design concept was observed to have properties of being fixed, yet flexible and mobile. However, unlike STS that observe material objects the design concept is an immaterial phenomenon. For this reason and because I draw from a range of epistemic arguments in the last phase, the categorisation of the conceptual, immaterial objects of design is highlighted as an area for further investigation.

ACKNOWLEDGEMENTS

The discussion of this research with the participants at the DTRS7 symposium was constructive and has informed developments made to this paper, together with additional communications with Ben Matthews, Martin Stacey and Eric Blanco.

14
Collaborative Negotiation in Design: A Study of Design Conversations between Architect and Building Users*

Janet McDonnell

Analysis of the conversational exchanges taking place during two meetings between architects and their building-user clients at an early stage in a building design project reveals blurred boundaries between architects' and building users' (argumentation) positions and the potential of these as conversational moves. As we might expect, we see conversational turns in which each person contributes from their own territory of expertise and we see them respond to each others' conversational invitations to supply information. However, we also see tentative excursions where one party invokes the position or knowledge of the other to propose or justify a design decision, provoking, in turn, an expert response. Incidents such as these, revealed as conversational exchanges, show us how design progression is negotiated collaboratively. They provide us with a way of grasping concretely some of the subtleties of how shared ownership of a design is established within a non-participatory design setting.

The architectural meeting data provides an opportunity to analyse the interaction between designers and client-users as they meet to refine design details at two early points in an architectural design project. In the video recordings of the two meetings we see that outstanding issues with the current design are introduced, elaborated and explored in conversation. In talking about each topic, information is exchanged and a course of action is negotiated to move the project forward. Although the meetings are not set within a participatory design context, it is clear from participants' references to their prior interactions that the meetings are part of a setting in which the degree of consultation appears to be acceptable to all parties present and a variety of means of exchanging information has been established to mutual satisfaction. In this chapter some of the features revealed from analysis of the conversational exchanges during the two meetings are described. The work focuses on the verbal negotiation of what to do at two levels; namely, what the next actions are in the design process, and what to attend to next within each meeting – the next moves in the interaction.

In previous work which analysed naturally occurring conversations between an architect and building users[1], we classified the information exchanged conversationally in terms of representational elements of design (functional, perceptual, phenomenological and symbolic) originally proposed by Medway[2]. The objective there was to compare building users' and architects' facility in talking about

1.
Luck, R. and McDonnell, J. (2006) Architect and User Interaction: The Spoken Representation of Form and Functional Meaning in Early Design Conversations, *Design Studies*, 27, pp. 141–166.

2.
Medway, P. (2000) Writing and Design in Architectural Education in Pare, A. (ed) *Transitions: Writing in Academic and Workplace Settings*, Hampton Press, New Jersey, pp. 89–129.

*Reprinted from: CoDesign Vol. 5 No 1 (March 2009), pp. 35–50,
DOI: 10.1080/15710880802492862

these different design elements. Here, using the same perspective which focuses on conversation as a medium through which a design evolves, analysis of the meetings is framed in terms of sequences of interaction, initiated by the raising of a design issue to be resolved, where exploration of possibilities ensues, and through which the meeting participants work towards agreement about what needs to be done. Viewed in this way, the exchanges appear to allow progress to take place smoothly and effectively through skilled interaction. Particular attention is paid to some of the ways in which participants appeal to each other as audiences[3] during the conversations as design contributions are negotiated.

3. Perelman, C. and Olbrechts-Tyteca, L. (1971) *The New Rhetoric: a Treatise on Argumentation*, University of Notre Dame Press, Notre Dame, Indiana.

This inspection of the architectural meeting transcripts reveals blurred boundaries between architects' and building users' (argumentation) positions. We do see conversational turns in which architects and building users make assertions based on their own territory of expertise and knowledge and we see their territorial claims invited, and reinforced through acceptance by other parties to the conversation. We also see crossovers – tentative excursions – where one party invokes the position or knowledge of the other in making a case for some design decision and provoking, in turn, an expert response. These incidents, revealed through the conversational exchanges, provide us with a way of grasping concretely some of the subtleties of how shared ownership of the design is established between meeting participants. This shared understanding is created and revealed in conversation as the design is negotiated.

1 MOTIVATION

The DTRS data provides a rare opportunity to examine in detail conversational interaction during naturally occurring meetings at the pre-planning application stage of an architectural design project. These meetings are part of the normal work duties of those present, the interactions are taking place to move the design of the buildings forward, not to serve a research agenda; our intrusion as researchers is about as low as we can get if we are to collect video recordings. Thus, in some senses the data is immensely rich; in others it is extremely limited, as it comprises recordings of two working meetings from a single project. One motivation of the work presented here is to explore what can legitimately be concluded from close inspection of (only) two meetings between architects and client/users.

A second motivation is to explore how the everyday business of designing a building manages to take place more-or-less effectively; to examine some of the mechanisms by which the design progresses in a situation which is neither radically participatory, nor politically or architecturally high-profile. This particular study examines some of the ways in which constructive engagement, via conversation, takes place. Rather than focusing on short-comings, e.g. misunderstandings, communication failures, or the poverty of design representations, this study takes a closer look at an example of everyday architectural design practice to examine how stakeholders manage to communicate sufficiently well to move the design project forward to their mutual satisfaction. In this respect it tries to set assumptions aside by adopting the spirit of qualitative data interpretation which starts by examining 'how' before tackling 'why'[4].

4. Silverman, D. (2001) *Interpreting Qualitative Data (2nd Edition)*, Sage Publications.

2 METHOD

The approach was to work principally from the architectural meeting transcripts, A1 and A2. The video recordings were used firstly to get an overall picture of each event and secondly to clarify and disambiguate verbal references in the transcripts. The principal participants in the first meeting, A1, are the architect, Adam, and two clients who are at the same time building users, Anna and Charles. In the second meeting, A2, these same three participants are joined by a more junior member of the architectural team, Tony. The two client/users have slightly differing interests and perspectives on the project but there is much overlap. Although it is acknowledged that, for many purposes, clients and building users are entirely different stakeholder groups whose concerns cannot be conflated, here Anna and Charles will be referred to as building users because it is their roles as users of the building themselves and as spokespersons for other building user communities that are the focus of attention.

Following Fairclough[5], each meeting transcript was divided into phases, a phase being a large scale conversational structure within a meeting which imposes the social routine for interaction, setting the parameters of what interaction can take place. See Oak[6] for a description of the institutional setting within which the meetings analysed here take place. The two transcripts were segmented according to the topic or issue under discussion. Each topic or issue pertains to some element(s) of the design. These segments will be referred to as episodes. An episode begins and ends when a topic change is introduced. For each episode, the participant and conversational means by which a topic change occurred was identified (for example a question from Anna directed at Adam) and also the outcome(s) in terms of the implications for progressing the design (for example agreement that following the meeting Anna will provide information for Adam or that Adam will work on a design change).

5. Fairclough, N. (2001) *Language and Power* (2nd edition), Pearson Education, Harlow, England; p. 115.

6. Oak, A. (Chapter 17) Performing Architecture: Talking 'Architect' and 'Client' into Being.

Parsing the data in this way foregrounds firstly, the social interactional set-up which constrains and enables what can be discussed and how it can be discussed, and, secondly, the design contributions which result from what is said. A design contribution here is considered to be any outcome that contributes to moving the design (process) on, including exchange of information relevant to the design, and agreement to do or supply something, as well as changes or refinements to the architectural scheme itself.

It should also be noted at this point that, although the term conversation is used throughout, formally, according to Schegloff[7], conversation is: "talk which is not subject to functionally specific or context-specific restrictions or specialised practices or conventionalised arrangements"[8] and therefore, talk-in-interaction, which is a more general term perhaps would be a more accurate, if more awkward term to use here. The data is drawn from formal design meetings where work is taking place to develop the design through speech acts. There is therefore an organisational format[9] in place. It is for this reason, i.e. that there are some social routines at play, that before the detail of exchanges are presented, in Section 4, to illustrate how design contributions are negotiated at the micro-level, they are first set in the context of the macro-structure; the phases and episodes of each meeting. This is presented in Section 3. At both levels the focus of attention is on how the business of the meeting, which is to move the design forward, is negotiated i.e. what the outcomes of the talk are and how the participants collaborate to justify these for themselves.

7.
Schegloff, E.A. (1999) Discourse, Pragmatics, Conversation, Analysis, *Discourse Studies*, 1, pp. 405–436.

8.
ibid., p. 407.

9.
ibid., p. 410.

3 MACRO-STRUCTURE: PHASES OF THE CONVERSATIONS AND THEIR TOPIC-ORIENTED EPISODES

In meeting A1, Anna and Charles 'go along with' the context and goals for the meeting as explicitly articulated by Adam, namely: "hearing the feedback [on the design proposals so far] because that's the purpose of the meeting" (A1, 21) and what he terms the shared intention:

> "the purpose of the meeting as I see it is to obtain your feedback to the design ++ to update everybody as to where we've got to and to receive your feedback" (A1, 75–79).

Within this agreed agenda, movement between topics, a protocol with which all parties comply, is organised around a verbally enacted tour through the proposed physical spaces comprising the new building and its landscape setting (i.e. the virtual building[10]). This organisation of the talk is supported by a series of architectural drawings, particularly a plan of the scheme. Glock[11] describes in detail how Adam orients those present, focusing their attention for

10.
Medway, P. (1996) Virtual and Material Buildings: Construction and Constructivism in Architecture and Writing, *Written Communication*, 13, pp. 473–514.

11.
Glock, F. (Chapter 16) Aspects of Language Use in Design Conversation.

a particular reading of the design. At times discussion shifts from the proposed development to the current building on the same site (where the meetings are taking place) which fulfils a similar function to the proposed building – a crematorium complex incorporating a chapel, cremation facilities, and offices. Other documents and drawings temporarily replace the plan in smooth transitions according to what best supports the current topic under discussion. The phases identified for meeting A1 are given below in Table 1.

Phase	Description
0	Phatic exchanges; explicit establishment of common agenda, implicit establishment of the conversation's boundaries and its protocol
1	A conversational walk-through of the proposed building, led by the architect, during which issues are raised by either building users or the architect in a sequence of topic changes that flow from the current local context of the design proposals; predominantly dialogical
2	Similarly organised as for phase 1 but focused on the landscaping proposals; conversation oriented towards addressing third party input (an ecological appraisal)
3	Factual exchanges of information over costs
4	Reprise (summary) orchestrated by the architect during which outstanding topics – ones overlooked or which did not fit the organising protocols of phases 1 and 2 – are discussed

Table 1. Large scale structures (phases) of the conversation in meeting A1.

Table 2 gives a summary of the further breakdown of the event into topic, initiator, and outcome. (Note that column two in the table refers to the start line in the meeting transcript.) A brief inspection of Table 2 shows that Adam, Anna and Charles are able to initiate topics within the protocol introduced by Adam i.e. dealing with matters as they naturally arise, prompted by the verbal tour of, firstly, the proposed building itself and, thence, the landscaping of the site. Reymen, Dorst and Smulders[12] present a similar table describing topics. The two tables have points of correspondence, their differences are accounted for by our different foci of attention, theirs being episodes which may be candidates for instances where co-evolution of design 'problem' and 'solution' might be occurring.

12. Reymen, I., Dorst, K., and Smulders, F. (Chapter 4) Co-evolution in Design Practice.

The second meeting, A2, has been convened to answer: "all the items on the list" (A2, 10), a reference to matters raised during or in follow up to meeting A1. Initially, the sequence of the topics is organised round a 'tour' of the plans. In this respect, the early part of the conversation is similar to that in meeting A1. Adam sets this sequence in place: "I'll just start by going through the drawings from top to bottom" (A2, 20). However, Anna and Charles also have a series of issues that they wish to raise. In this meeting, in the phases where the virtual tour through the space is maintained, we see a more monological presentation by Adam, interspersed with phases in which topic changes are initiated more frequently by the building users.

Phase	Start Line	ISSUE or TOPIC	INITIATOR	OUTCOMES
1.1	90	Size of the waiting area	Question from Adam	Adam to revise design *[design refinement]*
1.2	204	Access for the hearses and unloading of coffin	Anna on behalf of others	Clarification of what is proposed by Adam *[information exchange]*
1.3	263	Accommodating multiple vehicles	Raised as an issue by Charles	Adam volunteers to revise design (despite Anna and Charles protesting that it is unnecessary) *[design refinement]*
1.4	343	Size of catafalque (as a space and 'altar')	Question from Adam	Discussion, revision (or not) unresolved (design is revised before meeting A2) *[design refinement]*
1.5	391	Matters associated with catafalque area seating, visibility, doors	Anna and Charles	Adam to revise design (reconsider) *[design refinement]*
1.6	472	Sanctuary and features associated with spirituality	Adam	Adam gets mandate for incorporating (stained glass) features into the design *[design refinement]*
1.7	550	Provision for AV facilities and operator	Anna and Charles	Adam requests brief from Anna and Charles *[information exchange]*
1.8	1006	Route to committal room	Question from Anna	Clarification of what is proposed by Adam *[information exchange]*
1.9	1079	Additional cremator facilities – cremulator room	Question from Adam	Clarification from Anna, follow up q. from Anna leads to general discussion about coffin storage and disposal *[information exchange and design refinement]*
1.10	1130	Route over the pond and alternatives	Anna on behalf of others	Clarification of what is proposed by Adam – discussion of alternatives *[information exchange]*
1.11	1208	Movement of hearse to wreath court	Anna on behalf of others	Clarification of what is proposed by Adam *[information exchange]*
1.12	1228	Size of wreath court- and the roofing of it	Tentatively raised by Anna	Clarification of what is proposed by Adam – leads into 1.13 *[information exchange]*
1.13	1263	Overall form of the building	Anna – by conveying general impressions of others	*[information exchange]*
1.14	1305	Materials	Adam	Explanation by Adam and clarification prompted by queries from Anna *[information exchange]*
1.15	1343	Use of glass, its properties, pattern, colour	Opinion volunteered by Anna	Sense of consensus over how to achieve 'spiritual' feeling *[information exchange and design refinement]*
1.16	1443	Roofing material	Adam	Explanation of possible alternatives by Adam – final choice unresolved *[information exchange]*
2.1	1545	Ecological appraisal: reducing impact of (landscape) design on wildlife	Adam	Consensus over the way appraisal recommendations should be addressed by Adam *[information exchange and design refinement]*

(continued)

(continued)

2.2	1719	Transport routes	Charles on behalf of others	Agreement that good signage is the 'solution' to the objection to the scheme raised *[information exchange]*
2.3	1893	Landscaping and paths	Charles – revisiting topic 2.1	Adam seeks and gets formal mandate to respond to 2.1 through design changes *[information exchange and design refinement]*
2.4	2000	Book of remembrance	Charles	Adam sketches, literally and verbally, some possibilities– Adam to develop these in a revised design *[design refinement]*
2.5	2055	Ecological impact mitigation	Topic revisited by Adam	Adam mandated to incorporate ecological mitigation *[information exchange and design refinement]*
3.1	2154	Cost of the proposal	Adam	Understanding of broad costs and what they include *[information exchange]*
4.1	2233	(lack of) Space in cremulator room	Potential problem raised by Charles	Clarification from Adam *[information exchange]*
4.2	2253	Materials (reprise)	Charles	Elaboration from Adam *[information exchange]*

Table 2. Meeting A1 Macro-structure: episodes within phases of the conversation.

Despite this, the conversation still flows in a way that reveals opportunistic movement between topics. Each new topic is introduced where there is a natural link offered by the current context. For example, during a discussion about audio-visual (AV) equipment, the matter of the acoustical properties of the building arises. This leads to a discussion about the merits and pitfalls of carpeting, before 'returning' to other technical components of AV provision. The phases identified for meeting A2 are given in Table 3. An analysis of this meeting into episodes was produced, similar to that shown for meeting A1. This is referred to below but a table summarising the data is omitted here for brevity.

In both meetings, Adam sets the protocol in place initially, implicitly ordering the topics via the 'tour' of the current plans, however, if we look at where the initiative lies for introducing a topic change, we see that, in meeting A1, Adam, Anna and Charles are able to do this interchangeably (Table 2, Column 4). They do so, keeping the topic change either linked logically with a very local conversational context or linked into the virtual tour. The protocol for the second meeting, again, as with meeting A1 is set up explicitly by Adam at its outset. However, in meeting A2, although we still see a balance in topic changes, i.e. no one party dominating the conversation viewed as a whole, there is a bias in each phase (see Table 3). Thus, a balance is still operating in A2 but at a coarser level of granularity. Each 'side', architects and building users, has its own agenda for this second meeting; essentially, each has a notional list of issues that they want to address and these lists differ. Although two of the phases in the second meeting (phases 1 and 4 in Table 3) are predominantly monological on the part of

Phase	Description
0	Preamble
1	A conversational walk-through the proposed building, led by the principal architect, during which design changes since meeting A1, and the issues they address, are described; largely monological on the part of the architect
2	A sequence of topics driven by building users' initiative
3	Initiative returns to the principal architect, walk-through of the design resumed dealing with features not addressed so far; a mixed initiative phase where topics are raised opportunistically using natural links suggested by local context of current topic
4	Short, predominantly monological phase driven by the principal architect, based on a verbal tour of the perspective drawings
5	Mixed initiative discussion about arrangements for next (plan presentation) meeting
6	Resumption of substantial discussion of unaddressed topics; driven by user concerns
0	At conclusion of formal business multiple simultaneous exchanges as meeting participants and observers disperse

Table 3. Large scale structures (phases) of the conversation in meeting A2.

Adam, these are not strong evidence of conversational control on his part. These two phases are presentational in nature to a greater extent than the walk-throughs of the first meeting and thus his extended turns might be reasonable, and willingly granted by the others present on that basis. (A comparison may be made with Wheatley's[13] work on the presentation of document design.)

In summary, in both meetings we see turn-taking (who speaks next) and topic change occurring by negotiation, a pattern associated with interaction among equals. We see no evidence of the control of the conversation via, for example, interruption or enforcement of explicitness, tactics that are sometimes characteristic of control of the conversation by one participant[14]. Although there is organisation of topics, topic introduction is not controlled by any one party. In particular, Anna and Charles are able to introduce their own agendas: in the first meeting as their issues naturally arise in the tour of the design plans; in the second principally by initiating a phase change on two occasions (phases 2 and 6 in Table 3). In the second case, this is achieved by overriding what might otherwise be construed as a controlling mechanism – Adam's attempt to summarise the meeting (A2, 1432).

Looking at Table 2, we see that the outcomes of conversation on a topic are of two types: refinements to the design and information exchanges. Once an issue has been discussed to the point where a change to, or development of, the design seems indicated, for example making a room bigger, it is usually agreed that the architect will work on this outside the meeting. Information exchange is a salient feature of the interactions between architect and building users. Information is exchanged through discussion and more directed provision of information: architect to users, particularly explication of current design or design intentions to supplement

13.
Wheatley, J. (1995) Locating Negotiation Activity within Document Presentations in Firth, A. (ed) *The Discourse of Negotiation*, Elsevier, Oxford, pp. 373–397.

14.
Fairclough (2001), *op cit.*

what the drawings show; and vice versa, users explaining technical and social aspects of the buildings uses. These types of outcome also account for what takes place in the second meeting, although in the second meeting there is a single case of joint design problem solving (A2, 451–537) (defined narrowly as engagement by both building users and architects in the redesign of building features during the conversation itself).

The segmentation of the meetings in the way outlined, and shown in summarised form for meeting A1 in Table 2, accounts for all the talk that takes place i.e. references to topic start are contiguous with previous topic end. It lets us see that, through the smooth flow of exchanges, the business of the meeting is achieved surprisingly economically. A course of action is agreed; stuff gets done. Perhaps it begins to account for the mutual satisfaction with progress the participants express. We also notice that, whatever their shortcomings as representations of the design, in both meetings the drawings play a prominent role in setting the routine for the interaction and in organising the topics in a way which can be made to work to fit the interests of all parties.

4 MICRO-STRUCTURE: SUBTLE NEGOTIATIONS OF CONTRIBUTIONS TO THE DESIGN

In this section some of the conversational moves that meeting participants use to collaboratively negotiate what is to be done (i.e. the outcomes identified above) are examined in more detail. In Section 4.1 the focus is on how the architect appeals to (the audience of) the building users through attempts to provide scenarios of building use. Conversely, in Section 4.2, attention turns to some of the ways in which the building users bolster the architect's position by assuming custodianship of the design concept. Among other issues, these examples reveal the blurred boundaries between building users' 'functional' requirements, and their other 'architectural' concerns ('perceptual, phenomenological and symbolic' to use Medway's[15] classification of design elements). In Section 4.3, further subtleties of the negotiation of design features are illustrated using examples from the conversation in which arguments drawn from the different backgrounds of the participants are brought to bear in support of the same course of action. Finally, Section 4.4 sets the preceding examples in the context of the interaction as a whole, briefly discussing how expertise is asserted and acknowledged by all parties to the events.

15. Medway (1996), *op cit.*

4.1 *Architect's tentative appeals to building users' perspectives*

Here we identify three episodes where an issue is introduced by the architect attempting to sketch out a scenario of building use.

Despite their limitations, the architect's scenarios present conversational openings for the building users to proffer their own scenarios. And by drawing on their hugely richer scenario repertoire, they are able to share their knowledge with the architect. The first episode, early in meeting A1, concerns the size of the waiting area. Both Oak[16] and Glock[17] discuss this episode in some detail. The topic is initiated by Adam with a question about whether the size of the waiting room should be increased: "the first query I have [...] about whether you wanted the size of the waiting room increased" (A1, 90–92). Anna's response shows that, looking at the plan, she cannot tell how many seats there will be in the area. The architect compensates for the representation (the plan) by outlining a skeletal scenario of use, focusing on the functional aspects of the space thus: "the room is doing so much it's allowing people through into the porch area so it's also allowing access to the loos" (A1, 115–117). Anna gives a functional affirmatory response to Adam's question, i.e. in terms of number of seats. (It is worth noting here Glock's[18] more fine grained analysis of this part of the exchange where he interprets Anna's hesitancy as a possible disagreement.) But then she moves on to give a rich account of what people might be waiting for, different kinds of 'waiting', and she demonstrates a requirement to be able to separate waiting people for a whole series of reasons: "we might get people waiting for the eleven o'clock funeral erm and people at the ten o'clock perhaps arrive [...] they don't want to mix [...] and we do have problems with families during funerals" (A1, 132–137). After further rich elaboration, she moves on to say: "people like to smoke at funerals [...] and the seat that we've got out by the car park [...] even if it's cold and not very nice [...] people feel more happier out there than they do sometimes in the waiting room" (A1, 164–168).

16.
Oak (Chapter 17) ***op. cit.***

17.
Glock (Chapter 16) ***op. cit.***

18.
Glock (Chapter 16) ***op. cit.***

This episode shows how the architect's non-expert notions about waiting at funerals, once expressed, offer a conversational entry point through which the building user's extensive understanding of what waiting is about can be explained, and thus emerges what adequate provision for waiting might entail. The information she volunteers gives the architect the opportunity to understand 'waiting' in the crematorium context and thereby the requirements to cater for the practical, social and psychological needs associated with 'waiting' and separation.

The second episode, late in A2, is focused around the use of carpet in chapels. Here we see a similar pattern. Prompted by an apparently straight-forward question from the architect: "what was your view on a carpeted floor" (A2, 1133), Anna gives a collection of reasons, functional and aesthetic, saying: "number one it's the design of them pretty crap [...] number two they get filthy and dirty [...] it would have to be something more in keeping" (A2, 1135–1138) and later: "it needs to match up with other things"

(A2, 1157). She also refers to fading and fraying. The architect appears to agree to no carpet using a jocular intonation he says: "Ok I'll minute that as no carpet" (A2, 1164). Anna retracts her position slightly in response, and the architect takes up the opportunity for renegotiating the position it implies. Anna says: "it would be dependent on what sort of material I'm not [...] anti-carpet" (A2, 1168) eliciting the response from the architect of: "but you can get some extremely good carpets" (A2, 1169). Now, instead of pursuing an opportunity to persuade Anna to agree to carpets, he offers a scenario to support the building user's position conceding: "but you always have the problem of somebody somehow spilt something [...] I don't know what they'd spill in a chapel" (A2, 1171–1172). He tries to take on a building user perspective but is at a loss for experience. This presents the opening for Anna and Charles to pitch in with multiple illustrations of what can be spilt on carpets and how the spills arise. In a series of light-hearted exchanges candles, wine, vomiting and other spillages are reported in a very compact sequence of turns, amongst which the architect concedes: "I get the picture" (A2, 1184) and: "Ok" (A2, 1191). The last words on the topic lie with Charles and Anna respectively saying: "I think it would be possible not to carpet it" (A2, 1192) and: "I was just thinking what else you could use" (A2, 1193).

In both these examples, the architect's scenarios are attempts to take the building users' perspective, to make a case for a design decision from a scenario of use. His two scenarios are not drawn from his own direct experience. Despite their shortcomings, they act as prompts; their inadequacy is compensated for by the responses they elicit from the building users who are able to offer more convincing material from their rich repertoire of relevant experience of use. A third exchange, where the architect draws on his own personal experience of attending a funeral occurs during episode seven in A1 (1.7 in Table 2). Referring to a previous conversation, Anna gives Adam his opportunity to tell his story, she says: "but what you don't want them to do is play the music wrong like they did for your friend's funeral" (A1, 930–931). Adam then relates briefly (principally for the benefit of the observer) the tale of having the dramatic effect of a piece of music compromised by a technician inappropriately adjusting its volume. This scenario, in contrast to the two above, is a convincing personal testimony, and is therefore recognised by the building users as plausible and adequate as it stands, they do not embellish it. However, it does little work in argumentation terms, other than reinforcing a design decision that all the participants already are agreed upon.

These episodes show that even the most rudimentary attempts to appeal to an audience's position can be valuable as conversational openings which invoke in response explanations, via scenarios of use, which are valuable inputs in negotiating design requirements.

These episodes also show how functional, perceptual, phenomenological and symbolic 'level' issues can be intertwined in the construction of an argument about a design decision at any level (cf. the provision for 'waiting'), raising questions about the value of any attempts to establish simple correspondences between 'requirements' and features of design 'solutions'. Reymen, Dorst and Smulders[19] touch on these same issues from a different stance, identifying use plans and use scenarios as important 'bridging activity' between the parties and commenting on the failure of 'problem' and 'solution' as concepts to explain co-evolution in design.

19.
Reymen, Dorst and Smulders (Chapter 4) ***op. cit.***

4.2 *Building users as custodians of the design concept and of design integrity*

In the data there are many examples of where the building users, particularly Anna, talk about their aspirations for the building which are not readily expressed in functional terms (some examples are given later below). In some cases, their advocacy of design integrity sets them in opposition to other building user colleagues. Noticing this reminds us of the dangers of over-simplifying perspectives in general, and over-generalising about building users as though they are a homogeneous group (see for example the cautions offered by Luck[20] in an entirely different context). In episode fifteen in A1 (1.15 in Table 2) there is a series of exchanges between Adam and Anna over how to achieve a spiritual feeling using light. Anna argues the case for the proposed design, explicitly setting herself in opposition to other building users. Talking of patterns in stained glass that might achieve the effects she and the architect seek, she says: "it can't be sort of too twee because it just wouldn't go" (A1, 1376) and "I mean it won't work with a design like this" (A1, 1381). This supporting argument is offered at a moment when the architect's own reason for the same design decision is weakly grounded in appeal to (his own) authority: "it's not in my nature to do that sort of thing" (A1, 1380). Anna sets herself firmly in the architect's camp saying: "you know that's [where other building users] might spoil it for us" (A1, 1384). Her use of 'them' – the other building users – and 'us' to position herself is a rare use of this relational modality cue in the two meetings.

20.
Luck, R. (2003) Dialogue in Participating Design, *Design Studies*, 24, pp. 523–535.

In the second meeting, during the episode where provision of space for wreaths is discussed (starting at A2, 190), Adam concludes the exchanges by checking whether what he and his colleague Tony have proposed as a design revision is acceptable, saying: "so you you'd like us to do that put in some additional bays in there" (A2, 234–235). There is a query from Charles about whether the bays will be covered, at which point Anna comes in with concerns over compromising the design: "well it would be nice but I don't know whether that would cause problems with the design at all" (A2, 239–240). There follows a brief exchange between the architects over what is and is not covered, culminating with Adam reaffirming: "we could put a cover over that I think we could extend this line and cover that" (A2, 246–247). The use of 'I think' reduces the categorical strength of his reassurance, so again, Anna returns to architectural concerns, seeking further assurance with: "would that cause any problems at all for your design at all" (A2, 248). Here again, she shows concern to maintain the integrity of the design, this time not by conflating her interest with that of the architect (no 'we' here, she refers to the design as 'your' design), and by this question she opens up the possibility for the architect to (re)introduce a counter-argument to the introduction of more bays, on the grounds of design integrity. Luck's analysis[21] also recognises that Anna is 'attentive to not compromise the design' when raising concerns which imply possible modifications to the design.

21.
Luck, R. (Chapter 13) 'Does this compromise your design?' Socially Producing a Design Concept in Talk-in-Interaction.

There are many examples throughout the data where building users express 'requirements' as properties of the design 'solution', not as design 'problems' or requirements to be 'solved'. To give just a few examples:

- Users' concerns with spatial relationships (for example with lines of sight within the chapel (A1, 397–403) and the visibility of entrances (A1, 371–375));
- Concern over the balance in scale between landscape design and the proposed new chapel: "there's got to be a certain amount of balance between the building and the (landscape) design features" (A1, 1611);
- Integration of the existing buildings with the new one (A1, 2038–2043);
- Concerns with light and its contribution to perceptual, phenomenological and symbolic elements of the scheme;
- The role of clean lines, simple and uncluttered interiors in contributing similarly.

All of the above are used by the building users in their own arguments for design elements. They use these arguments unselfconsciously as part of their legitimate concerns as clients and building users. This highlights again that arguments for design features draw on a complex web of functional and aesthetic concerns and spatial appreciation, reminding us not to mistake lack of fluency with formal architectural vocabulary for lack of architectural sensibility.

4.3 *Collaborating to justify agreed design decisions*

Negotiation does not necessarily imply opposing positions or a dispute to be settled, but an orientation towards shared objectives[22]. In Sections 4.1 and 4.2, we have seen some features of the conversation that show subtle negotiation of a justifiable course of action.

22.
Wheatley (1995), *op. cit.*

Here two examples are presented where the course of action is agreed (a conclusion is reached) and the collaboration centres on arriving at robust justifications of the decision which satisfies each participant. In both of these examples what we see is the architect's professional knowledge of space combining with the building users' lived understanding of place.

In phase three of the second meeting during a monological presentation of the sectional perspective of the building (see Table 3), Adam gives a technical description about the lighting effects realised by the sectional profile of the chapel:

> "the chapel itself will be top lit err and there will be a roof light over the top of the central strip of natural day lighting [...] daylight will bounce its way in through the roof and be reflected off the underside of the vault [...] this diffuser is also doing a number of other jobs [...] the artificial light source will end up coming from [...] the same direction as the natural light" (A2, 786–800).

Anna takes the opportunity to ask about the colours of the artificial light. Adam and Anna exchange ideas about the use of coloured glass. During this exchange Anna volunteers her experience of the site itself in supporting the case for using coloured glass. This is a decision to which the architect is not opposed; they are already in agreement. Adam says: "we're with you one hundred percent" (A2, 816). Nevertheless, Anna volunteers support for his abstract argument from her concrete experience of working in the existing building on the same site as the one being designed. Her offer to justify what is proposed comes from her intimate knowledge of that particular place. She says: "the sun comes up this way and sets this way, so it would be sort of erm that would be you know quite nice" (A2, 817–818).

In the second example, the architect again draws on his professional expertise, this time not with technical input but via reference to an exemplar[23]. He draws on the work of Le Corbusier to support his case: "if you want to we could puncture the wall with some more ++ holes if you like what I'm thinking of is like Le Corbusier's chapel at Ronchamp" (A1, 492–493). After an exchange for clarification and further description by Adam, Anna contributes a different supporting argument for the same outcome, again based on her experience at the site, with: "yes that would also be nice [...] the water [...] would also make the light move [...] through the glass" (A1, 503–506).

23. Lawson, B. (2003) Schemata, Gambits and Precedent: Some Factors in Design Expertise in Cross, N. and Edmonds, E. (eds) *Design Thinking Research Symposium 6:* Expertise in Design, Sydney, pp. 37–50.

4.4 *Deferring to expertise and assertion of expertise*

In Sections 4.1 to 4.3, a series of examples have been given to illustrate the collaborative way design progression is negotiated.

Only a small selection has been given, it is offered in the context of the interaction as a whole (Section 3 above) and it is important to make it clear that neither building users nor architects are reticent about asserting their own expertise and assuming their professional role in the meetings. Their interactions explicitly draw out the roles of professional architect and user/client expert in each other. As Oak[24] expresses it, 'the participants competently perform as 'architect' and 'client' through talk'. The most obvious form that information exchange takes is simply through asking questions of each other. In the first meeting in particular, this agenda for interacting is set out at the outset by the architect (as we have noted above in Section 3: "to obtain your feedback to the design" (A1, 75–76)). During that meeting Adam raises issues about the design and makes it plain that he is seeking both instruction and requirements expertise from his client-users.

24. **Oak (Chapter 17) *op. cit.***

Exchanges of information (drawn to attention in column 5 of Table 2 for meeting A1) show the prominence of questioning and answering, and the assumptions about authority which these exchanges create and reinforce. An example from A1, episode 4.2 (Table 2) proceeds as follows. Charles, looking at a sample of stone, asks: "is this a facing material" (A1, 2253). This initiates an exchange about material for external and internal walls, in which the decisions are clearly claimed by the architect unapologetically – and unchallenged. He says: "for internal I haven't quite made up my mind [...] I might want a plain white block" (A1, 2299–2300) and: "that's how I see the interior as a smooth block face so it might be smooth internally and coarse externally" (A1, 2311–2312). Adam does not use any masking devices such as an impersonal tense, for example. There are many further examples in both meetings where the building users defer to the architect in this way. Towards the end of meeting A2, for example, again on the topic of materials, Anna asks: "can I ask do we have any thoughts on what the actual outer material will be or is that still too early at the moment" (A2, 1359–1360). Despite the use of 'we' here, the question shows deferral to the architects' expertise on both process issues (when to decide things), as well as technical and aesthetic matters – here the visual and functional properties of the external materials.

Here we have only drawn on the more obvious markers indicating willingness and expectation of both architect and building users to defer to the other's expertise. There are potentially more subtle indicators. Oak[25] observes of these meetings that the language in which information is sought does not necessarily correspond to the form in which it is provided. She interprets this sort of evasive response as a possible deferral strategy (giving as an example the apparent unwillingness of Anna to stipulate room dimensions). Glock[26] proposes that deferral may also be signalled by some cases of giving the preferred answer, but with a hesitation.

25. **Oak (Chapter 17) *op. cit.***

26. **Glock (Chapter 16) *op. cit.***

5 CONCLUSIONS

The data analysed comes from the 'swampy ground' of everyday professional design practice[27]. It is not drawn from the architectural practice that belongs in glossy architectural magazines or art galleries, where architects design to impress and compete against each other[28]. Neither does it belong in the equally rarefied atmosphere of those collaboratively designed projects where the political stakes are so high, the benefactor is so wealthy, or the design is so unconstrained that the comprehensive participatory engagement of emancipated parties is possible[29].

What does a close reading of the material of two meetings from the mundane world of designing buildings for clients invite us to conclude? The first thing we notice is how immensely rich the material is as a resource for design research. There are many avenues for analysis, as the chapters in this volume show. In the work described here only a few limited aspects of the events have been explored. Given the direction taken, the data shows that when we examine the talk in these meetings, we can see that *a priori* designations of the roles of building user, client, designer, and so on play their part, but that they are also to some extent continually negotiated during conversation; to some degree they are emergent features of the social interaction. Oak[30] makes this phenomenon the centre of her attention. As studies in other situations have shown, roles are both an outcome of and a rationale for various social situations; roles are something that are assumed, they are not inherent properties of an individual[31].These roles and their relationship to one another are complex and their epistatic nature is revealed in verbal interactions. Negotiating positions, and contributions to the negotiations, shift during conversation as meeting participants collaboratively establish what is to be done and how collectively they will justify what they agree to do.

Other simple polarities are also suspect. We would do well neither to over-privilege nor to under-rate expertise wherever it resides (in building users or building designers). Recognition of the expertise of others, and assertion of expertise when appropriate, is a practical way to get things done; it can be viewed as a consensual act without implying power inequality. Building designers have experience of using buildings. In interaction, opportunities for expressing what little they do know may act as a starting point for drawing out information from 'expert' building users that can then inform design. Conversely, building users are capable of symbolic, aesthetic and further 'other-than-function' appreciations of the properties of space, and here we have seen that they can assume a custodianship role towards both a design concept and preservation of design integrity. The ability to distinguish clear boundaries between functional elements of the design and others should be

27.
Schon, D. (1987) *Educating the Reflective Practitioner*, Jossey-Bass, Oxford.

28.
Mitchell, C.T. (1993) *Redefining Design: From Form to Experience*, Van Nostrand Reinhold, New York.

29.
See for example Day (2003), Blundell Jones, Petrescu and Till (2005), and Sanoff (2007)
Day, C. (2003) *Consensus Design: Socially inclusive process*, Architectural Press, Oxford.
Blundell Jones, P., Petrescu, D. and Till, J. (eds) (2005) *Architecture and Participation*, Spon Press, Oxford.
Sanoff, H. (2007) *Design Studies* (Special Issue on Participatory Design) 28(3).

30.
Oak (Chapter 17) *op. cit.*

31.
West, C., Lazar, M., and Kramarae, C. (1997) Gender in Discourse, in Van Dijk, T. (ed) *Discourse as Social Interaction 2*, Sage Publications, pp. 119–143.

questioned as should notions that we might be able to associate functional, perceptual, phenomenological and symbolic user needs with distinct features of the design and the arguments made for them. To understand the complex web of relationships we need to take account of the social expectations of all those contributing to a design.

Design representations, such as plans, sectional drawings, perspective drawings, sketches and so on, have received the attention of many researchers as objects for design practitioners to think with individually and communicate with collectively. Their values and shortcomings for communicating with clients, users and other non-designer groups have been studied extensively. Here, with a focus on conversational interaction, not withstanding any shortcomings as a means of communicating design intentions they may have, the drawings provided an external, common reference point for organising systematic consideration of issues that needed to be talked about, without imposing stringent control over what might be discussed.

This study focused on the verbal negotiation of what to do next in a design process which made no claims to be collaborative or participatory. However, by paying attention to what the talk in the meeting achieved, the study reveals collaboration occurring at different levels of granularity oriented towards deciding how to move the design along. A sense of satisfactory engagement in the design process and 'ownership of the design' by building users may accrue from collaborative negotiation of design contributions and collective agreement over what the justifications are for these decisions. In everyday, non-participatory architectural practice understanding the nuances of these more subtle engagements with designing may be a way to account for effective meetings between architects and those who will inhabit the buildings they design.

15
The Function of Gesture in an Architectural Design Meeting

Willemien Visser

This chapter presents a cognitive-psychology analysis of spontaneous, concurrent speech gestures in a face-to-face architectural design meeting (A1). The long-term objective is to formulate specifications for remote collaborative-design systems, especially for supporting the use of different semiotic modalities (multi-modal interaction). According to their function for design, interaction, and collaboration, we distinguish a number of gesture families: representational (entity designating or specifying), organisational (management of discourse, interaction, or functional design actions), focalising, discourse and interaction modulating, and disambiguating gestures. Discussion and conclusions concern the following points. It is impossible to attribute fixed functions to particular gesture forms. 'Designating' gestures may also have a design function. The gestures identified in A1 possess a certain generic character. The gestures identified are neither systematically irreplaceable, nor optional accessories to speech or drawing. We discuss the possibilities for gesture in computer-supported collaborative software systems.

This chapter presents a cognitive-psychological analysis of a design meeting, focusing on the function of gestures in collaborative design. We are interested by all the gestures that designers may use to elaborate on their project. This choice only excludes from our analysis signed gestures, commonly used instead of speech by deaf or hard of hearing people to convey meaning.

To examine the function of the gestures we analysed them in the context of other semiotic modalities[1]. However, for reasons of space here, we do not discuss either gesture-related expression forms such as posture, facial expression, gaze, or paralinguistic and prosodic aspects of speech, nor hand-made actions such as drawing. Given the scarce knowledge about gesture in collaborative design, this selection already provides us with a considerable amount of material.

Our investigation is the first stage of a larger cognitive-ergonomic analysis. Its long-term objective is to identify the consequences that knowledge about gesture in face-to-face design meetings has for remote collaborative-design environments, especially with respect to different interaction modalities and their support.

The present analysis continues the approach of our work in cognitive design research since the 1980s[2]. We don't use in our analysis

1. Goodwin, C. (2002) Multi-modal Gesture in *The Living Medium: First Congress of the International Society for Gesture Studies*, The University of Texas at Austin, http://tinyurl.com/4vv8cu (accessed September 2008).

2. Visser, W. (2006a) *The Cognitive Artifacts of Designing*, Laurence Erlbaum.

pre-existing theoretical entities such as the design 'task' or 'process' invoked in design methods and prescriptive models of design. Instead, we focus on the dynamic aspects of design activity during designers' actual work on a design project. Glock[3] adopts a similar focus, looking in detail at the way in which design is done in practice.

3.
Glock, F. (Chapter 16) Aspects of Language Use in Design Conversation.

1 GESTURE IN COLLABORATIVE DESIGNING

1.1 *Gesture research*

Gesture studies have been conducted in the fields of semiotics, ethology, dance and choreographic research, psychology/psycholinguistics, and pragmatics. These have focused mainly on the use of gesture in everyday conversation. The use of gesture in goal-oriented, professional activities has received far less attention. There are some notable exceptions, however, such as studies by the Work, Interaction and Technology research group at Kings College in London led by Christian Heath[4,5,6,7,8]. In the cognitive ergonomics of design, however, the analysis of gesture has only just begun.

4.
Heath, C. and Luff, P. (2007) Gesture and Institutional Interaction: Figuring Bids in Auctions of Fine Art and Antiques, *Gesture*, 7, pp. 215–240.

5.
Hindmarsh, J. and Heath, C. (2000) Embodied Reference: A Study of Deixis in Workplace Interaction, *Journal of Pragmatics*, 32, pp. 1855–1878.

6.
Goodwin, C. (2003) Pointing as Situated Practice in Kita, S. (ed) *Pointing: Where Language, Culture and Cognition Meet*, Laurence Erlbaum, pp. 217–241.

7.
Koschmann, T., LeBaron, C., Goodwin, C. and Feltovich, P. (2006) The Mystery of the Missing Referent: Objects, Procedures, and the Problem of the Instruction Follower in *Proceedings of the 20th Anniversary Conference on Computer Supported Cooperative Work*, pp. 373–382.

8.
Mondada, L. (2002) Describing Surgical Gestures: The View from Researcher's and Surgeon's Video Recordings, in *The Living Medium: First Congress of the International Society for Gesture Studies*, University of Texas at Austin, http://tinyurl.com/534s2a (accessed September 2008)

1.2 *Gesture in design meetings: previous studies*

In an early collaborative design study of a remote control handset, Tang[9] identified that the 'hand-made actions' of writing, freehand drawing, and gesturing had three functions: to 'store information', to 'express ideas', and to 'mediate interaction'. A full third of such hand-made actions were gestures. Only a limited number of these actions were classified as serving an information-storage function, while expression of new ideas was performed as much by gesture as by drawing and writing. However, more than half of all gestures served interaction mediation, considerably more than the other hand-made actions.

9.
Tang, J.C. (1991) Findings from Observational Studies of Collaborative Work, *International Journal of Man-Machine Studies*, 34, pp. 143–160.

Bekker, Olson, and Olson[10] distinguished four types of gesture in the design of an automatic post office: kinetic (action execution – which we do not consider as gesture), spatial (indication of distance, location, or size), pointing, and other (with an emphasis on verbal utterances and attracting the attention). They identified design, management, and overall conversation regulation as the purposes for which gesture was used. All combinations of types of gestures and functions occurred.

Studying a team of architects, Murphy[11] analysed their designing in terms of 'collaborative imagining'. He noted the use of gesture for imagining characteristics of design entities – particularly their motion, structure, and functioning – and of user experience – actions performed on the design entities themselves.

In our own gesture-related research we have developed a description language for graphico-gestural design activities[12,13,14] We have used this language, combined with COMET, our method for analysing collaborative design[15], to describe and analyse the verbal and graphico-gestural dimensions of interaction in situations such as an architectural design meeting. We have interpreted co-designers' graphico-gestural actions according to their functional roles in the project or the meeting[16]. We have also examined different forms of multi-modal articulation, between graphico-gestural and verbal modalities, in parallel interactions between the designers. Activities taking place concurrently revealed either alignment or disalignment between the designers regarding the focus of their activities[17,18].

We wish to emphasise two functions of gesture shown by these previous studies. First, gesture offers specific possibilities to render the spatial (especially 3D) and motion-related qualities of entities, and to embody action sequences through their mimicked simulation[19]. Second, gesture plays an important interactional or, as we define it below, organisational role.

2 DATA ANALYSIS: USE OF GESTURE BY DESIGNERS IN AN ARCHITECTURAL MEETING

In order to continue and broaden our previous work on gesture in collaborative architectural design, we chose the architectural meetings from the DTRS7 data. Given the explorative nature of our study, and the time-consuming character of gesture analysis, we selected only one meeting, the first architectural design meeting (A1).

2.1 *Data analysed: architectural meeting A1*

The three participants in the meeting were Adam, the architect in charge of the project, and two clients, Anna, registrar of the crematorium, and Charles, whose precise function was not explicit. A DTRS7 organiser (Peter) was also present. The project concerned the design of a crematorium to be built on a site alongside an existing crematorium.

Even if different roles can be distinguished among the participants, we consider that all three participants are 'doing' design. We analyse design not only as the activity of somebody whose profession is a 'designer', but also as a more general cognitive activity.

> "Design consists in specifying an artefact (the artefact product), given requirements that indicate—generally neither explicitly, nor

10.
Bekker, M.M., Olson, J.S. and Olson, G.M. (1995) Analysis of Gestures in Face-to-Face Design Teams Provides Guidance for how to use Groupware in Design in *Proceedings of Designing Interactive Systems 1995: Processes, Practices, Methods and Techniques*, Ann Arbor, MI, pp. 157–166.

11.
Murphy, K.M. (2005) Collaborative Imagining: The Interactive Use of Gestures, Talk, and Graphic Representation in Architectural Practice, *Semiotica*, 156, pp. 113–145.

12.
Détienne, F. and Visser, W. (2006) Multimodality and Parallelism in Design Interaction: Co-designers' Alignment and Coalitions in Hassanaly, P., Herrmann, T., Kunau, G. and Zacklad, M. (eds) *Cooperative Systems Design: Seamless Integration of Artifacts and Conversations-Enhanced Concepts of Infrastructure for Communication*, IOS, Amsterdam, pp. 118–131, http://tinyurl.com/5tkc99 (accessed September 2008).

13.
Détienne, F., Visser, W. and Tabary, R. (2006) Articulation des Dimensions Graphico-Gestuelle et Verbale dans l'analyse de la Conception Collaborative, *Psychologie de l'Interaction (Numéro spécial 'Langage et Cognition: Contraintes Pragmatiques')*, 21–22, pp. 283–307.

14.
Visser, W. and Détienne, F. (2005) Articulation Entre Composantes Verbale et Graphico-gestuelle de L'interaction dans des Réunions de Conception Architecturale in *Actes de SCAN'05, Séminaire de Conception Architecturale Numérique: Le Rôle de L'esquisse Architecturale dans le Monde Numérique*, Charenton-le-Pont, France,

completely—one or more functions to be fulfilled, and needs and goals to be satisfied by the artefact, under certain conditions (expressed by constraints). At a cognitive level, this specification activity consists of constructing (generating, transforming, and evaluating) representations of the artefact until they are so precise, concrete, and detailed that the resulting representations—the 'specifications'—specify explicitly and completely the implementation of the artefact product."[20]

2.2 Description scheme: The function of gestures in design meetings

Not every hand or arm movement is a gesture. There is a difference between gestures and actions[21]. Actions have a practical aim: they transform the material state of the world and are, in principle, without any meaning (even if bodily actions can also be analysed as contributing to the construction of meaningful structures). Gestures, on the contrary, convey meaning: they have a symbolic function and transform the cognitive state of the addressee.

The gestures analysed here are firstly 'gesticulations'[22], spontaneous, speech-accompanying gestures[23]; and secondly, 'emblems', quasi-linguistic, lexicalised gestures, with conventional forms and meanings, and which don't necessarily accompany speech.

There are many different classifications of gestures. The most well-known ones, adopted by many authors, are those proposed by Kendon[24] and by McNeill[25]. These classifications combine the form and the use of gestures. Two classical examples are deictics – pointing, surrounding, or covering gestures, generally considered to designate entities, and iconics – gestures that bear a physical resemblance to an entity that are generally considered to convey the entity. Our analysis is guided by the functions of the activities that the gestures are performing or supporting, not by their form. As we will show, a particular type of gesture may take various forms, and a particular gestural movement may fulfil different functions. A gesture may be 'iconic' in McNeill's[26] classification, but we also ask ourselves: why does a designer use this gesture in this meeting? Table 1 presents the five gestural families and the sub-families distinguished in this chapter as a result of our analysis (detailed in the following section).

Since we analyse design as the construction of representations[27], our representational gestures correspond to Bekker, Olson, and Olson's 'design'[28], and Murphy's 'imagining'[29] activities. Naively, one might expect that, in order to develop the representation of the artefact, designers would elaborate on the artefact's qualities. They do indeed do this, but our observations also show that they often explicitly introduce the design entity concerned by their

http://tinyurl.com/6z73pp (accessed September 2008).

15.
Darses, F., Détienne, F., Falzon, P. and Visser, W. (2001) *COMET: A Method for Analysing Collective Design Processes* (Research report INRIA No. 4258), Rocquencourt, France: INRIA, http://tinyurl.com/3panbc (accessed September 2008).

16.
Détienne, Visser and Tabary (2006), *op. cit.*

17.
Détienne and Visser (2006), *op. cit.*

18.
Visser and Détienne (2005), *op. cit.*

19.
Tversky, B. and Lozano, S.C. (2006) Gestures Augment Learning in Communicators and Recipients in Coventry, K., Tenbrink, T. and Bateman, J. (eds) Spatial Language and Dialogue, Oxford University Press, http://tinyurl.com/46r3nl (accessed September 2008).

20.
Visser (2006a), *op. cit.*, p. 116.

21.
Grosjean, M. and Kerbrat-Orecchioni, C. (2002) Acte Verbal et Acte Non-verbal, Ou: Comment le Sens Vient aux Actes, paper presented at *Les relations intersémiotiques Conference*, Lyon, France.

22.
Kendon, A. (2004) *Gesture: Visible Action as Utterance*, Cambridge University Press.

23.
McNeill, D. (ed) (2000) *Language and Gesture*, Cambridge University Press.

24.
Kendon (2004), *op. cit.*

25.
McNeill (2000), *op. cit.*

26.
ibid.

27.
Visser (2006a), *op. cit.*

Gestural families	Sub-families	Sub-sub-families
Representation	Designation	Identification
		Qualification
		Comparison
	Specification	
Organisation	Discourse and interaction management	Management of one's own discourse
		Management of others' interaction
	Functional design-action management	
Focalisation		
Modulation		
Disambiguation		

Table 1. Families of gestures distinguished in this chapter.

representational activities (object, process, action, state, or one of its components). We distinguish such designational activities from the representational activities that (further) specify the entity.

Designating gestures point out an entity (the designatum) to one's co-participants. Their form is most frequently deictic. The designation may be more or less global: deictics vary from pointing with a finger to waving with a hand. Designata can be various: static physical objects, directions or movements in space, or moments in time. Deictics are generally considered to be the most typical gestures and have received much attention, both in pragmatics and in computer-system and HCI research (this is also the main type of gesture discerned by Glock[30] in his analysis of meeting A1).

In order to specify gesturally an entity, people physically display one or more of its qualities. They may use illustrative gesticulations[31] or emblems. Contrary to the deictic designating gesture, an illustrative gesture resembles its designatum: one must look closely at the gesture, because the designatum is, at least partially, 'in' it.

As most interaction between co-participants in a design meeting is, or serves, design in a more or less direct manner[32], we do not treat interactional activities separately, we only distinguish their management. Different authors have analysed and qualified management gestures in different ways: Tang[33] refers to 'interaction mediation', while 'management' and 'overall conversation regulation'[34] have also been used. We use the term 'organisational' to refer to gestures that cover, on the one hand, the management of one's own discourse and of the interaction between the co-participants, and on the other hand, the planning and organisation of the functional design actions (for example to propose to proceed first to brainstorming and only afterwards to the formulation of critique).

28.
Bekker, Olson and Olson (1995), *op. cit.*

29.
Murphy (2005), *op. cit.*

30.
Glock (Chapter 16), *op. cit.*

31.
Cosnier, J. and Vaysse, J. (1997) Sémiotique des Gestes Communicatifs, *Nouveaux Actes Sémiotiques*, 52, pp. 7–28.

32.
Visser (2006a), *op. cit.*

33.
Tang (1991), *op. cit.*

34.
Bekker, Olsen and Olsen (1995), *op. cit.*

To organise their own discourse, people often use 'beats'. Beats are gestures whose form is shaped by one or two hands that move along with the rhythm of speech (as a music director conducting her orchestra). They structure verbal discourse by accentuating certain parts of it. What distinguishes beats from both pointing designating and illustrative gestures is that they tend to have the same form regardless of the discourse content.

Three types of gesture are not specific to collaborative design, and exist in almost any interaction: focalisation, modulation, and disambiguation. Focalisation serves to make one's addressees focus on a fraction of the world that one considers critical – be it 'outside' or within a particular discourse. Focalisation always combines with one or more other functions. Each representational gesture focuses. We show below how it combines with designation and with organisation.

Modulation has an expressive function. Discourse or interaction components can be modulated, or emphasised, generally with an emotional loading.

Gestures are not simple illustrations of verbal discourse. They are generally associated with speech, but have their own contribution. They can also help to disambiguate a discourse or interaction component expressed using another semiotic system.

Our description scheme is preliminary with respect to the range of gestures used in meeting A1. We did not describe all gestures, but rather tried to identify the different types of gestures performed during A1, grouping them according to their function for design, interaction, and collaboration. As a result our analysis is neither an exhaustive list of gestures, nor a quantitative indication of gesture.

3 RESULTS: GESTURES MADE BY THE DESIGN PARTICIPANTS

3.1 *Formalism used in the meeting extract examples*

We will describe the physical realisation of certain gestures ('gesture movements'), relating them to the spoken communication in the meeting by adding two or three additional lines under each transcription line. The first of these lines of annotation indicates, between square brackets, the identity of the gesturer and the duration of the gesture with reference to that of the corresponding portion of the transcript, for example [g_Anna_____]. The second and, if necessary, a third line provide a brief description of the gesture movement (see Extract 1).

Extract 1, A1, Formalism used for describing gestures.

352 Anna they can just go in side by side but it's difficult to squeeze in to
353 put the coffins on at the moment even because you've also you've got the
[g_Anna________][g_Anna________
draws apart her hands
opens each hand
354 two catafalques in side by side and you need to have four routes for
________________________][g_Anna____________________
into a ball-enclosing shape *spreads apart two fingers of each hand*
355 people to go either you need the one in the middle for both people to go
____________]
and advances them over two separate parallel tracks

In Extract 1, when saying: "even because you've", Anna draws apart her hands. Then, during: "also you've got the two catafalques in side by side", she opens each hand in a grasping movement, as if she encloses a ball with each hand; finally, saying: "and you need to have four routes for people to go", she spreads apart two fingers of each hand and advances with these fingers over two separate parallel tracks.

3.2 *Representational gestures: Designation and specification*

3.2.1 *Designation: An identificatory, qualificatory, or comparative orientation, combined with a focusing goal*

We observed that designating gestures are generally identificatory: gesturers designate entities in order for their co-participants to be able to identify them. Gestures were used in a qualificatory manner – even if qualification generally relied more on intonation, facial expression, or posture than on gesture. Gestures were also used comparatively: in contrast to: "that chapel" (first pointing gesture), "the original concept of this chapel" (second pointing gesture) "was that the bier … would actually come out and meet the hearse" (A1, 245–246). Such a comparative function was generally combined with an identificatory function. All three functions combined with a focusing use of the gestures in question.

3.2.2 *Spatial deictics*

In order to confirm our observation, when first viewing A1, that spatial deictic gestures were relatively frequent, we took a closer look at the first hour of the meeting and found that out of the 215 verbally expressed spatial deictics more than 200 seemed to be accompanied or completed by a deictic gesture.

We observed the use of spatial deictics in various ways: indirectly designating an entity through pointing to the location where the entity was situated, or directly designating a direction or a sequence of things.

For example in Extract 2, Adam indicates three possible routes for entering the cremator area, tracing with his finger, over the plan, three lines indicating three different paths.

Extract 2, A1, Spatial deictic gesture designating three itineraries.

1134	Adam	obviously there are numerous ways of getting into this accommodation
1135		that's route number one the second route is round the end of the pond
		[g_Adam______________________________________
		tracing with a finger consecutively three lines over the plan
1136		and the third route is through the chapel so there's numerous ways of
		_____________________________]

3.2.3 *Specifying an entity: Displaying its qualities*

Some examples of qualities used to gesturally specify entities were: size, movement, luminousness, boldness, importance, intimate character, private character, calmness, and spiritual atmosphere. Both physical and abstract qualities were specified, the former through 'iconic' gestures, the latter using 'metaphoric' illustrative gestures; this distinction is a relative one, gestures being more or less iconic[35]. Furthermore, it is not the form of a gesture, but its relationship with the designatum that makes the gesture iconic or metaphoric. Iconic gestures bear a physical resemblance, metaphoric gestures a metaphoric relation to the entity they are meant to convey.

35. Krauss, R. M., Chen, Y. and Gottesman, R.F. (2000) Lexical Gestures and Lexical Access: A Process Model in McNeill, D. (ed) *Language and Gesture*, Cambridge University Press, pp. 261–283.

Adam's 'tracing of line' gestures in Extract 2 were both iconic and metaphoric; both the routes and the perspective being represented as lines.

In Extract 3, Adam moves apart his arms to ask Anna about the width necessary to make two coffins pass into the chapel side by side.

Extract 3, A1, Gesture specifying a physical quality.

368	Adam	OK so my question for you is how wide would it need to be for two
369		coffins or if we're going for two it would need to be

In Extract 4, Adam proposes a new door, and possibly a new window, for the audio-visual technicians to be able to monitor the funeral services that precede and follow the one in progress. He points to the location where he plans this door and draws it. Charles, following with his finger the direction of view, gesturally 'sees' what becomes possible.

Extract 4, A1, Gesturally 'seeing' a possible direction of view.

635	Adam	the answer is then to have a door there
636	Charles	a door
637	Adam	maybe a window
638	Charles	a window
639	Adam	and they can
640	Charles	and a window this way

3.2.4 *Asserting a quality or asking for one*

Instead of gesturally attributing a quality to an entity, a participant may also gesturally enquire about an entity. In Extract 3, Adam asks his client about the required width of a passage. A request has been identified in combination with a specification. Obviously, other representational and organisational expressions can also be combined with these two enunciative modalities (for example, Charles questioning his co-participants concerning seating, referred to below).

3.3 *Organisational gestures*

Anna, in her answer to Adam, who asks whether she sees any need for a particular alteration (A1, 299), Anna organises – and underlines – her discourse with beats (Extract 5). To do this, she performs a sequence of hands opening and hands closing movements, accompanied with a hands rising and/or hands advancing movement. Considering a sequence of hands opening and closing, combined with a rising and/or advancing movement as one beat, Anna performs some 13 beats during this 5 second utterance.

Extract 5, A1, Organising one's discourse using beats.

300	Anna	not no as long as there's a that's what they're I mean if we have a sort of
301		consultation meeting with them that's what they will be interested in

Interaction management gestures may moderate co-participants' turn-taking by using a stop-sign (an emblem) or call co-participants' attention by waving a hand, or lifting a finger. An example of this is Charles who, in order to focus the attention of his co-participants on a question concerning seating, points to the seats asking: "are we going to have fixed seating" (A1, 392).

The following example illustrates several qualities that may be encountered in gestural (or verbal) expression – though not necessarily coupled as they are presented here. After a long intervention by Anna (A1, 174–194), Adam refocuses the topic: "yes + so having got this this far we've got a bigger waiting room" (A1, 197).

While saying this he points with his pen to the waiting room on the plan: that is, to the design entity concerned in his following discourse. In doing this he establishes a transition from Anna's discourse to the next topic that he presents: "having got this far" (A1, 197). This leads to a number of observations.

- Adam uses two associated deictic expressions (gestural and verbal) to 'point' where they have reached in the meeting, that is, to make a designation in the notional domain.
- Adam's gesture precedes the associated verbal deictic expression "this far" (A1, 197), gestures do not necessarily follow their associated verbal expressions.
- There are other occurrences of organisationally used deictic expressions in the meeting. For example Adam saying: "this leads us to the first query I have" (A1, 90–91) and "OK so having got this far" (A1, 323).
- Adam's gesture is also an emphatic modulating gesture – something useful for an expression with an organisational purpose.
- Besides its interaction-organisational function, Adam's gesture also serves a design-action management function. By refocusing the topic, he aims to make his co-participants change their collective design action.

3.4 *Focalisation*

Various examples of the focusing use of gestures have been presented above, in combination with designation and with organisation.

3.5 *Modulating gestures*

Extract 6 shows how Anna presents a book in which the author ("she" (A1, 34)) mentions the already "existing" crematorium. Anna had wanted to order the book, but encountered many administrative problems, so bought it herself. In saying this, she both highlights content elements (partly through using beats) and expresses their emotional loading, using different types of gestures (deictics, iconics, metaphorics). Pointing to the book (A1, 34), Anna uses the deictic gesture to highlight that: "[they]'re mentioned in the book". Through frequently made long continuous hands opening, advancing metaphoric gestures (A1, lines 36, 37, 40, 41, 42, 43, 45–46), and a sequence of detached short beats (A1, 40–41), she underlines discourse components. Enacting how one holds a car steering wheel (A1, 46) and suddenly changes direction (A1, 47), she uses iconic gestures to highlight her idea that people, normally, do not want to visit cemeteries during their holidays.

Extract 6, A1, Modulating discourse components using various forms of gesture.

34 Anna yes she's been to have a look at our existing we're mentioned in the
[g_Anna__________
points to the book

35 book quite a bit with the existing chapel she was quite impressed
_____]

36 because we've also got quite a lot of photographs and other things and
[g_Anna____________________]
long continuous hands opening, advancing gestures

37 plans like the forward plan of extending the chapel originally the original
[g_Anna____________________________]
long continuous hands opening, advancing gestures (LCHOAG)

38 idea so that was quite forward thinking in nineteen eighty or seventy

39 whenever seventy eight when they decided on that + so I've sent off for

40 that but you try and get them to raise a SAP order for it and you get how
[g_Anna__________________________] [g_Anna___
LCHOAG *sequence of*

41 often will we need the SPIRE BOOKS COMPANY I said well we'll never use
_________________________________] [g_Anna_______]
detached short beats *LCHOAG*

42 them again probably why do you need to do that oh for god's sake
[g_Anna__________]
LCHOAG

43 so I just went out and bought it and the one above as well a history
[g_Anna_________________]
LCHOAG

44 of cremations that's what we have at home on the bookshelf

45 cheerful reading like you when you go on holiday with your part- your
[g_An
LCHOAG

46 husband and wife + you see a sign for crematorium straight away
na_____________] [g_Anna______]
enacting how one

47 ignoring them whether you're abroad or in this country the
[g_Anna__________________________________]
holds a car steering wheel
enacting how one suddenly changes direction

48 first thing you do straight away oh not another cemetery they go oh dear

3.6 *Disambiguation gestures*

While the co-participants are discussing cars and how coffins are off-loaded from the cars, Anna refers to: "the doors" (A1, 290; A1, 315).

Given the car-centred character of the interaction sequence and the absence of any clue signalling that a new topic is being introduced, the doors of "the cars" are probably a reference expected by the other co-participants. Yet Anna's gesturing over the doors of the waiting area representation on the plan probably disambiguates their actual reference.

4 DISCUSSION

4.1 *Attributing fixed functions to particular gesture forms*

In this chapter *designation* was presented as mostly performed through deictics, and *specifying* mostly through iconic and metaphoric illustrative gestures and emblems, however we also saw that deictics can have other uses and specification can be expressed differently. Indeed, there is no simple correspondence between gesture forms and functions.

Besides their identificatory, qualificatory, or comparative orientation, designating gestures also have a focusing function. Other instances of this multifaceted nature of gesture were present in our analysis as follows.

- Designating representational gestures having an interactional function.
- Deictic designating gestures also functioning metaphorically.
- Various types of gestures used to modulate the discourse.

4.2 *Designating or designing?*

Our presentation of designating gestures may have seemed to presuppose that the entities designated already existed before the meeting, designed by the architect who 'simply' presents them to his clients during the meeting. The architect, however, enunciates, gestures, and draws entities that he may be designing while he is conveying them to his co-participants. It is difficult, if not impossible, especially for an external observer, to distinguish between entities that already 'exist' and entities that are being designed while they are being designated. Designing is taking place continuously; the meeting is not an event where a finished project is being reported.

Regarding the use of the artefact by different users (crematorium personnel, funeral directors, family of the deceased), especially in space and in motion (people sit here, pass by there), architectural plans are not explicit, and thus provide no fixed instructions. It is entirely possible, for example, that the itineraries that Adam indicates as possible ways to enter the crematorium area (in Extract 2) are designed by him in response to, and interacting with, or in collaboration with, Anna, when she formulates the funeral directors' worries concerning access routes.

Our observations concerning gesture should also be situated in a wider context. We have stated previously that an essential part of collaborative design occurs in and through interaction:

> "the different forms that interaction may take in collaborative design—especially, linguistic, graphical, gestural, and postural—are ... not the simple expression and transmission (communication) of ideas previously developed in an internal medium (such as Fodor's 'language of thought'). They are more and of a different nature than the trace of a so-called 'genuine' design activity, which would be individual and occur internally, and which verbal and other forms of expression would allow sharing with colleagues."[36]

36.
Visser (2006a), *op. cit.*, p. 199

Several other authors in this volume adopt comparable positions. According to Luck[37], talk *is* action: it brings about 'real' actions in the world. McDonnell[38] focuses on conversation as a medium through which a design project advances and Oak[39] analyses how design-meeting participants create their roles through their talk.

37.
Luck, R. (Chapter 13) 'Does this Compromise your Design?': Socially Producing a Design Concept in Talk-in-Interaction.

38.
McDonnell, J. (Chapter 14) Collaborative Negotiation in Design: A Study of Design Conversations between Architect and Building Users.

39.
Oak, A. (Chapter 17) Performing Architecture: Talking 'Architect' and 'Client' into Being.

Concerning gesture specifically, recent research shows that it not only conveys information to one's addressees, but also affects the gesturer. In short it may help a person think[40] and learn[41].

40.
Lozano, S.C. and Tversky, B. (2006) Communicative Gestures Facilitate Problem Solving for both Communicators and Recipients, *Journal of Memory and Language*, 55, pp. 47–63.

41.
Goldin-Meadow, S. and Wagner, S.M. (2005) How our Hands Help us Learn, *Trends in Cognitive Sciences*, 9, pp. 234–241.

4.3 *The generic character of the gesture functions identified in A1*

In order to appreciate the generic nature of our results with respect to the possibly idiosyncratic nature of the analysed meeting, we also viewed engineering design meeting E1, a different type of design and at another stage in the design process. As in A1, both representational and organisational gestures were used in E1. Yet the gestural enactment of motion seemed particularly important in E1, more so than in A1. Interaction around the possible motions of E1's artefact, the digital pen, involved sequences of gesturally performed actions, often many small ones. Even if the use of both a building and a pen imply action, it may be harder to formulate verbally and easier to express gesturally the many, small movements involved in manipulating a pen, compared to the different ways of passing through a building.

The correspondence with the results of previous studies on design meetings (see Section 1.2) also substantiates the generic character of our results regarding at least two points: activities performed using gesture, and their objects. Gesture plays a role in both representational and organisational activities. It is used to specify both the characteristics of the design entities as well as aspects of user experience.

4.4 *The irreplaceability of gesture relative to other semiotic modalities*

Previous work on gesture in collaborative design confirmed the conclusion of classical gesture research[42, 43] that gesture has a unique role and often cannot be replaced by other semiotic modalities. Anticipating the next stage in our longitudinal project, we start a discussion about the degree to which gesture is irreplaceable relative to other semiotic systems.

42.
Kendon (2004), *op. cit.*

43.
McNeill (2000), *op. cit.*

Certain gestures in A1 seem to add nothing to the verbal expression. For example, Adam's hands moving apart when he asks how wide the passage would need to be for two coffins (Extract 3). Sometimes, a gesture has a function that might also, or better, be performed by a drawing. The possibility for people, when seated in the chapel, to see the coffin on the catafalque is a question of measuring on a drawing, even if this can be approximated by simulating the orientation of gaze using a finger. This does not imply, however, that questions concerning orientation of gaze can systematically be dealt with graphically (Extract 4).

There are, however, many situations in which the contribution of gesture is not redundant relative to verbal or graphical expression. In the example presented in Section 3.6 it is only due to Anna's gesturing that her co-designers properly understood 'the doors' as being the waiting area doors. If one only listens to her talk, one might interpret them as the car doors.

Our preliminary conclusion is that, among the families of gestures identified in our study, gestures are neither systematically irreplaceable, nor simple an optional accessory to speech or drawing. Furthermore, even if certain gestures might seem optional relative to other forms of expression, their discretionary character would require empirical confirmation.

The usefulness of gesture for the gesturer introduces another viewpoint about gesture's irreplaceability. A gesture that may be of little or no particular value for an addressee may play a unique role for the gesturer.

4.5 *Gesture in computer-supported collaborative software (CSCW) systems*

How should gestures that have been identified as irreplaceable in non-mediated collaboration, be handled in computer-supported collaborative software (CSCW) systems? CSCW-related research on gesture, and gesture-substitution devices, mostly focuses on technological aspects, without considering empirical research on human gesturing in face-to-face interaction. When gesturing is supported, it is mostly deictic manipulation through pen-based devices, generally not other types of gesture, such as specifying.

In face-to-face interaction, designating gestures are often particularly useful associates of verbal expressions. If a verbal expression alone ('this…', 'over there') is incomprehensible without a deictic gesture, is it possible to use another mode, even if the substitute – typically pen-based devices or telepointers[44] – is not as easy and efficient as a deictic gesture?

44. Gutwin, C. and Penner, R. (2002) Improving Interpretation of Remote Gestures with Telepointer Traces in *Proceedings of CSCW'02*, New Orleans, pp. 49–57.

Both analysis of face-to-face collaborative work, and specification and evaluation of technological approaches, still require much endeavour in order to develop support systems that will be useful and useable in collaborative design meetings.

4.6 *Contribution to gesture studies*

The study of gesture in collaborative professional activities has been little studied (Section 1.2 notwithstanding). Most research has been conducted in interactional conditions that were not goal-oriented – generally in conversational situations.

The present study has examined the function of gesture in a goal-oriented, professional work context; a collaborative design meeting. It has shown how the 'classical' gestures (such as deictics, iconics, metaphorics, and beats) are used with specific functions related to the nature of the design task.

We have identified representational and organisational gestures, corresponding to the design activities' main functions. Organisational gestures occur in each activity, but representational gestures are, in our view, typical for design[45]. Among the representational gestures, we identified designation and specification. Specification gestures may seem to be the distinctive design gestures however, if designation occurs in all interaction, it does not do so with the specific design function identified here. With regards to specification, all interaction involves specifying meaning for others in order to establish common ground, but, in design, specification aims not only at understanding, but also agreement, a distinction that was introduced in our analysis of 'common ground'[46].

45. Visser (2006a), *op. cit.*

46. Visser, W. (2006b) *Design: One, but in different forms*, INRIA, http://tinyurl.com/6dec7x (accessed September 2008).

4.7 *Contribution to design cognition research*

The role of gesture as an irreplaceable semiotic system for expression is not restricted to design. People not only always gesture, be they alone or in discussion, blind or sighted, visible to their addressees or not, they would not be able to express themselves

47.
McNeill (2000), *op.cit.*

satisfactorily without gesture. For speaking and hearing people, speech and gesture operate as an inseparable unit, reflecting different semiotic aspects of the cognitive structure that underlies both[47]. What is under discussion here is the specific contribution of gesture to collaborative design.

This study contributes to our knowledge on design thinking in respect to both the activities of design and the bodily realisation of those activities. Regarding aspects related to space (especially 3D), motion, and (sequences of) actions, our study points out (1) the particular nature of the way in which they are realised through designing and (2) how the semiotic system of gesture facilitates the implementation of these activities. Indeed our results point to the essential role of gesture in both the design of, and the communication around 3D and dynamic design aspects (such as motion and action), which are difficult, if not impossible, to elaborate using verbal or graphical expression.

Until now, cognitive design studies have paid little attention to designational activities and our analysis has shown the extent of these activities in collaborative design. We have known about gesture's appropriateness to render signalling through deictics. What our results have added to this knowledge is (1) the importance of designational activities with a design function in collaborative design, and (2) the implementation of this function through gesture.

Even if it is not specific to design, we also wish to underline the importance of gesture – a 'silent' form of expression – for organisational matters. This is a result to note as it confirms previous work on gesture in design meetings. The present study has shown this organisational role of gesture to have wider application than attributed in other research, which also includes the management of design actions.

ACKNOWLEDGEMENTS

The author gratefully acknowledges Françoise Détienne's remarks concerning initial versions of this chapter, and the comments and suggestions by two anonymous colleagues concerning later versions of this chapter.

16
Aspects of Language Use in Design Conversation*

Friedrich Glock

This chapter investigates some conversational episodes in an architectural meeting (A1). The analysis adopts an interpretative approach to design research and is guided by a qualitative research strategy. Designing is conceived as a social, interactive, interpretative process. Sociological and sociolinguistic concepts and research results are deployed to analyse design conversation and designing in terms of contexts and frames. The aim of the analysis is to reconstruct how participants interactively construct meaning in the design process and to describe practices they employ in the process.

Studies of work as it occurs in practice have gained interest in several fields such as science and technology studies and design research field studies[1]. Ethnographic research methodologies are increasingly employed to investigate the culture of the work of scientists, engineers, and other professionals. Rather than attempting to determine and prescribe how practitioners *ought* to do their work, these studies investigate how work is *actually* done. Many models of designing conceive of designing as a goal-directed process, with the specification of requirements prescribed at the beginning of a design process[2]. By contrast, the work presented here adopts an interpretative approach and suggests a methodological reorientation[3].

An interpretative approach is oriented to a social constructionist perspective wherein social life is analysed in terms of how individuals construct meanings and identities and so make sense of their everyday lives and interactions. Such an approach assumes that design goals are more or less incomplete and vague at the beginning of a design process. Requirements (such as the needs, wishes, etc. of clients, users, stakeholders, and others) and constraints are of-course crucial to designing. However, design goals are interpreted in contexts and in part are created by designers in the design process on the basis of their experience, embodied skills, practices, and tacit knowledge. These are acquired by participants in design through enculturation in a community of practice. From this perspective, design goals give orientation but cannot serve to direct each move in the design process. Designing is conceived as goal-oriented rather than goal-directed. From this view 'models of technical rationality'

1.
For an introductory overview see for example Heath, Knoblauch, and Luff (2000). Heath, C. Knoblauch, H. and Luff, P. (2000) Technology and Social Interaction: The Emergence of 'Workplace Studies', *British Journal of Sociology*, 51, pp. 299–320.

2.
One example of these models is that of Pahl and Beitz (2006)
Pahl, G. and Beitz, W. (2006/1984) *Engineering Design*, Design Council, London.

3.
Glock, F. (2003) Design Tools and Framing Practices, *Computer Supported Cooperative Work*, 12, pp. 221–239.

*Reprinted from: CoDesign Vol. 5 No 1 (March 2009), pp. 5–19,
DOI: 10.1080/15710880802492870

might serve as idealizations but appear inadequate to account for actual design work.

The interpretative paradigm in design research seeks to observe, investigate, and describe practices that designers use in the process of designing. An interpretative approach assumes that designers' 'skilful coping'[4,5] with situations does not require the goals to be fully represented in detail. Rather than using social factors to explain observable deteriorations of practice from some idealized model, design work is conceived as a social process. Studies of work have turned attention to the practitioners' use of language embedded in their practices, with talk at work gaining increasing interest for researchers[6,7]. In design processes participants use various sign systems such as drawings, formal as well as natural language, gestures, etc. It can be observed, for instance in the recording of meeting A1, that the use of natural language is pervasive in design conversation. The participants' use of natural language in designing can be taken as an argument for adopting an interpretative approach in design research. Since vagueness is an inherent feature of natural language and relies on context for appropriate understanding, sociolinguistic concepts and research results are useful for an interpretative approach to design research.

4.
Dreyfus, H. (2002) Refocusing the Question: Can There be Skillful Coping without Propositional Representations or Brain Representations? *Phenomenology and the Cognitive Sciences,* 1, pp. 413–425.

5.
Dreyfus, H. (2007) The Return of the Myth of the Mental, *Inquiry*, 50, pp. 352–365.

6.
Drew, P. and Heritage, J. (1992) Analyzing Talk at Work: An Introduction, in Drew, P. and Heritage, J. (eds) *Talk at Work*, Cambridge University Press.

7.
Sarangi, S. and Roberts, C. (eds) (1999) *Talk, Work and Institutional Order*, de Gruyter, Berlin.

Participants in the design process are conceived as interpreters of their own and each others utterances, actions, or 'moves'. These moves "are *context shaped* … (and) *context renewing*. 'Context' is used to refer both to the immediately local configuration of preceding activity in which the utterance occurs, and also to the 'larger' environment of activity"[8].

8.
Drew and Heritage (1992), *op. cit.*, p. 19.

1 THE SETTING

The architectural meeting A1 is one in a series of meetings in a design project. The task of the project is the design of a crematorium in a town in the UK. The meeting takes place in a room at the site of the proposed building. The architect (Adam), two client representatives (Anna and Charles) and an observer (Peter) participate in the meeting. The architect takes the initiative for the meeting: "we thought it would be a very good idea to have another meeting" (A1, 77–78). He suggests a definition of the situation: 'another meeting'. Goffman[9] has introduced the concept of frame to focus on the definition which participants give to their current social activity and the roles which the participants adopt within it. In 'frame analysis', behaviour, including speech, is interpreted in the context of participants' current understandings of what frame they are in.

9.
Goffman, E. (1986): *Frame Analysis: An Essay on the Organization of Experience*, Northeastern University Press, Boston.

> "When the individual in our western society recognizes a particular event, he tends ... to imply in his response (and in effect employ) one or more frameworks or schemata of interpretation of a kind that can be called primary... a primary framework is one that is seen as rendering what would otherwise be a meaningless aspect of the scene into something that is meaningful"[10].

10.
ibid. p. 21.

Goffman's[11] related notion of 'footing' addresses the fluctuating character of frames, together with the moment-by-moment realignments which participants may make in moving from one frame to another. As expected by frame analysis, it can be observed in the video-recording of the meeting and the transcript of it that further framings are introduced particularly in the beginning of the meeting, mainly by Adam. He initiates the transition from the gathering and initial small talk to the meeting proper (A1, 68) and states the purpose of the meeting: "the purpose of the meeting as I see it is to obtain your feedback to the design" (A1, 75), and "to update everybody as to where we've got to and to receive your feedback" (A1, 78–79). Adam "starts" (A1, 87) the meeting with a presentation of (his) "the design" (A1, 85). Framing provides a 'definition of the situation', an understanding of 'what is going on', and establishes some expectations of (types of) upcoming activities.

11.
Goffman, E. (1981) *Forms of Talk*, University of Pennsylvania Press, Philadelphia.

2 DESIGNING THE SIZE OF THE WAITING ROOM – A SNIPPET OF DESIGNING

The episode discussed in this chapter can be conceived of as a 'snippet' of the design process[12] concerned with the (re)design of the size of the waiting room. Several other authors in this volume also provide analysis of this episode, in particular the chapters from McDonnell[13] and Oak[14]. A brief summary of the first episode is as follows. Adam (the architect) presents the (drawings of the) new design; he starts to explain the new design and raises an issue: the size of the waiting room. The waiting room is designed according to Anna's (the client's) brief but Adam evaluates the waiting room as too small; he asks the clients whether he should increase its size. Extract 1 shows the conversation which leads to the design of a bigger waiting room. (It should be noted that for the analysis presented here gestural information and fine grained attention to overlapping talk is needed. Extracts 1 and 2 in this chapter therefore include additional notation beyond the DTRS7 basic indicators. Temporal alignment of gestures with speech are show by | and the starts of overlapping talk by /. Significant durations of gestures occurring in parallel with talk are indicated by ----------.)

12.
Schön, D. and Wiggins, G. (1992) Kinds of Seeing and their Functions in Designing, *Design Studies*, 13, pp. 135–156; p. 136.

13.
McDonnell, J. (Chapter 14) Collaborative Negotiation Design: A Study of Design Conversations between Architect and Building Users.

14.
Oak, A. (Chapter 17) Performing Architecture: Talking 'Architect' and 'Client' into Being.

Extract 1, A1, Designing the size of the waiting room.

		| [*leafs through drawings*]
85	Adam	| well the design has developed / a wee bit since / y- you saw it last time
	Anna	/ okay / okay
		| [*points to drawing*]
86	Adam	erm the design | obviously is still in exactly the same place
	Anna	[*nods*] yes [*low voice*]
		| [*takes drawing*]
87	Adam	but the the design is extended | to include the actual cremator facility
	Anna	yes
		|[*presents dr*] | [*gesture*]|
88	Adam	so if I can start with this particular drawing | err you've |seen |
		| [*Anna nods*]
		a version of this drawing before
		| [*gesture over drawing r to l*] | [*gesture left to right-*] | (see Figure 1)
89		| [*Anna nods-------*] |
		| basically we're arriving (in) | the new car park |
		| [*directional gesture with pencil*]
		+ / in this area | and
	Anna	/ yeah
		| [*surrounding gesture.Anna* | *nods*]
90	Adam	from the car park we'll enter the building | through a waiting area | this this
		| [*points to waiting room*]
91		| + lea- leads us to the first + query I have because um there was some discussion about whether
		| [*gazes to Anna* *Adam gazes to drawing* | *Anna nods slightly*]
92		you | wanted the size of the waiting room increased | at the moment it's
93		| [*gazes to Anna* *gazes to Charles* | *Charles nods slightly*]
		| exactly on brief / (but) it does look + kind of | + small to my eye
	Anna	/ yes [*nods*]
		| [*gazes to anna*]
94	Adam	in relation to the | + / size of the project
95	Anna	/ yes I'd be interested to know in a sense what sort of + seat- er how many
		| [*Adam takes ruler*]
96		seats in a sense you could get | in a sense that most people / arrive probably at least
97	Adam	/ ok [*takes measure with ruler*]
98	Other	[*knocks on the door*]

We enter the conversation as Adam leafs through a pile of drawings and says: "well the design has developed a wee bit" (A1, 85). The utterance informs the recipients about the current state of the design, announces some recent changes, and implies an assessment of the design. He does not mention who has developed the design and who is responsible; this might be a form of modesty or an aspect of the design process that is taken for granted[15]; but the agent-less development of the design may also indicate internal necessity of the development. Anna approves: "okay" (A1, 85).

15.
See Luck on 'ownership of the design concept'.
Luck, R. (Chapter 13) 'Does this Compromise your Design?' Socially Producing a Design Concept in Talk-in-Interaction.

The utterance: "since y- you saw it last time" (A1, 86) relates 'the design' to the design 'you saw'. This reminds the recipient of a previous situation 'last time' and suggests common ground for understanding. The pronoun 'it' refers to a previous design as an entity which 'has developed'. Adam continues: "erm the design obviously is still in exactly the same place" (A1, 86). As he says 'obviously' he points at a drawing; the pointing gesture directs the recipients' attention to the part of the drawing where the content of the statement 'obviously' can be seen (the use of modal adverbs is discussed later in Section 3.1). Anna nods and acknowledges: 'yes'.

Figure 1. Adam says: "basically we're arriving (in) the new car park".

Adam performs a series of activities which suggest how to understand the situation; he invokes framings for upcoming activities and makes types of activities expectable and relevant. The context 'so' defined, Adam announces: "if I can start with" (A1, 88) and presents "this particular drawing" (a plan) he has selected out of the pile. He 'start(s)' a presentation of the design 'with this particular drawing' to 'update' for 'feedback'. Adam introduces the object with the proximal deictic reference 'this' and directs the recipients' attention to 'this particular drawing' – the participants look at the drawing. 'Err' marks the insertion 'you've seen a version of this drawing before'; he does not continue the presentation but reminds the referent(s) that 'you' – Anna (and Charles) saw a previous drawing and presupposes that the drawing is known in some part. As Adam says 'you've seen' only Anna nods. The utterance relates the presented drawing to previous drawings and situations and implies common ground (just as does 'you saw last time'). The utterance provides the reader(s) of the drawing with an interpretation cue: to see the drawing as a version of a previous drawing she/they has/have 'seen before'.

Architectural drawings (just as technical drawings generally) are drawn up (in contrast to most sketches) to be read by every competent reader. Each reader may read the drawing, perceiving its parts and relating them to each other in a particular order to understand and judge the design, according to their 'motivational relevances'[16]. Adam's presentation guides the participants' focus of attention and suggests a particular sequence of reading; he provides a point of view, a perspective, a framing for the understanding of 'the design'.

16. Goffman (1986), *op. cit.*, p. 64.

2.1 *Situations of designing – designing of situations*

The adverb 'basically' (A1, 89) marks a change. Adam suggests an understanding of the design as he invokes contexts of use: "we're arriving in the new car park" (A1, 89), "we'll enter the building through a waiting area". Analytically, the actual situation of the meeting – the speech event in which design conversation takes place – can be distinguished from the transformed, (sketched, reported, imagined, etc.) situations or virtual worlds[17] the participants talk about.

17. Schön, D. (1983) *The Reflective Practitioner*, Basic Books; p. 157.

18.
Jakobson, R. (1971, 1957) *Selected writings*, Vol. 2, The Hague, Mouton.

This distinction is inspired by linguist Jakobson's[18] distinction between the speech event – the interaction among participants in the conversation – and the narrated event – what speakers are talking about. Designing can be conceived as 'keying' where the key refers to:

> "the set of conventions by which a given activity, one already meaningful in terms of some primary framework, is transformed into something patterned on this activity but seen by the participants to be something quite else"[19].

19.
Goffman (1986) *op. cit.*, p. 43.

20.
Goffman (1986) *op. cit.*, p. 48.

In design processes many basic keys employed in our society[20] can be observed such as planning, simulating, experimenting, practicing, dramatic scripting, fantasy, mock-ups, rehearsals, etc. Like most of these keys, designing is performed under conditions in which there is no actual engagement with the real world setting and events are thus "decoupled from their usual embedment in consequentiality"[21].

21.
Goffman (1986) *op. cit.*, p. 59.

Participants use natural language in design conversation. Pronouns and spatial and temporal adverbs are examples of deictic or indexical expressions whose interpretation depend on context. Deictic or indexical expressions such as: 'I', 'you', 'it', 'we', 'this', 'that', 'there', etc. are pervasive in the design conversations in A1. This chapter seeks to demonstrate that paying attention to the use of indexical language in design conversation is a means of investigating contexts in design processes.

The referents of 'we' in "we're arriving" (A1, 89) appear ambiguous; that is, the pronoun might be used impersonally or it may refer to the present participants. The utterance 'basically we're arriving' in the context of Adam's presentation of the drawing and gesture over the drawing locates the referents of 'we' virtually in the drawing. Through gesturing in reference to the drawing of the building, Adam asks the recipients to virtually arrive at and enter the building. The use of the present continuous and future tenses in the utterance emphasizes ongoing (virtual) activities, 'we're' renders 'us' as a kind of figures 'arriving' in the drawn world. The pronoun 'we' just like the first person singular pronoun 'I' seems to have properties "...bridging between the scene in which the talking occurs and the scene about which there is talking, for it refers both to a figure in a statement and to the currently present, live individual who is animating the utterance"[22].

22.
Goffman (1981) *op. cit.*, p. 150.

Adam's activities are constructed through the simultaneous use of different kinds of semiotic practices in different media (language, gesture, and the drawing) which mutually elaborate each other. As Adam says: 'basically we're arriving', he simultaneously performs a gesture over the drawing; the hand-gesture moves from the drawn entrance over the drawn road to the drawn car park. As he says 'the new car park', he opens his hand and makes a gesture over the part of the drawing where the car park is drawn. According to Visser's typology[23] Adam performs a spatial directional gesture followed by a spatial surface gesture. Visser also recommends that "one must look closely at the gesture, because the designatum is – at least, partially, – 'in' it"[24]. Consider Adam's spatial directional gesture as he traces the itinerary on the drawing where 'we're arriving' (see Figure 1), – presumably by car: 'the new car park'. Adam forms his hand as if he moves a toy car; this supports the interpretation that a keying is occurring, that is, there is an 'as if' scene on the drawing, a kind of stage with figures and an imaginary car: "we're arriving (in) the new car park" (A1, 89).

23.
Visser, W. (Chapter 15) The Function of Gesture in an Architectural Design Meeting.

24.
ibid.

The gesture that immediately follows has a different function. Here, Adam opens his hand and moves it to the right; the gesture does not describe a direction of movement but surrounds the area of 'the (drawn) new car park'. Notice that the spatial surface gesture accompanying the utterance: 'the new car park' is for the participants viewing the drawing from a bird's eye view, and not so much for the figures in the described scene. The figures might be somewhere 'in this area' where 'this' refers to the area indicated by the preceding gesture. Anna's responses concur with the units described. Anna nods after Adam's first gesture and her nodding gesture is as long as Adam's spatial surface gesture and then she verbally confirms understanding: 'yeah'. Adam continues: 'and' – returning to the virtual scene and restates where the figures are in it with: 'from the car park we'll enter the building'. This new change in perspective is also expressed through a change of 'physical realization of (the) deictic gesture'[25]. Adam uses a pencil to trace the way the figures will go to enter the building. Different perspectives can be discerned within Adam's turn of talk. However, it is not his talk alone, let alone isolated utterances, but the analysis of the activities in different modalities (talk, gesture, visual field) which elaborate each other to jointly establish context.

25.
ibid.

Adam announces a query: "this this lea- leads us to the first query I have" (A1, 91). The meaning of the deictic expressions: 'this', 'us', 'I' depends on contexts; contexts of the virtual situations and contexts of the speech event. Proximal deictic 'this' might refer to both the virtual, keyed scene that the figures experience and to the waiting room on the drawings that the participants of the meeting look at. Adam utters 'this' twice and he points at the drawing having said 'this' for the second time. Analogously: 'us' – just like 'we' above – might refer to both the figures in the scene talked about, and to the present participants.

The first person pronoun 'I' refers to Adam as the present speaker and to the figure which has encountered the question in a previous situation and has prepared to ask it now; the present speaker reports or cites the query. Goffman[26] has suggested going beyond

26.
Goffman (1986), *op. cit.*, p. 516.

analysis of roles to also consider footing[27]. In terms of his footing Adam acts more as an animator (physical speaker), and not so much as a principal or originator i.e. "someone who is committed to what the words say"[28], on the basis that it is as easy to cite what someone else said as to cite oneself[29]. He is a reporter, he says: "because um there was some discussion". As a reporter, the animator of the query in the present speech event, he is less responsible for the impact of the query: "whether you wanted the size of the waiting room increased" (A1, 92). Adam's next utterance supports this interpretation.

27. Goffman (1986), *op. cit.*

28. *ibid.*

29. *ibid.*

The discourse marker "because" (A1, 91) connects the query with a previous situation in which 'there was some discussion'. Through the invocation of a previous situation Adam accounts for putting the query. Rather than asking, he is stating a 'causal' consequence of previous events. Adam does not state the first query but reports the query as a citation or summary of 'some discussion about whether you wanted the size of the waiting room increased'. As Adam says 'you', he gazes towards Anna who nods slightly. Adam adds a description: 'at the moment it's exactly on brief'.

2.2 *A design goal created by the designer in the design process*

Adam closes his turn-at-talk with an assessment of the described waiting room: "(but) it does look + kind of + small to my eye in relation to the size of the project" (A1, 93–94). Using Schön's terminology[30], Adam makes a visual judgment; just as in Schön's famous example "Petra … sees that … the classrooms are too small … (and) Petra's judgment is … a subjective judgment"[31]. Likewise Adam's judgment is subjective: 'small to my eye'. Small "…is a quality of spatial configuration … Whether or not a given configuration (the waiting room) is … significant (big) enough, depends, at least in part, on its relation to other configurations around it in some context (the size of the project) … We can say that on the basis of … (Adam's) initial appreciation of the small (waiting room) …, (he) form(s) the intention of changing … (it) to a more significant layout" (bracketed text added by author)[32]. Adam's assessment implies and suggests a design move to increase the size of the waiting room.

30. Schön (1983) *op. cit.*

31. Schön and Wiggins (1992), *op. cit.*, p. 137.

32. Schön and Wiggins (1992), *op. cit.*, p. 139.

The design goal – to increase the waiting room – is created by the designer in the design process based on his interpretation. Adam's judgment is just like Petra's, (the student whose work Schön discusses). That is "Petra's judgments are embodied in acts of seeing"[33]. This supports the assumptions made by an interpretative approach as indicated in the introduction to this chapter. Since "the size of the waiting room" (A1, 92) is "exactly on brief" (A1, 93) Adam's assessment of the waiting room also

33. Schön and Wiggins (1992), *op. cit.*, p. 137.

may imply a judgment of the brief and disagreement with Anna. "Assessments reveal not just neutral objects in the world, but an alignment taken up toward phenomena by a particular actor"[34]. The designer assesses and criticizes the client's brief. Rather than presenting "requirements, defined in the clearest possible terms"[35] Adam states the design goal indirectly through his assessment or critique. Notice how the design move is suggested by the conversational practices that Adam uses. The practice of indirect formulation of design goals in the form of critique has also been observed in engineering design processes[36].

34.
Goodwin, C. and Goodwin, M.H. (1992) Assessment and the Construction of Contexts, in Duranti, A. and Goodwin, C. (eds) *Rethinking Context: Language as an Interactive Phenomenon*, Cambrifge University Press; p. 166.

35.
Pahl and Beitz (2006), *op. cit.*, p. 53.

36.
Glock, F. (2001) Design work in contexts: contexts in design work, in Lloyd P. and Christiaans, H. (eds) *Designing in Context*, Delft University Press, pp. 199–217.

Studies in Conversation Analysis have revealed preference formats of some types of action; for an assessment action, the preferred format is agreement, the dis-preferred format is disagreement[37]. "(T)he standard components for a dis-preferred action (are) – delay, mitigation, accounting"[38]. This is the format in which Adam produces his utterance (see Extract 1 lines 93–94): which exhibits all three of these, delay (+), mitigation 'kind of' and accounting 'in relation to the size of the project'. Goodwin and Goodwin[39] have investigated assessment activities in more detail. They have called utterance constructions such as (A1, 93–94) post-positioned assessments because the assessment occurs after the assessable has been made available. The Goodwins observed

37.
Heritage, J. (1984) Garfinkel and Ethnomethodology, Polity Press, Cambridge; p. 269.

38.
ibid., p. 267.

39.
Goodwin and Goodwin (1992), *op. cit.*

> "...that by moving to the assessment the speaker shows that though her talk is continuing, a marked structural change has occurred in it... (W)hen the speaker begins the assessment she is no longer describing events, but instead commenting on the description already given. Such a shift from description to assessment of described events in fact constitutes one of the characteristic ways that speakers begin to exit from a story"[40].

40.
Goodwin and Goodwin (1992), *op. cit.*, p. 161.

As Adam shifts from description to assessment, Anna attends to the assessment as marking a move toward exit and she takes the turn in overlapping talk.

3 VAGUENESS AND NATURAL LANGUAGE IN DESIGN CONVERSATION

Although the precise size of the waiting room can be read from the plan, participants in this episode do not discuss or negotiate the size of the waiting room in terms of precise numbers but in rather vague, qualitative terms: "small", "relatively small"(A1, 112), "big enough"(A1, 117). Moreover the vagueness of the expression is even enhanced: "kind of + small" (A1, 93). Whereas the statement: "at the moment it's exactly on brief" (A1, 93) conveys some state of affairs which constitute the truth conditions of the proposition, the vague expression used in the assessment relies on a resemblance relation: 'small … in relation to the size of the project'.

Vague, natural language terms replace precise numbers and introduce interpretative flexibility[41] and room for negotiation. This supports the idea that "[a]mbiguity is essential to design process…"[42]. Natural language is also the language which is shared among the participants, designer and clients. The (re)designing of the size of the waiting room begins with a qualitative judgment, expressed in vague terms such as: 'kind of small'. In contrast, it could be said that returning to exact terms later in the conversation (e.g. "two metre extension" (A1, 161)) marks the closure[43] of the design episode. Just as with any indexical term the meaning of terms such as 'kind of small', or 'big enough' depends on context. The use of indexical, natural language requires contexts for understanding and may serve several functions.

41.
Pinch, T. and Bijker, W. (1987) The Social Construction of Facts and Artifacts: Or how the Sociology of Science and the Sociology of Technology might Benefit Each Other, in Bijker, W., Hughes, T.P. and Pinch, T.J. (eds) *The Social Construction of Technological Systems*, MIT Press, pp 17–50; p. 40.

42.
Bucciarelli, L.L. (1994) *Designing Engineers*, MIT Press. p. 178.

43.
Pinch and Bijker (1987) *op. cit.*, p. 44

> "(V)ague expressions may convey different meanings, compared to exact numbers… Vague expressions are not just poor but good-enough substitutes for precise expressions, but are preferable to precise expressions because of their greater efficiency… (They) may convey more relevant meaning than would a precise number. For example it may do a better job at conveying the lack of significance of a quantity itself…. The use of the vague quantifying might be maximally relevant in that it may yield more contextual assumptions than a precise number would. Speakers appear to exploit the inherent vagueness of these expressions for particular communicative purposes. Vagueness may then be seen as a deliberate function of these quantifying expressions rather than as a defect in them. The degree of attitudinal meaning might collocate with their increased degree of vagueness. That is, the higher the degree of vagueness of the expression, the more emphasis seems to be put on the interpretive function of the quantifier … they seem to involve a higher degree of interpretive resemblance"[44].

44.
Jucker, A.H. Smith S.W. and Ludge T. (2003) Interactive Aspects of Vagueness in Conversation, *Journal of Pragmatics* 35, pp. 1737–1769; p. 1755.

The turn from considering precise, technical, formal language to the investigation of natural, more informal language in design conversation is to be reflected, I suggest, in design research that itself moves away from considering formal 'models of technical rationality'[45] to instead using interpretative methodologies. In design research vagueness in reference is often seen as a deviation from clarity, as it is more generally in 'models of technical rationality'. But, as Jucker Smith and Ludge[46] have pointed out, vagueness is not only an inherent feature of natural language, it is crucially an interactional strategy.

45.
Schön (1983) *op. cit.*

46.
Jucker, Smith and Ludge (2003), *op. cit.*

3.1 *Dealing with vagueness*

A particular kind of expression which can be observed recurrently in the meeting participants' utterances, including expressions such as: 'sort of', 'kind of', or 'a bit'. These expressions have been called downtoners; they increase the degree of vagueness of an utterance. "They may serve as politeness strategies, softening implicit complaints and criticisms. They also provide a way of establishing a social bond"[47].

47. Jucker, Smith, and Ludge (2003) *op. cit.*, p. 1766.

Anna, in overlapping talk (see Extract 1), responds to Adam with an inserted question. She acknowledges Adam's assessment and concern about the size of the waiting room with: 'yes' and addresses an aspect relevant to this issue: "I'd be interested to know ... how many seats in a sense you could get" (A1, 95–96). Anna accounts for her question as she invokes the context of building use: 'in a sense that most people arrive probably at least'. Her talk is interrupted as the secretary knocks on the door.

After the secretary leaves the room, Extract 2, Adam takes the floor. He explicitly refers to Anna's question as he announces: "to answer your question" (A1, 112) and responds to the conditional relevance of Anna's question. Adam begins his answer with a preliminary evaluative assertion which reformulates his assessment: 'it's a relatively small room'. He also provides an interpretation cue for the vague expression: 'relatively small' with a resemblance relation: "the area is similar to the existing waiting room" (A1, 112–113). Anna confirms: 'right'. On the basis of this understanding, 'and as such', he estimates: "you are really unlikely to get more than about eight to ten maximum twelve seats in there" (A1, 113–114). The construction of the utterance seems to suggest that it is *not* actually 'unlikely to (need) more than'. The modal adverb 'really' amplifies the relevance of the utterance. Adam qualifies the statement as an estimate with "I would have thought" (A1, 114), this is indicated also through 'unlikely' and 'about', and he gives an approximate number of seats. In her study of the use of deictics in planning meetings between planners and developers Glover[48] found that proximal references such as 'this' or 'here' introduce objects into the talk, while distal ones like 'that' or 'there' orient to some aspect(s) which have been established[49]. Adam refers to the established waiting room with distal deictic: 'there'.

48. Glover, K. (2000) Proximal and Distal Deixis in Negotiation Talk, *Journal of Pragmatics*, 32, pp. 915–926.

49. *ibid.*, p. 918.

Having answered Anna's question, Adam adds a further argument: "erm probably because..." (A1, 115) before he repeats his query. He marks a change from answering Anna's question with 'erm' to outlining his next activity: 'because' announces a connection of upcoming utterances. Here, there seems to be two uses of 'because'. The first is a speech-act connection: 'because ... the room is doing so much ... ' I am asking 'my question'. The second is a causal connection, in effect: 'because the room is doing so much ... so ... increase it'. The inferences appear in the talk as reasonable consequences.

Extract 2, A1, Designing the size of the waiting room continued.

111 Other [*leaves room*]
| [*Adam leafs through drawings*]
112 Adam | to answer your question it's a relatively small room the area is similar [*takes plan A1_4*]
| [*presents drawing* | *points*]
113 to the existing waiting room / | + and as such + erm you are really | unlikely
Anna / (right)
Adam to get more than
114 about eight to ten maximum twelve seats in there I would have thought +
115 erm probably because the building well the room is doing so much its
116 allowing people through into the porch area / it's also allowing access to the loos
Anna / yeah [*low voice*]
| [*gazes to Anna* | *short gaze to Charles*]
| | [*looks at drawing* | | *gazes to Anna* | *Anna headshake*]
117 Adam | so + | my first question to | you | is is the waiting room | big enough
Anna / no
118 Adam / and would you like us to increase it [*puts drawing aside*]
119 Anna I would say although the time spent sitting and waiting might not be
120 very long and eight seats + seems enough at some stages even our waiting
| [*Adam slightly nods*]
121 room is + too big so it's slightly | I'm also thinking of the fact that if we've
122 got a flow of people walking through that then restricts us we can't put
| [*Adam nods*] ( ) | [*Adam draws*]
123 | seats through that because in a sense we need to keep an access |
124 / open / and so the seating will be against / the wall
125 Adam / yeah / yeah / yeah
what I'd recommend is that we look at doing something like that
126 extending it which will give you seating areas here seating areas here
127 seating areas here as well as here and here
Anna yes
Adam which effectively will double
128 the seating / capacity from what I was just saying
Anna / yes
129 Anna yes I mean people waiting for cabs or for people waiting to be picked
130 up as well for services you know it-it might not sort of eight might be more
131 than enough for funerals for the majority of the time but I would think
132 it's nice to give them a bit more space as well / because we might get
Adam / yes
133 Anna people waiting for the eleven o'clock funeral erm and people at the ten
137 do have problems with families like that during funerals
| [*gazes to Charles*]
138 | [*Charles changes posture, smiles, gazes to Anna*]
| [*Adam smiles*]
| they don't want to be [*laughs*] th / |
| [*gazes to Peter*]
139 Charles | [*Adam takes drawing*]
/ | police attendants [*laughs*]
140 Anna police attendants quite often you know you'd think it would bring
141 them together but it actually makes it worse
| [*gazes to Peter*]
142 Adam really | gosh
143 Anna yeah and they sit separately in the chapel as well it's all to do with
144 money and you know they've left someone something wonderful
145 that's most of the time what it is or the other family are cross because
146 one family has arranged it and they used they never visited her while she
147 was alive and how dare they get involved with this and it all escalates
148 Peter it's like east EASTENDERS [*all laugh*]

The use of modal adverbs seems to serve a similar function as vague expressions in the speech events of the design conversation, that is, to present the viewpoints as negotiable. Notice that Adam initiates the utterance with the modal adverb 'probably' and uses several modal adverbs in his turn. Hoye[50] has found that modal adverbs occur:

> "in precisely those areas where speakers have something to gain or lose by their addressee's acceptance or rejection of what they are saying... Modality is non-factual and is concerned solely with the moment of utterance, the 'action of speaking'. It involves the speakers' needs and wants to convey their opinions and attitudes..."[51]

50.
Hoye, L.F. (2005) "You may think that; I couldn't possibly comment!" Modality Studies: Contemporary Research and Future Directions. Part II, *Journal of Pragmatics*, 37, pp. 1481–1506.

51.
ibid., p. 1484.

Adam has repeatedly indicated his preference for increasing the size of the waiting room before he asks his first question[52]. He explicitly announces: "my first question to you is" (A1, 117) and addresses the question 'to you'. The referents of you are Anna and Charles. As Adam utters 'you', he gazes at Charles and as he continues speaking he gazes at Anna. The utterance contains two consecutive questions, these are, firstly: "is the waiting the room big enough"? and secondly: "would you like us to increase it"? (A1, 118). These phrases are linked with the connective 'and'. Adam puts the drawing aside, implying that the next conditionally relevant activity is not expected to take place with reference to the drawing. The first question might be understood to ask for agreement with his repeatedly stated evaluation; the second question asks for a decision; the decision seems to be a consequence of the evaluation.

52.
See Oak's consideration of questions and answers, **Oak (Chapter 17)** ***op. cit.***

Anna projects the question as she starts to (give the preferred) answer through a headshake before Adam has fully articulated the first question. Anna's answer supports the interpretation of the two questions. Anna delivers the preferred answer: "no" (A1, 117) to Adam's first question in the format for preferred actions that includes both simple acceptance and no delay. Anna agrees with Adam's judgment and acknowledges his competence and authority[53]. This sequence exemplifies that "the design of actions can contribute to the maintenance of social solidarity ... the preferred format responses to requests ... and assessments are uniformly affiliative actions which are supportive of social solidarity"[54].

53.
See also **Luck (Chapter 13)** ***op. cit.*** and **McDonnell (Chapter 14)** ***op. cit.***

54.
Heritage (1984) *op. cit.*, p. 268.

Anna begins her answer to the second part of the question with a hedge, 'I- I would say'. A hedge is normally used to express consent and indicate consequences of the imagined, talked about situation (cf. Section 2.1). The speaker presents herself

> "...through the office of a personal pronoun 'I' and it is thus a figure – a figure in a statement – that serves as the agent, a protagonist in a described scene, ... someone, after all, who belongs to the world that is spoken about, not the world in which the speaking occurs. And once this format is employed, an astonishing flexibility is created.

For one thing, hedges and qualities introduced in the form of modal verbs (I 'think', 'could' etc.) introducing some distance between the figure and its avowal"[55].

55. Goffman (1981) *op. cit.*, p. 147.

In their study of 'I-marked-utterances' in work-meeting conversation Fasulo and Zucchermaglio[56] investigate utterances such as 'I would say' which they call 'decisionals':

56. Fasulo, A. and Zucchermaglio, C. (2002) My Selves and I: Identity Markers in Work Meeting Talk, *Journal of Pragmatics*, 34, pp. 1119–1144.

"Decisionals, especially in the conditional, are typical of proposals of solutions... Speakers not only reveal their stance..., but they offer an immediate line of action ... Some linguists have considered every declarative sentence to be a dependent clause of the performative 'I say to you'. But when the sentence is actually spoken, it assumes an ethical nuance ... Such uses are crucial interactional resources; these uses constitute part and parcel of the moral dimension of social interaction"[57].

57. *ibid.*, p. 1128.

Therefore, when Anna says: "eight seats + seems enough at some stages even our waiting room is + too big" (A1, 120) the delay she makes before the assessment might indicate disagreement with Adam.

Adam nods or uses verbal acknowledgments at those parts of her utterances which seem to conform to his position: "it's slightly" (A1, 121), "we can't put seats through" (A1, 122). As Anna says: "we need to keep an access open" (A1, 123), Adam confirms 'yeah' and starts to sketch to increase the waiting room in the drawing. It is not clear to a viewer of the video recording of the meeting whether Anna's requirement to 'keep an access open' and to arrange 'the seating ... against the wall' actually reduces the possible number of seats in the waiting room; the extended waiting room will: "double the seating capacity" (A1, 127–128).

Designing does not appear to be solely the realm of the designer; the client also designs. Anna's utterances might be conceived as designing[58]. Thus, although the requirement mentioned by Anna: "keep an access open" (A1, 123) might not necessarily be required to increase the size of the waiting room, Adam fixes his suggested design move in the drawing and makes it visible and more enduring than transient words. In his next utterance Adam refers to the sketched waiting room with the distal deictic expression "that" (A1, 125), the now visually established design[59] and spins out consequences of the move (A1, 126–128). Anna confirms: 'yes' and continues to consider consequences in the context(s) of use.

58. See **McDonnell (Chapter 14)** ***op. cit.***

59. See Glover (2000), *op. cit.*

3.2 *Vagueness, social bonds, and common ground*

At one point, Anna says: "I mean people waiting for cabs or for people waiting to be picked up as well for services you know it – it might not sort of eight might be more than enough for funerals for the majority of the time but I would think it's nice to give them a bit more space as well" (A1, 129–132). Here, there appears to be an observable pattern in the sequences of interaction under discussion. That is, Anna follows Adam's judgment (A1, 93–94), provides preferred answers to his questions (A1, 117), and confirms his design move (A1, 127). Further, she consistently uses talk that begins with a hedge (e.g. "I'd be interested" (A1, 95), "I would say" (A1, 119), "I mean" (A1, 129 and 163)). Anna's concerns appear to be in a loose relation to the size of the waiting room. Her use of hedges seems to serve several functions: for instance, the phrase 'I mean' signals a relationship of the form 'something like' between the qualified construction (people waiting for cabs...) "and an item in absentia"[60] – here the criteria or requirement for judging the waiting room. 'I mean' also functions as a marker of 'imprecision', an expression of '*like*-ness'[61] or resemblance relation, that is, a vague or loose use of expressions which involve using interpretive rather than descriptive language[62]. 'I mean' also seems to function as a compromiser or as a mitigator of the strength of her evaluative statement by making her less committed. The phrase also acts as cajoler, a device by which harmony between interlocutors is established, increased, or restored[63]. In this function the hedges are metacommunicative that is, they modify her intentions.

Notice that Anna uses several markers and downtoners in her utterance (A1, 129–131) including 'I mean', 'you know', 'sort of', 'I would think', and 'a bit'. In their studies of pragmatic devices Stubbe and Holmes[64] have found that 'I mean' is often used to mark informality, as a politeness strategy which reduces social distance, working in a similar way to 'you know' and 'sort of'[65]. Anna's use of hedges might be interpreted as an attempt to reduce social distance, establish harmony, and contribute informal conversation as a common ground and as a basis for understanding her imprecise concerns.

A simple count of the frequency of the use of some pragmatic devices (without further differentiation of their function in context in meeting A1) is shown in Table 1. This indicates that Anna more often uses pragmatic devices such as: 'you know', 'I mean' and 'sort of' which tend to emphasize solidarity. Conversely, Adam more often uses the pragmatic particle: 'I think', which is more explicit in the way it draws attention to perceived gaps in shared understanding between speakers.

Returning to the conversation (Extract 2) at Anna's contribution at 129–132, we see Adam acknowledges with an overlapping: "yes" (A1, 132) and Anna continues with: 'because' and tells him about some people's behaviour and her organization's problems with it: 'we do have problems with families like that anyway

60. Brinton, L. J. (2003) I Mean: The Rise of a Pragmatic Marker, Paper presented at *GURT, Cognitive and Discourse Perspectives*, Georgetown University Round Table on Language and Linguistics, Washington DC; p. 2.

61. *ibid.*, p. 2.

62. Jucker, Smith, and Ludge (2003) *op. cit.*, p. 1742.

63. Brinton (2003), *op. cit.*, p. 3.

64. Stubbe, M. and Holmes, J. (1995) 'You Know, Eh and Other Exasperating Expressions': An Analysis of Social and Stylistic Variation in the Use of Pragmatic Devices in a Sample of New Zealand English, *Language & Communication*, 15, pp. 63–88.

65. *ibid.*, p. 70.

during funerals'. As she adds: 'they don't want to be' she gazes towards Charles and laughs. Charles inserts: "police attendants" and also laughs (A1, 139). He contributes to the jovial mood that Anna has contributed to with her utterance. Charles gazes at Peter, the observer, when he makes the insertion. In her next utterance: "quite often you know you'd think it would bring them together but it actually makes it worse" (A1, 140–41), Anna repeatedly uses the pronoun 'it' which is the vaguest possible expression to refer to a variety of entities. In the introduction to this section (Section 3) the idea that vagueness has strategic uses has been introduced. The vagueness we see here is neither bad nor good, "[W]hat matters is that vague language is used appropriately"[66].

66.
Channel, quoted in Jucker, Smith, and Ludge (2003) *op. cit.*, p. 1738.

67.
Lloyd, P. (Chapter 5) Ethical Imagination and Design.

68.
See Matthew's remarks on method in **Matthews, B. (Chapter 2) Intersections of Brainstorming Rules and Social Order.**

69.
Hoye (2005), *op. cit.*, p. 1484.

Through his response: 'really gosh' Adam appreciates Anna's account as a story – a story about people's behaviour with ethical implications[67], this interjection displays (to us) his interpretation of Anna's utterances[68]. Rather than feedback on the design suggested by Adam, he treats Anna's utterances as going beyond the purpose of the meeting, to be 'out-of-frame', as a story, or gossip addressed to the individuals as participants in informal conversation. As the interjection: 'really' indicates, Adam: "…is concerned solely with the moment of utterance, the action of speaking"[69]. As he says: 'gosh', he gazes at Peter. Adam includes the observer – who is not (to be) addressed regarding design issues in the conversation – as a legitimate addressee of his utterance. His activities (re)frame Anna's utterances as more informal conversation.

Peter affirms this framing of the situation of Anna's account. He contributes to an informal conversation as he jokingly resumes the stories about people's behaviour with his comment: "it's like EAST – EASTENDERS" (A1, 148). Anna continues her talk which is responded to by Adam with: "ok" and: "yeah well" (at A1, 160) before he closes the episode by saying 'so shall we agree a two metre extension'. From this we can observe how the client's concerns and perspectives are interpreted in the conversation.

Pragmatic device	Anna	Adam	Charles
You know	96	13	5
I mean	82	26	36
Sort of	90	22	8
I think	22	31	14

Table 1. Frequency of the use of some pragmatic devices by the participants in meeting A1.

4 CONCLUDING REMARKS

Anna has provided her concerns as 'feedback to the design' and in response to Adam's suggested design move. Her depictions might be understood as her attempt to describe requirements for the architecture that are relevant from her perspective. This investigation of the use of discourse markers has indicated that Anna seems to appeal to informal social relationships and their implied common ground as the basis for understanding her concerns. This appears to be her strategy to manage the problems of providing feedback from: "everybody that's come in" (A1, 83). The episode, largely encompassed by the two extracts which have been discussed in detail here can be construed as an example of the problem of how to specify requirements in conversations between designers and clients. This episode is also an example of how clients' perspectives are interpreted or translated into design considerations. "(W)hat turns out to count as 'requirements' is what emerges in the course of negotiation ... Requirements are the upshot of a process of (often protracted) social interaction"[70].

70.
Woolgar, S. (1994) Rethinking Requirements Analysis: Some Implications of Recent Research into Producer-Consumer Relationships in IT development, in Jirotka, M. and Goguen, J. (eds) *Requirements Engineering: Social and Technical Issues,* Academic Press, London; p. 203.

In this chapter, an interpretative approach to design research has been used to investigate design conversation. An episode from the architectural design meeting A1 concerning the design of the size of the waiting room, viewed as a snippet of the design process, has been investigated and some practices such as presenting the design and making assessments have been described.

The use of natural language is pervasive in the design conversation investigated. Natural language and its inherent vagueness introduce interpretative flexibility into design conversation. Such interpretative flexibility appears crucial in design processes. Vague expressions perform important functions in conveying meaning which require context for an appropriate understanding. Sociological and sociolinguistic research results and concepts for the investigation of the pragmatics of natural language – such as the functions of personal pronouns, use of hedges, downtoners, modal adverbs, discourse markers, and so on are helpful for the analysis of design conversation in terms of contexts and frames.

The episode investigated in this paper indicates that the client in this particular case tends to appeal to informal social relationships as a common ground to get her concerns understood. This strategy appears to receive selective treatment by the designer.

Part 6

Constructing Roles

17
Performing Architecture: Talking 'Architect' and 'Client' into Being*

Arlene Oak

This chapter explores how the roles or social categories 'architect' and 'client' are performed by participants as they meet to talk about the design of a crematorium. The analytic framework through which the interaction is studied is Membership Categorisation Analysis (MCA). By attending to the participants' talk through the perspectives of MCA, we can see how questions and answers, attributions of building ownership, and assessments of the building are enacted in ways that enable the participants to competently perform as 'architect' and 'client'. Thus, as well as the participants' interaction helping to shape the actual form of the building, it also helps to shape and perpetuate ideas concerning what it is to 'do' architecture.

This chapter presents a study of how the roles of 'architect' and 'client' are performed in a meeting between an architect and two clients (though primarily it considers the interactions that occur between the architect and one client). Rather than view the roles of 'architect' and 'client' as unproblematic categories that may be applied to the meeting's participants, this study views these roles as the participants orient to and perform them through their talk. What do the participants say to adequately and convincingly perform as an architect and a client and how do they say it? How does their interaction help to create the 'doing' of architecture? (A meeting about designing a building is part of the architectural process and therefore is part of 'doing' architecture.)

While the participants perform other role-specific behaviours (such as gesturing at drawings) this work focuses solely on the participants' talk, to explore how interaction allows the speakers to organize their social roles and get things done. In this case, one of the things that is done is the designing of a crematorium but also the participants accomplish the presentation of themselves as relevant participants in the design process. Accordingly, this chapter deals with talk, not as the straightforward reporting of events, but as 'discourse' that involves participants in networks of social and moral obligation[1]. Therefore, underpinning this work is a Wittgenstenian[2] perspective in which: "the participants [in talk] play to some tacit set of rules about what is permissible and obligatory in the linguistic interaction"[3]. In terms of its affiliation with design research this work relates to prior scholarship in areas associated with design and collaboration[4]. It also relates to work done on design's relationship to interaction[5,6,7,8,9,10,11,12,13,14,15,16].

*Reprinted from: CoDesign Vol. 5 No 1 (March 2009), pp. 51–63,
DOI: 10.1080/15710880802518054

1.
Shotter, J. (1981) Telling and Reporting: Prospective and Retrospective Uses of Self-Ascriptions in Antaki, C. (ed) *The Psychology of Ordinary Explanations of Social Behaviour*, Academic Press, London.

2.
Wittgenstein, L. (2001) *Philosophical Investigations*, Blackwell, Oxford (first published in 1953).

3.
Antaki, C. and Leudar, I. (1992) Explaining in Conversation: Towards an Argument Model, *European Journal of Social Psychology*, 22, pp. 181–194; p. 183.

4.
See for example Cross, Christiaans, and Dorst, Scrivener, Ball, and Woodcock, and the journals *Co-Design* and *Design Studies*, especially *Design Studies* 28 (3), a special issue on participatory design.
Cross, N., Christiaans, H., and Dorst, K. (eds) (1996) *Analysing Design Activity*, Wiley, London.
Scrivener, S., Ball, L., and Woodcock, A. (eds) (2000) *Collaborative Design*, Springer Verlag, London.

5.
Cross, N. and Clayburn Cross, A. (1995) Observations of Teamwork and Social Processes in Design, *Design Studies*, 16/2, pp. 143–170.

6.
Fleming, D. (1998) Design Talk: Constructing the Object in Studio Conversations, *Design Issues*, 13 (2), pp. 41–62.

The perspectives of this study primarily stem from conversation analysis[17] and ethnomethodology,[18] which are concerned with how members orient to each other and make sense of each other's actions in specific contexts. This work is especially influenced by the ethnomethodologically-informed work of sociologist Harvey Sacks and his modes for analysing conversation, particularly his approach now known as Membership Categorization Analysis (MCA)[19]. Membership Categorization Analysis considers both the details of talk's structure (such as turn-taking) while also considering how participants' talk may be associated with particular social roles or 'membership categories'[20,21]. Broadly speaking, membership categories refer to how people are named in ways that give relevance and coherence to a particular situation of interaction. Membership categories may be broad (e.g. man/woman) or more narrowly construed; what is at issue in MCA is how categories have relevance for the interaction under investigation. Therefore, in the meeting between the architect and client discussed here, the first architectural design meeting (A1), when the architect says: "it's a dream come true for an architect to do such a project so I'm very excited by it" (A1, 73–74) he positions his role as 'architect' as relevantly related to action ('to do such a project'), enthusiasm ('dream come true'), and emotion ('excited').

As an analytic approach, MCA has especially influenced scholars who study talk at work, wherein speakers are in institution-based roles that may impact upon the topics and orderliness of the interaction[22,23,24]. Since the talk between the architect and the client in meeting A1 takes place in the context of their work-based roles, it is pertinent to consider their interaction through MCA in order to better understand how the categories of 'architect' and 'client' are performed in the specific setting of their meeting. What is investigated here are some of the ways in which their interaction allows the roles of architect and client to be '*ultimately and accountably talked into being*'[25]. The main issue discussed in this chapter is how participants orient to the categories of architect and client and how this orientation impacts upon the structure of their talk, especially their performance of questions and answers. Also, though more briefly considered are: how some attributions of building 'ownership' are made; and, how the delivery of some negative assessments is managed. Finally, the chapter outlines how the participants' talk may relate to some generally-held beliefs about the practice of architecture.

1 BACKGROUND INFORMATION: THE ROLES OF 'ARCHITECT' AND 'CLIENT'

The materials provided to the DTRS7 analysts did not include recordings or transcriptions of the first meeting(s) between the

7.
Lloyd, P. and Busby, J. (2001) Softening up the Facts: Engineers in Design Meetings, *Design Issues*, 17 (3), pp. 67–82.

8.
Luck, R. (2007) Learning to Talk to Users in Participatory Design Situations, *Design Studies*, 28, pp. 217–242.

9.
Oak, A. (2000) "It's a Nice Idea, but It's Not Actually Real": Assessing the Objects and Activities of Design, *The International Journal of Art and Design Education*, 19 (1), pp. 86–95.

10.
Oak, A. (2001) *Identities in Practice: Configuring Design Activity and Social Identity Through Talk*, unpublished PhD dissertation, King's College, University of Cambridge.

11.
Oak, A. (2006) Particularizing the Past: Persuasion and Value in Oral History Interviews and Design Critiques, *Journal of Design History*, 19 (4), pp. 345–356.

12.
Stumpf, S. and McDonnell, J. (2002) Talking About Team Framing: Using Argumentation to Analyse and Support Experiential Learning in Early Design Episodes, *Design Studies*, 23, pp. 5–23.

13.
Glock, F. (Chapter 16) Aspects of Language Use in Design Conversation.

14.
Luck, R. (Chapter 13) "Does this compromise your design?" Socially Producing a Design Concept in Talk-in-Interaction.

15.
Matthews, B. (Chapter 2) Intersections of Brainstorming Rules and Social Order.

architect and his clients. However a video recording of a short, informal interview with the architect conducted by a researcher was made available. In this the architect outlined the background to the crematorium design project. Some of the points that he makes in this 'background-information' video are relevant to his perceptions of his role as architect and so will be briefly outlined here, prior to a more in-depth discussion of the architect/client interaction that occurred in meeting A1.

In the background-information video, the architect talks about the early stages of the crematorium project. For instance, he expresses some frustration with the initial brief for the project, noting that it was: "very simple" (5.37) and that: "we really want a lot more information" (5.39). (Numbers in brackets indicate the time in minutes and seconds at which the talk occurred in the interview.) The architect also mentions that he went on to develop a more successful brief with the clients, though he also states that, within the collaborative process, he made several decisions himself (e.g. "I was determined that this should be a concrete building" (21.39)). As the architect talks about the early stages of the crematorium's design, it is apparent that he was personally highly invested in the project. For instance, he states that: "of-course it's every architect's dream" (2.53) to work on such a building. Also, he notes that, for him, the project: "was a dream come true" (2.59) and: "amazing" (3.03). Further, he says from: "very early on" (9.16), he had a particular idea for: "what sort of form the building should take" (8.17); that of Louis Kahn's Kimbell Art Museum in Texas (which the architect had visited in 1993). In the background-information video, the architect also talks about how he showed images of the Kimbell Art Museum to his clients at an early meeting and he: "couldn't believe it they were sold on the concept virtually straight away" (9.41).

The architect's vision of a link between the Kimbell Art Museum and the crematorium is accepted by his clients and so he is able to follow his inspiration which, as the "starting point for the architect, [as] a way in to the problem"[26] could be considered a 'primary generator' for the project. A primary generator is "an article of faith on the part of the architect, a designer-imposed constraint"[27] that is "strongly valued and self-imposed"[28]. That the architect has created a personal link between the crematorium and the Kimbell is evident from how he associates the project with his memories of visiting the Kimbell ("the whole feeling of it [the Kimbell] had never left me since visiting" (11.40)). The architect thus indicates that it is appropriate for him to associate his professional work with memory, emotion, and embodied experience[29,30,31]. Thus, the architect's talk in the background video establishes his decision-making capabilities and personal engagement with the building, as well as his recognition of the significance of the clients since, as he notes, if

16.
McDonnell, J. (Chapter 14) Collaborative Negotiation in Design: A Study of Design Conversations between Architect and Building Users.

17.
Sacks, H. (1992) *Lectures on Conversation Vols. 1 and II*, Blackwell Publishing, London.

18.
Heritage, J. (1984) *Garfinkel and Ethnomethodology*, Polity Press, Cambridge.

19.
op. cit., Volume 1, p. 40.

20.
Housley, W. and Fizgerald, R. (2002) The Reconsidered Model of Membership Categorization Analysis, *Qualitative Research*, 2, pp. 59–83.

21.
Psathas, G. (1999) Studying the Organization in Action: Membership Categorization and Interaction Analysis, *Human Studies*, 22, pp. 139–162.

22.
Boden, D. and Zimmerman, D. (eds) (1991) *Talk and Social Structure: Studies in Ethnomethodology and Conversation Analysis*, Polity Press, Cambridge.

23.
Drew, P. and Heritage, J. (eds) (1997) *Talk at Work: Interaction in Institutional Settings*, Cambridge University Press, Cambridge.

24.
Housley, W. (2006) Membership Categorisation Analysis, Sequences, and Meeting Talk, *Working Paper 84*, School of Social Sciences, Cardiff University.

25.
Heritage (1984) *op. cit.*, p. 290, emphasis in original.

26.
Darke, J. (1984) The Primary Generator and the Design Process in Cross, N. (ed) *Developments in Design Methodology*, John Wiley & Sons, Chichester, pp. 175–188; p. 181.

they had not been 'sold on the concept' he may have had to design differently. Having outlined the architect's perspective on some issues pertaining to the roles of 'architect' and 'client' as expressed in the background video to the crematorium's design, let us now turn to some talk from meeting A1 to consider how the architect and his client (and, less so, the other client present at the meeting) use talk to manage their relationship(s) and perform architecture.

27.
ibid., p. 181.

28.
ibid., p. 186.

29.
Downing, F. (2000) *Remembrance and the Design of Place*, College Station, Texas A&M University.

30.
Gero, J. (1999) Constructive Memory in Design Thinking in Goldshmidt, G. and Porter, W. (eds) *Design Thinking Research Symposium: Design Representation*, MIT, Cambridge, pp. 29–35.

31.
Solovyova, I. (2003) Conjecture and Emotion: An Investigation of the Relationship Between Design Thinking and Emotional Content. http://tinyurl.com/5z8xte (accessed 22 July 2008).

2 MEETING STRUCTURE

As outlined above, this study focused on membership categories and how they constrain and afford certain forms of behaviour, including certain forms of talk. For example, the data indicates that the architect perceives his role as one in which it is acceptable behaviour for him to define the nature of the meeting ("I'll look forward to hearing the feedback because that's the purpose of the meeting" (A1, 21–22)). Indeed, as we see below, the architect frames the meeting as a way to gather feedback from several project stakeholders, including funeral directors and other building users. However, although the client indicates that she has consulted others widely ("everyone that's come in has been dragged in" (A1, 83)) she does not actually tell the architect what others have said (this issue is considered later in this chapter).

Extract 1, A1, Defining the meeting.

Arch we thought it would be
a very good idea to have another meeting to update everybody as to
where we've got to and to receive your feedback because the last time
you and I met – you said you were interested in talking to funeral
directors and so on and so forth obtain feedback from people who'd be
using the building
Client which is everyone that's come in has been dragged in to show even
my mum's had a look [*laughs*]

3 QUESTIONS AND ANSWERS IN THE PERFORMANCE OF ARCHITECTURE

Although the architect seizes the initiative in defining the purpose of the meeting as a feedback session, feedback from others is not forthcoming and the architect instead seeks information through asking questions of the clients at hand. The architect draws attention to the appropriateness of his role as inquisitor by saying: "the first query I have" (A1, 91), thereby implying that he will have several others. From this point on a pattern of interaction develops in which questions and answers are significant to the conduct of the meeting (particularly its first half). In MCA (as in Conversation Analysis more broadly), questions and answers are understood as a type of adjacency pair; that is, a form of talk wherein one part (a question) is followed by a second part (an answer)[32,33]. While apparently straightforward in terms of their structure, the asking and answering of questions is significant in interaction, particularly in institution-based talk, because the speaker who does the asking constrains the person asked to answer, thereby creating a potential difference in power. Indeed, much of the way that talk can be identified as occurring within an institutional setting is through considering how questions and answers are managed. For instance, interaction between doctors and patients or teachers and students is identifiable as occurring within 'medicine' or 'education' in part because certain parties can and do ask questions of the other in particular ways (and in ways that differ from the usually more equally-distributed asking and answering of questions that characterizes ordinary conversation between peers)[34,35]. In meeting A1 the manner in which the architect asks questions and the topics that he asks about indicate that he seeks particular types of answers that he can use in designing the building; accordingly, meeting A1 proceeds within the institution of 'architecture' or, more broadly, 'design practice'. However, as we will see, the sort of information that the architect seeks is not always directly forthcoming from his client.

32.
Sacks (1992), *op. cit.*

33.
Hutchby, I. and Wooffitt, R. (1998) *Conversation Analysis: Principles, Practices and Applications*, Polity Press, Cambridge.

34.
Heritage, J. (2005) Conversation Analysis and Institutional Talk in K. Fitch and R. Sanders (eds.) *Handbook of Language and Social Interaction*, Routledge, pp. 103–148.

35.
Hutchby and Wooffitt (1998) *op. cit.*, pp. 149–154.

3.1 *Questions and answers: Clarity and ambiguity*

In the following extract, the architect seeks information concerning room size, data that can be translated into a drawing and then, ultimately, into the building's final form. Despite the clarity of his first question, and despite it being rephrased as a recommendation (A1, 125), the client's replies do not clearly answer the architect's questions about the appropriate dimensions for the waiting room:

In this extract the architect makes a clear request for information about room size (A1, 117–118), but the client's answer leaves open for interpretation exactly how big the new waiting room should be. Later, when the architect suggests possible dimensions for the room (A1, 126–128) the client's reply is also ambiguous: the current room may or may not be big enough (A1, 129–141). Eventually the client says that she would like the waiting room: "a little bit bigger I think + not hugely because … it is a wasted space most of the day" (A1, 158–159). The architect responds to the client's ambivalent comments by choosing dimensions ("a couple of metres" (A1, 160)) and writing on the drawing, minuting their talk as a design decision made (see the second meeting, A2: "one of the items on the minutes last time was to increase the size of the waiting room" (A2, 49–50)). However, despite the architect's decision

Extract 2, A1, Questions and answers.

Arch	so my first question to you is is waiting the room big enough and would you like us to increase it
Client	I would say although the time spent sitting and waiting might not be very long and eight seats seems enough at some stages even our waiting room is too big so its slightly I'm also thinking of the fact that if we've got a flow of people walking through that then restricts us we can't put seats through that because in a sense we need to keep an access open and so the seating will be against the wall
Arch	what I'd recommend is that we look at doing something like that extending it which will give you seating areas here seating areas here seating areas here as well as here and here which effectively will double the seating capacity from what I was just saying
Client	yes I mean people waiting for cabs or for people waiting to be picked up as well for services you know it might not sort of eight might be more than enough for funerals for the majority of the time but I would think it's nice to give them a bit more space as well because we might get people waiting for the eleven o'clock funeral erm and people at the ten o'clock perhaps arrive and so they keep in their little groups they don't want to mix with other people so the feeling of keeping them segregated just because they don't know the other people might also be there and we do have problems with families like that during funerals
Client	you know you'd think it would bring them together but it actually makes it worse
Arch	really gosh
Client	so I'd like it a little bit bigger I think + not hugely because there is it is a wasted space most of the day really
Arch	yeah well I would have thought another couple of metres on there [*writes on drawing*] would do the trick so shall we agree a two metre extension yes or thereabouts hmm
Client	I mean the other suggestion that perhaps I could make at this stage would be perhaps for a small amount of outside seating…

to extend the space, the client has not actually made a definitive response about her preferred room size. Instead, she answers his last question about waiting-room size (A1, 161–162) by changing the topic to suggest seating outside (A1, 163–164).

In the meeting a turn-taking format arises in which the architect tends to ask questions about the building's formal qualities, such as room dimensions, spatial arrangements, and/or amenities with the client's responses delivering somewhat ambiguous accounts in reply. This especially happens in the first half of the first meeting. For example, the talk in Extract 3, below, follows a similar pattern to that outlined in Extract 2, wherein direct questions are answered by information-rich accounts[36].

36.
See also **McDonnell (Chapter 14)** ***op. cit.*** who discusses the client's rich descriptions of building use.

Extract 3, A1, Further questions and answers.

342	Arch	so my next two questions are are the sanctuary and the catafalque big enough
345	Arch	there might be the possibility of a double funeral in which case would this be
346		wide enough for two coffins ++
347	Client 1	it wouldn't probably I don't know
348	Arch	it's just over three metres diameter it's about three point one metres
349		diameter
350	Client 2	how wide is the existing er trolley
351	Client 1	I don't know I think I would say it might just I mean at the moment they
352		can just they can just go in side by side but it's difficult to squeeze in to
353		put the coffins on at the moment even because you've also you've got the
354		two catafalques in side by side and you need to have four routes for
355		people to go either you need the one in the middle for both people to go
356		and the ones at the end for them to drop the coffins off erm but even two
357		catafalques isn't always enough we've had three or we've had car
358		accidents you know we've had three coffins and we've not been able to
359		accommodate all the you know I mean if we can do two that's the
360		majority of them put them side beside or in the sense perhaps have the
361		catafalque so it can expand to accommodate two I don't know one
362		catafalque that spreads out like a sort of a table or something I don't know
368	Arch	OK so my question for you is how wide would it need to be for two
369		coffins or if we're going for two it would need to be
370	Client 1	we'll have a measure up on that

Here, the architect first asks the client (in this extract, noted as Client 1) a clear question about the size required for the sanctuary entrance (A1, 342): "would this be wide enough for two coffins", but Client 1 is not sure. This is soon followed by a question from the second client who was present at the meeting. He asks about the size of the trolley on which a coffin is carried (A1, 350), implying that if this can be established then from it may be estimated the space needed for two trolleys. Despite the specificities of the questions raised by the architect and the second client, the replies of the first client express doubt over her ability to answer (A1, 347 and 351: "I don't know"). Eventually, rather than answering their questions about the required size of the sanctuary or the actual size of the trolley, she describe some of the problems associated with a multi-coffin funeral (A1, 351–363). Whilst the event of a three-coffin funeral might be rare, her reference to it implies that it should at least be considered (A1, 357–359). Eventually the architect again seeks clarification of the specific dimensions of the entrance space ("my question for you is how wide would it need to be" (A1, 368)), with the first client's answer ("we'll have to measure up on that" (A1, 370)) indicating that she does not know how wide the area would need to be, and that she, alone, is unwilling to estimate its size. Her reply puts off into the future a decision about the dimensions of the sanctuary, a decision that she proposes should be made by more than one person ('we').

3.2 *The management of clarity and ambiguity*

From these two representative extracts we can see that the architect performs his role as architect through actions such as defining the terms of the meeting and asking questions that are designed to elicit a certain type of information. In contrast, Client 1 performs her role as client through replying in ways that tend not to directly answer the architect's questions; that is, she does not reply using the kind of terms in which the questions are presented (terms associated with spatial dimensions or room measurement). Not all clients would respond as she does but her replies are significant because they are relevant within the context of the design process for this particular building. In effect, the interaction between this architect and this client creates a framework for design in which questions about the specific formal qualities needed in the future building are answered with detailed accounts of behaviour and descriptions of events as they occur in the present building. For instance, Extract 2 deals with the client's answer to the architect's first question ("is the waiting room big enough" (A1, 117)). Here, her response does not mention room size but she does provide a rich description of the behaviour of people who wait. She mentions the usual numbers of people, and notes that their waiting is an ordered activity whose management needs to be sensitively handled ("we might get people waiting for the eleven o'clock funeral and people at the ten o'clock perhaps arrive" (A1, 132–134)). Also, the client notes the potentially fraught emotional climate of those who wait: "they don't want to mix with other people" (A1, 134–135); "we do have problems with families" (A1, 136–137); "you know you'd be thinking it would bring them together but it actually makes it worse" (A1, 140–141)). As McDonnell[37] also notes in her discussion of this portion of the meeting, the client's account of others' activities shows the complexity of what goes on in the building and enables the client to perform effectively as a 'building expert'. However, by evading a direct answer in which she stipulates room dimensions, the client suggests that she does not consider herself to be best suited to decide precisely what the space should be like.

37. ***op. cit.***

Extract 3 also features the client offering answers that seem somewhat evasive, this time in response to the architect's question about the size of the sanctuary entrance and to the second client's question about trolley size. Yet, again, the client actually does provide information that is relevant to the discussion of the appropriate size for the sanctuary, though the significance of her answer does not seem to have been recognized in this sequence of interaction. That is, the architect's question: "would this be wide enough for two coffins" (A1, 345–346) is actually met by the client with a precise answer as she first says: "it wouldn't" (A1, 347). However, she immediately follows this statement with: "probably" and "I don't know" (A1, 347). That is, she couches the clarity of her initial answer in terms that suggest doubt. Eventually, she provides a detailed description of a multi-coffin funeral that actually supports her initial answer of: "it wouldn't" (i.e. the entrance wouldn't be wide enough for two coffins). This description indicates how cramped the space is now ("it's difficult to squeeze in" (A1, 352)) and upgrades this negative assessment to raise the extreme example of a three-coffin funeral (A1, 357–359). Through referring to such a rhetorically-hyperbolic case the client's answer can be understood as a form of 'disagreement management'[38] in which an unusual claim (such as a three-coffin funeral) may act as an expression of doubt. Thus, the client both implies that she disagrees with the dimensions suggested by the architect while also evading a personal recommendation of what those dimensions should be.

38. Antaki and Leudar (1992) *op. cit.*, p. 190.

4 PARTICULARIZATION, EXPERTISE, AND OWNERSHIP

The clients answers to the architect's questions offer descriptive stories that contain examples of particularization, or, details that enable the client to perform the specific knowledge and proficiency she has as an 'expert' on the building[39,40,41]. The rhetorical performance of such knowledge enables her to contribute to meeting A1 and thus be deemed relevant as 'a client'. However, the complex and nuanced information that she includes in her stories about the building's use does not always seem to be understood as relevant by the architect who, for instance, repeats his questions to her (A1, 342, 345–346, 368–369), perhaps seeking an answer more in accordance with the information he seeks (e.g. room dimensions). This level of apparent miscommunication does not appear to trouble the client; indeed, as we have seen, her answers seem designed to achieve a certain level of ambiguity. In effect, by offering descriptions rather than straight answers, the client casts the architect into the roles of 'client-interpreter' and 'decision-maker'. Her talk of the behavioural nuances of room use, rather than of room size, puts the architect into a position from which he is constrained to make decisions about interior spaces without clear direction from her. That the client believes the architect is responsible for making such decisions is further suggested by some passages of talk in which changes to the building's plans are suggested, and in which the client is sensitive to how the architect may perceive such changes; for instance:

39. Billig, M. (1996) *Arguing and Thinking: a Rhetorical Approach to Social Psychology*, Cambridge University Press, Cambridge.

40. Oak (2006), *op. cit.*

41. Wiggins, S, and Potter, J. (2003) Attitudes and Evaluative Practices: Category vs. Item and Subjective vs. Objective Constructions in Everyday Food Assessments, *British Journal of Social Psychology*, 42, pp. 513–531.

Extract 4, A1, Building ownership.		
802	Client	I don't want to compromise your design

Extract 5, A1, Building ownership.		
1152	Client	compromising your design all the time [*laughs*]

Extract 6, A1, Building ownership.		
1174	Client	what I don't want you to do is sort of come back in five years after
1175		we've done all this and then find that we've mucked everything up

Extract 7, A1, Building ownership.		
813	Arch	we might be able to get it to work it does go slightly against the
814		grain for me to do that but it does satisfy what you wanted and it means
815		that we could link this up to it actually so- ++++++
816	Client	OK is that too heartbreaking for you [*all laugh*]
817	Arch	well it's not as pure a summation as I was looking for but I mean
818		maybe there's another way of doing it maybe if I keep my thinking cap
819		on because you can see I'm trying to keep the spaces pure the
820		purer the space the more spiritual I think…

In Extracts 4 and 5 the client is concerned with not: "compromising your [the architect's] design" (A1, 802, 1152)[42]. In Extract 6, above, the client worries that if the architect returned to the building in the future he would find that subsequent changes they made on their own might have: "mucked everything up" (A1, 1175). In Extract 7 the client upgrades her concern with not compromising the architect's design by joking that such compromises may be: "too heartbreaking" for him (A1, 816). While all participants laugh at such an affect-related term, her words suggest that she recognizes aspects of the architect's emotional investment in the building (as was established earlier through his use of terms such as: "excited" and "dream come true" (A1, 73–74)). Extracts 4 to 7 thus suggest that the client accedes ownership of many qualities of the building to the architect, a position that the architect seems to accept since, when a compromise to his proposal is suggested he does not readily agree with it (as indicated by comments such as those in Extract 7: "it does go slightly against the grain for me" (A1, 813–814) and: "it's not as pure a summation as I was looking for … because you can see I'm trying to keep the spaces pure" (A1, 817–820)).

42. An issue also dealt with by **Luck (Chapter 13)**, *op. cit.*

5 THE PERFORMANCE OF ASSESSMENT

Finally, in Extract 8 below, we see how the client indicates an awareness of the architect's personal engagement with the building in relation to the negative judgments of others. Here, the client both delivers bad news while also apparently seeking to protect the architect from it.

Extract 8, A1, Building assessment.

1263	Client	because I think what [*funeral directors*] can't quite see from the drawings
1264		obviously the first drawings that we've got there is the fact that some of
1265		them have mentioned the feeling that they get from those sort of what they
1266		think is some of the comments that have been made about
1267	Arch	the aircraft hangar
1268	Client	the aircraft hangar or a chicken hut or-
1269	Arch	[*makes a sound with his lips*]
1270	Client	I'm just pre-warning you what they might use as a comment so I don't
1271		want to make you feel you know that's what they might mention but they
1272		can't as I've said to them
1273	Arch	chicken hut
1274	Client	I said what you're not looking at is the sense of what the roof will be
1275		covered in in a sense how it will look as we drive as you said to me I
1276		said to them what you've got to remember is you're looking at it from
1277		this way you won't be looking at it when you drive in that way ...
1278		which is why I've done the photographs sorry to point
1279		the photographs of the actual building itself so they could see the actual
1280		sort of you know the feel of how the roof shape is from this angle in a
1281		way so I've tried to explain that to them and pre-warn them so they don't
1282		pick on yer [*laughs*]

In this extract, it is notable that the client is not the first to report a specific, derisive term that others have applied to the crematorium; instead, the client forecasts[43] that bad news is imminent through terms such as: "some of them have mentioned the feeling that they get" (A1, 1264–1265). Structurally, in conversation, such a forecast or 'preannouncement' acts as a "device by which a news giver can discover whether a recipient already knows some news-to-be-told"[44,45]. Indeed, the architect's reply "aircraft hangar" (A1, 1267) indicates that he does know the specifics of some of the negative judgments. In keeping with the characteristic structure of delivering bad news, the client then 'elaborates'[46] on the negative term by adding another, even more negative term ("chicken hut" (A1, 1268)). This sequence indicates the manner in which the participants are able to negotiate a series of social relationships and follow conventions for politeness in language. That is, 'polite' conversation restricts a participant from making a report that threatens another participant's 'face' (the positive, public self image that they project for others)[47,48,49]. In this extract, the client's use of a preannouncement maintains her 'face' and that of the architect (by not bluntly reporting the negative assessments of others she can distance herself and the architect from the judgments of the funeral directors and so maintain a cordial personal relationship with the architect). Further, the architect's utterance of 'aircraft hangar' indicates that

43.
Maynard, D. (1996) On 'Realization' in Everyday Life: The Forecasting of Bad News as a Social Relation, *American Sociological Review*, 61, pp. 109–131.

44.
ibid., p. 115.

45.
Maynard, D. (2003) *Bad News, Good News: Conversational Order in Everyday Talk and Clinical Settings*, University of Chicago Press, Chicago, pp. 88–119.

46.
ibid., p. 94.

47.
Goffman, E. (1999/1967) On Facework: An Analysis of Ritual Elements in Social Interaction in Jaworski, A. and Coupland, N. (eds) *The Discourse Reader*, Routledge, London, pp. 306–321; p. 306.

48.
Brown, P. and Levinson, S. (1987) *Politeness: Some Universals in Language Usage*, Cambridge University Press, Cambridge.

49.
Mills, S. (2003) *Gender and Politeness*, Cambridge University Press, Cambridge.

he is aware of the client's impending news and so, by reporting this negative assessment himself, he saves the client's face (by removing from her the responsibility to report the negative comments of the funeral directors). Additionally, he saves his own face, by indicating that the bad news is not a surprise.

The manner in which the participants manage the delivery of the negative assessments of the funeral directors is interesting in part because the sequencing of these utterances demonstrates how the client meets her responsibilities to the funeral directors to communicate their perceptions to the architect, while also meeting her own need to maintain an amiable relationship with the architect. In effect, the client manages to: inform the architect of the negative opinions of others; establish herself as the architect's protector ("I'm just pre-warning you" (A1, 1270)); and, present herself as a client who has gained knowledge through her previous interactions with the architect. Indeed, not only has she gained knowledge of architecture, she reports that she has used it in an attempt to educate the architect's critics ("as you said to me I said to them" (A1, 1275–1276); and, "which is why I've done the photographs … so they could see the … feel of how the roof shape is … I've tried to explain that to them" (A1, 1279–1281)).

50.
Silverman, D. (2006) *Interpreting Qualitative Data* (3rd edition), Sage Publications, London, pp. 208–209.

The issue of how the participants jointly manage the reporting of the funeral directors' negative assessments returns us to the early part of this chapter wherein Extract 1 shows the architect proposing that the meeting should be a feedback session; a proposal that was evaded by the client, since she did not offer him direct feedback from others. Given that we now know the negative qualities of some of this feedback, the client was constrained not to report it, especially not early in the meeting. That is, conversation's structure sees certain types of response as 'preferred',[50] with a dispreferred response being delayed or hedged. Thus, we can see how the characteristics of preferred responses and politeness conventions help to structure meeting A1 so that it proceeded as an occasion for the architect to ask questions of the client (rather than as a session where the client would readily report the feedback of others). Accordingly, as meeting A1 becomes an architect/client question-and-answer session, the talk enables the client to perform as someone who is knowledgeable about the activities that occur in the existing building, but as someone who is disinclined to state precisely how the new building's spaces should be arranged.

6 CONVERSATION, CATEGORIES, AND DESIGN

An interesting aspect of the way this meeting unfolds is that, together, seamlessly, and without prior arrangement the participants in meeting A1 perform their roles within the membership categories of client and architect in ways that help to perpetuate everyday perceptions of architectural practice. That is, although at several points the architect draws attention to the collaborative nature of design (e.g. "well it's your building you know" (A1, 1177)), nevertheless as we saw in Extracts 4 to 8, the talk of the participants accedes ownership of the building's form to him, as architect. This is further supported through the architect's claiming 'ownership' of the structure of the meeting, and through his inclination to determine the kind of topics that should be asked about (e.g. the appropriate dimensions of rooms). Such interaction subtly supports a popular view that architects are likely more concerned with building form than with its function and that they may be inclined to fulfil their own vision rather than that of the clients[51,52,53]. Such talk also supports a professional perspective wherein architects are presented as people who use "their unique creative skills to advise individuals" and as people who "can be extremely influential as well as being admired for their imagination and creative skills"[54].

In the interactions discussed here the architect was placed in a decision-making role partly in response to the client's talk, in which the client herself avoided making design-related judgments about, for instance, the measurement of rooms. Yet, perhaps under the circumstances, the client would prefer not to be responsible for the configuration of a building that she may have to justify to others for years to come. That is, the client's everyday life brings her into frequent contact with the funeral directors, some of whom have stated that they believe the proposed building looks like an aircraft hangar or a chicken hut. In effect, the client's disinclination to precisely answer the architect's questions about building form and space allocation may allow her, in the future, to save face with colleagues who could question her about such aspects of the finished building. This is not to suggest that the client's somewhat evasive answers to the architect are the result of cowardice. Instead, in the context of an analysis of the interaction of meeting A1, her talk suggests the complexities of her role as client, and indicates that she manages to communicate the needs of a range of building stakeholders (from bereaved visitors to critical funeral directors) while also skilfully maintaining a genial relationship with the architect (and her colleague, who was also present at meeting A1). Here, evasiveness in talk can be seen as something of an interactional accomplishment and demonstration of the diverse requirements that may be associated with her role as client.

51. Jenkins, S. (2006) The Gherkin is Magnificent but it Should Have Been Built Elsewhere, *Guardian Unlimited*, http://tinyurl.com/6oporw (accessed July 22 2008).

52. Mawer, N. (2007) Crystal Vision Blurry: ROM's Addition Does Little to Brighten City's Streets, *The Star.com*, http://tinyurl.com/5s3x5y (accessed 22 July 2008).

53. Morrison, R. (2007) For Ever Thinking Outside the Boxy: A New Design Museum Show Pays Tribute to Zaha Hadid's Remarkable, if Often Impractical, Vision, *Times Online*, http://tinyurl.com/6bkn2c (accessed July 22, 2008).

54. RIBA: Royal Institute of British Architects Website, *Becoming an Architect*, http://tinyurl.com/6eplau (accessed 22 July 2008).

7 CONCLUDING REMARKS

This chapter's discussion of questions and answers, attributions of ownership, and the management of negative assessment in a meeting about architecture has shown how an architect and a client constrain and afford each other's design-related behaviour through their talk. While the structures and topics of their interaction are ordered, the specificities of their performance ensures that this design process is a singular and nuanced context from which a singular and nuanced building will emerge. Despite the orderliness and specificity of this process, some generalized comments may be made. First, participants come to design-related contexts with pre-existing knowledge and beliefs about what design (architecture) is and how it happens. It is possible that designers may have reflected upon what constitutes appropriate design-related knowledge and behaviour, but it is likely that many clients have not. While the impact on the design process of participants' knowledge and beliefs can perhaps be traced, it is difficult to anticipate, given that one participant will elicit a context-specific performance from another. Nevertheless, it may be worthwhile for architects and clients to occasionally discuss the nature of their dialogue, particularly early in the design process, so that participants may become more aware of how they may be talking in ways that could perhaps limit aspects of the design process.

Second, and related to the first point, is that roles that become associated with membership categories may unconsciously constrain participant behaviour. That is, even if each party comes to the design process with a willingness to collaborate, it may be difficult to achieve, given that collaboration occurs partly through moment-by-moment interaction. Thus, as we have seen, a client's talk may help an architect perform initiative and ownership, while an architect's talk may help a client perform ambiguity and relative acquiescence. Third, although the roles of architect and client are performed in the present, they are also carried into the future. Therefore, a client may manage their role in the design process by trading off a high degree of participation in the present with the ability to save face with colleagues in the future. Given this possibility, perhaps as the design process unfolds, architects and clients could discuss how a client's relationships with other stakeholders may impact upon their decisions concerning a future building's form and function. Finally, in terms of how the topics discussed here may be useful in the context of design education: students could be made more aware of how the orderly performance of (polite) interaction may itself contribute to the practice of design, and that clients may demonstrate ambivalence within the design process for reasons that could range from their lack of design-related knowledge to the nature of the ongoing relationships they have with colleagues.

To summarize then, we have seen how an architect and a client perform their roles in an accountable and competent manner, in part, through the structures and topics of social interaction. In effect, their communication is constitutive of an object (the crematorium), a process (design), and of their mutual roles in that process. Through their talk, the architect and client draw upon categories of action and so together they design a building, but their talk also helps to create and perpetuate the customs, attitudes, beliefs, and behaviours that form and inform the social practice of design.

18
Behind the Scenes of the Design Theatre: Actors, Roles and the Dynamics of Communication

Gabriela Goldschmidt & Doron Eshel

All the world's a stage,
And all the men and women merely players:
They have their exits and their entrances;
And one man in his time plays many parts...
(William Shakespeare, *As you like it*)

This chapter analyses architectural design meetings from the point of view of the actors who participate in the design 'theatre': their formal and informal roles in the design team, the specific contribution of designers and clients, and the nature of the communication between them. We look at the data in terms of adherence to purely professional decision making along with other dimensions of discourse that cushion the design process and provide the necessary social scaffolding that ensures successful team collaboration.

In this chapter we inspect the meetings concerning the architectural design of the new crematorium, as documented by video recordings and transcriptions of the verbalisations. At first sight the data seem straightforward and factual, but a second look reveals that they are telling in terms of the roles that all participants, professionals and non-professionals alike, play in these kinds of design meetings. The main participants in the meetings are Adam and Tony, the architects; and Anna and Charles, the clients. Two other people appear briefly: Sally, the customer service supervisor, and Peter, the researcher, who monitors and records the meetings. Our interest lies in the complex architects-clients interaction in these meetings, from which we learn a lot about how design is actually 'done' in practice[1].

1.
See also: Oak, A. (Chapter 17) Performing Architecture: Talking 'Architect' and 'Client' into Being.

The phase of the design process that we witness is intermediary. The design concept has already been set, the main features of the program planned and implemented in the preliminary design, and the functions well allocated; plans, sections and elevations have been drawn. These meetings, A1 and A2, are ordinary, run-of-the-mill sessions in which the design is re-visited, aiming at the refinement of the design; the tweaking of the plan and section so that the building will better fit its surroundings and the needs of the clients. The meetings in question are part of a series of meetings between the parties – architects and clients – held at various stages of the design, with other people joining in as needed. Together, the participants constitute a team whose objective is to refine the design of the new crematorium; they should be regarded as a team because they fit a widely accepted definition of a team: a small number of people with complementary skills who are committed to a common purpose, set of performance goals, and approach, for which they hold themselves mutually accountable[2]. We think of this team's meetings as somewhat resembling theatrical rehearsals, in which each actor plays a role and the group become an ad-hoc team that works to improve the outcome from one rehearsal to the next until the show opens (and in parallel, the building begins to be constructed). The theatre metaphor is not a structural one and we do not plan to offer a discourse grounded in performance theory. However, it reminds us to view the participants in the design meetings as actors (Section 2) who perform roles (Section 3), and the dynamics of turn-taking in the conversation as a joint effort to 'resolve' a situation. This effort turns out to involve an intricate pattern of communication and interaction among actors who respond to one another in a goal-oriented exchange, as in a play (Section 4).

2.
Katzenbach, J.R. and Smith, D.K. (1992) *The Wisdom of Teams*, Harvard Business School Press.

Our main aim in this chapter is to expose, analyse, and illustrate the ingredients of design-talk in ordinary design sessions, and to show that they are social settings in which achieving design outcomes requires a fair amount of conversation that does not pertain to 'hard-core' design topics only. We tie this analysis to the verbal behaviour of the actors in the design team in terms of the roles they play. A simple, straightforward and detailed analysis is what we aim at and it is for others to relate our findings to the rich literature in social psychology that studies the role of talk in various group settings.

The DTRS7 data are divided into video and text. While the visual data in the video files may hold vital information about behavioural aspects of the meetings, for us the transcriptions of the meetings hold a better body of data regarding the participants, the way they perceive their roles in the design team, the way they interact with each other, and the decision making process they go through throughout the meetings. The lack of annotated drawings from the meetings themselves and sketches made *in situ* precludes any attempts to analyse the graphic output. The sole role of the drawings has therefore been illustrative. In this capacity they helped understand the project. Our analysis methodology is spelled out in Section 1.

1 METHODOLOGY

The method chosen for an analysis of these meetings is text-based protocol analysis. What we sought to analyse was the explicit and implicit actors' roles in the meetings, and their chosen mode of interaction, in order to best move the design forward. To that end, we first divided the transcriptions into individual segments: 17 in the first meeting and 20 in the second meeting. Segments were later grouped into a number of categories. The individual verbalisations in each segment were then divided into utterances. An utterance could be as short as a single word (such as "yeah") or as long as a full comment composed of several sentences, by one of the actors. Utterances were then coded – there are seven different codes (see Section 1.2). The first four are directly related to the features of the design; the latter three have to do with interaction strategies and asides. The quantitative results of these analyses serve us throughout the rest of this chapter. The segmentation, division into utterances, and the coding of these utterances was done separately by two researchers who later discussed and resolved (a small number of) incompatibilities where they existed.

1.1 *Segmentation and categorisation*

The rationale behind segmentation is that every discourse is naturally driven by a succession of topics. Team members discuss different design issues and topics, and the transition from one to another is identifiable by a change of topic or subject matter; each segment deals with one subject and is categorised accordingly. As noted above, two researchers carried out the segmentation. There are of course overlaps and sometimes we find brief 'implants' of topics also dealt with in previous (or subsequent) segments. We view segmentation as important because we are interested in both the topics that are being discussed and the actors who initiate the discussion. Our approach to segmentation is akin to that of McDonnell[3], who parsed the transcriptions into phases which were then further divided into episodes.

3. McDonnel, J. (chapter 14) Collaborative Negotiation in Design: A Study of Design Conversations between Architect and Building Users.

Segmentation has further advantages. By looking at the dominance of specific actors in certain segments, we may infer insights regarding those actors' special interests and expertise. Segmentation also enables us to exclude from the discussion segments which pertain to irrelevant subjects, for example segment 16 in A2 which focuses on scheduling the next meeting.

The segments were grouped into the categories listed below. Categorising the segments gives us a handle on the main concerns in the particular project at hand, and the way it unfolds. The following is a brief description of the categories:

OVERVIEW
: A general overview of the project, the design process, a general introduction or a summary of a meeting.

DESIGNED ELEMENTS
: The actual elements of the project. In these segments the design of the building's various spaces and their functions, as well as adjacent open spaces, are discussed and evaluated.

MATERIALS

The materials the building will be made of, in particular the roof and the cladding of external walls.

EQUIPMENT

These segments are technical in nature and deal with the equipment to be installed in the building.

ENVIRONMENT

The site, the landscape, and other environmental issues.

REPRESENTATION

The way the project is represented to the client, and the types of representation used (plans, elevations, perspectives etc.).

Four segments were left out of the categorisation scheme: A1 segment 16 (A1.16) which deals with cost estimates; A2.16 which deals with scheduling of the next meeting; A2.17 which deals with the customer service department in the municipal administration and its role in the project, and A2.20 which is a farewell ceremony[4].

4.
There is no single scheme of design categories, and different researchers use different schemes with partial overlaps. The purpose of a study and the level of detailing are an important factor in determining the codes. See for example Broadbent (1973), Goldschmidt (1983), Heath (1986), Hillier and Leaman (1972), Schön (1983).
Broadbent, G. (1973) *Design in Architecture*, Wiley, London.
Goldschmidt, G. (1983) Doing Design, Making Architecture, *Journal of Architectural Education*, 37, pp. 8–13.
Heath, T. (1986) *Environment-Behaviour Research Inputs to Design Processes*, Queensland University of Technology, unpublished manuscript.
Hillier, B. and Leaman, A. (1972) A New Approach to Architectural Research, *RIBA Journal*, December, pp. 517–521.
Schön, D.A. (1983) *The Reflective Practitioner*, Basic Books, New York.

1.2 *Utterances and utterance coding*

As mentioned above, after the transcriptions were divided into segments which were each assigned a category, a finer parsing system was applied and segments were divided into utterances. An utterance is a word, a sentence or a phrase that was verbalised by one of the actors. One utterance may be distinguished from its predecessor either by a shift of turn to another actor, or by a change of subject within a single actor's turn. Segments vary in terms of the number of utterances they comprise. The longest is segment A2.04 which holds 184 utterances; the shortest is A2.13 with just 8 (the number of utterances does not necessarily correlate with duration of time). It is important to stress that utterances may have more than one code; quite often two or even three codes are assigned, therefore the number of codes does not equal the number of utterances. The quantitative data pertaining to utterances in this chapter refers to coded utterances, i.e. an utterance with two codes is counted twice. The breakdown of segments into utterances and their coding is illustrated in the example given in Table 1, a detailed description follows.

An utterance is thus defined as the basic building block for the analysis of the transcriptions. The scheme of codes for analysis was developed on the basis of transcription content in terms of issues and topics that are typical of architectural talk: we write this chapter from the perspective of architectural practice (both authors are architects). The choice of codes also facilitates the division into 'gross' and 'net' discourse (see Section 4). Descriptions of the codes are as follows:

Utterance No.	Actor	Transcript Lines	DESCRIPTION	ATTRIBUTES	BEHAVIOUR	PRECEDENT	QUESTION	CONFIRMATION	OTHER	Content
1	Ad	326–328	x							OK + from this point you go through a lobby into the chapel and the chapel layout is quite similar to the existing chapel
2	Ad	328–329	x							where it gets different is that we have what I call a sanctuary to one side or an ante chapel for the very small funerals
3	Ad	329–330	x	x						and we have a much more dominant catafalque design at the end
4	An	331						x		thank you yes
5	Ad	332–333		x						erm last time we spoke Anna you thought that that the original catafalque design just wasn't bold enough

Table 1. Example of segment parsing into utterances and their codes (A1 segment 4).

DESIGN DESCRIPTION

Spaces in the building and around it, movement through the complex, details and particular elements of the layout, openings, and so on. An example is the following utterance by Adam: "there are three possible routes they either come through here one or they go through here two or they can go through there three" (A1, 1013–14). When a speaker says, 'here' or 'there' or uses similar indications, he or she usually also points to the appropriate spot in one of the drawings that are present in these meetings.

ATTRIBUTES

This refers to a quality of the design, or rather of a particular feature or element that has been described. For example, Adam's remark: "I think it would be very light" (A2, 265), or Anna's: "yes one or two sort of quite nice oak trees that are quite big probably been there for a few years now and so it would be nice to keep those as well…" (A1, 1927–29).

BEHAVIOUR

An utterance that describes, or discusses the behaviour of various building users. These users may be the people working there or the public using it. The behaviour may be observed (past experience) or anticipated in the future. For example, Anna states: "I'm also thinking of the fact that if we've got a flow of people walking through that then restricts us we can't put seats through that because in a sense we need to keep an access open and so the seating will be against the wall" (A1, 121–24).

PRECEDENTS

Although rare, this is a very important code. It establishes an existing building as an exemplar. In design conversations precedents have the important role of creating shared images that help bring the mental models of designers to a common platform. Many non-professionals have difficulties reading technical drawings and consequently imagining what a building would look like when built. A precedent helps the architect convey the right kind of image. On the other hand, when trying to explain to the architects what they want, clients refer to buildings they know to explain their visions. A good example is a precedent Anna brings into the conversation early on: "actually that looks like the one [crematorium] they went to have a look at in NOTTINGHAM I think when they designed this one they went to quite a few places" (A1, 51–2).

The next two codes have to do with the flow of conversation and information within the team. They are QUESTIONS and CONFIRMATION. We shall have more to say about them later (Section 4). The last code is:

OTHER

Anything that does not fit in any of the previous codes.

Utterances by Peter and Sally were not coded, as they were not perceived as members of the design team.

2 ACTORS IN THE DESIGN THEATRE

The American Institute of Architects (AIA) acknowledges the great significance of clients' input in the success of building design. Cuff[5] cites an AIA handbook:

> "Architecture is a responsive art. Without a client, there is no architecture. A successful client-architect relationship constitutes the cornerstone of fine architecture."[6]

Lewis[7] also emphasises the importance of client contact in his guidebook to the architectural profession:

> "Client contact is among the most critical activities in architectural practice … the face-to-face interaction between architects and clients can disclose most clearly the differing agendas of each party, as well as offer the opportunity to reconcile those agendas."[8]

In what follows we try to fathom this relationship as revealed in common, on-the-job interaction. Our descriptions of the actors are based on the transcriptions, as well as on the posterior interview with the architect, which was included in the DTRS7 dataset.

2.1 *Architects*

The design of the building was entrusted to the municipality's own architectural design department. We do not know anything about this department, but of course we get to know the architect in charge of the design of this project – Adam, who is one of our

5. Cuff, D. (1991) *Architecture: The Story of Practice*, MIT Press, Cambridge, MA.

6. *ibid.*, p. 81.

7. Lewis, R.K. (1998) *Architect? A Candid Guide to the Profession (second edition)*, MIT Press, Cambridge, MA.

8. *ibid.*, p. 200.

principal actors. In the second design meeting we also meet Tony, another architect in the department, who was a member of the design team. Although present throughout the second meeting, his active participation is quite minimal. Also mentioned is Beatrice, to whom Adam refers as his boss. He reports many conversations with her in the early stages of the conceptual design, in which her input was apparently helpful.

2.1.1 *Adam: chief (municipal) architect*

Adam is an architect with the city municipality's design department. He was offered the assignment of designing the crematorium by the department head almost by accident, when he mentioned to her that he had almost run out of projects. He welcomed this assignment very much; in fact he refers to it in the interview as "every architect's dream", because it involved designing what he calls "spiritual spaces". By this he means first and foremost the chapel; the actual cremators were not included in the initial brief and were added only later.

Adam describes himself as very interested in spirituality and spiritual architecture, and this is indeed reflected in his comments during the design meetings. For Adam, spiritual, or sacred spaces, are characterised by a "cool and calm … feel of the architecture", which is achieved through the choice of form and materials, the way the building is laid out, and the way (diffused) natural light penetrates and illuminates the spaces. He also mentions symmetry and simplicity, spaces that are pure and not cluttered, as helpful in creating the right 'feel'.

When this assignment came along Adam looked for a precedent that would help him shape a concept for his design, and Louis Kahn's Kimbell Art Museum, built at Fort Worth, Texas in 1972, suggested itself as an immediate relevant candidate. Adam admires Kahn's work and knows it well; he owns books on Kahn and had visited the Kimbell museum in person. He has already used architectural principles derived from Kahn in previous designs. Kahn is known for his 'spiritual view of architecture' and it is therefore not surprising that Adam feels close to Kahn's concepts. The fact that the Kimbell is a museum, unlike the building Adam is designing, may be responsible for the image of the building Adam holds in his head when he says: "I mean I see it as a spiritual modern art gallery flavour sort of sp[ace]" (A1, 1389). In particular, Adam adopts Kahn's famous principle of served and servant spaces which seemed perfect to Adam who testified that this principle delivered what he was looking for.

However, Adam sees his building also as a compromise. There were many constraints regarding the precise location of the building at the site having to do with: wildlife and its conservation, the relationship to roads and parking and to the existing crematorium facility, and so on. There were also many major requirements that had to be reckoned with, which were revealed as the design process moved along. "The building is essentially a business", said Adam in his interview, and it has to be efficient and "function like a production line". Adam is not only a lover of spirituality but also an architect with a strong service orientation, very attuned to his clients and anxious to satisfy them as best he can (in addition, the city has a customer service agency which oversees the service he provides).

2.1.2 *Tony: associate (municipal) architect*

Tony, Adam's assistant, is a minor actor in the design meetings, of which he attends only the second. His involvement in the conversation is minimal, and he offers comments almost exclusively when asked. It is obvious from the discourse, though, that he is an active participant in the design of the building and is particularly involved in certain topics like the audio visual systems to be installed in the chapel. Tony participates more intensively in the discussion about the roof. At issue is the finishing material with which the vaults will be clad; pros and cons are brought up regarding copper and lead. Tony asks questions or offers local solutions (a folding partition to hide a kitchenette; a shelf for the display of a cross in the chapel, etc.), which demonstrates that he is versed in all aspects of the design.

Tony is also reported to have learned a new computer program (SketchUp) which is relevant to the production of presentations (walkthroughs, in this case), and we may infer that he is an indispensable, all-round assistant designer who knows the design well and can contribute at different levels. The hierarchy is very clear though: Adam is the project architect and Tony is a team member who assists him.

2.2 *Clients*

The clients are represented by Anna, the manager of the crematorium and registrar of the cemetery, and Charles, who heads the city office that runs the crematorium. They are both very involved in the design of the new facility and know a lot about the various aspects of this type of institution. Anna is thought of as the direct client who will have to live with consequences of all design decisions on a daily basis; Charles is seen as an indirect client who represents the interests of the community, in his capacity as the municipal officer in charge of the project.

2.2.1 *Anna: direct client*

Anna is a veteran; she joined the crematorium team in 1987. She seems to love her job and is deeply involved in every aspect of running the existing crematorium. The impression one gets is of a most dedicated worker who lives her job around the clock. Anna is sensitive to the needs of both her clientele and her colleagues and employees. She is aware of every possible scenario in the crematorium activity and knows how the physical environment may support or hinder people's activities. Quite surprisingly, for a non-professional, she is a good plan reader who can easily relate configurations and dimensions to requirements. She gives detailed and accurate descriptions of behaviours and feelings that the design must accommodate: she can tell how much space is needed for a small group of people in a sanctuary, where the minister should be able to see participants of the next funeral in order to walk out and greet them, and how best to accommodate a religious symbol or a book of remembrance. She knows when and where people like to gather, and who wants to sit in the back rows during the funeral service.

Apparently her long years of work in the crematorium, coupled with a natural sensitivity and empathy, have turned her into a first-rate expert in matters of funerals and cremation. She buys books on the subject and visits sites of other crematoria to widen her knowledge. She is also an appreciative and cooperative team-member who accepts and respects the architects' expertise. She has excellent communication skills in the team context, where her healthy sense of humour and self-deprecation make it very easy to relate to her and accept her comments, despite the fact that her remarks are often lengthy and she frequently repeats herself.

2.2.2 *Charles: indirect client*

Charles is mostly interested in technical issues, in budgetary concerns, in compliance with the ecological appraisal that had been prepared for the site, and in access and parking issues. He has less to say about the attributes of spaces which relate to the experience, or behaviour, of the building users, and appears to trust Anna to cover these aspects. At the same time he is not disinterested in all other design aspects, follows the conversation very closely, asks questions and voices opinions. He definitely sees himself as an integral member of the design team, while respecting Anna's role as the principal client whose requirements must be met. Charles and Anna appear to maintain a good working relationship based on mutual acceptance and respect.

In contrast to Anna, Charles is brief and hardly ever says more than a few words; he does not use humour and does not get personal. He is direct, focused and attentive, and makes sure nothing is forgotten or left unresolved.

2.2.3 *The users: absent clients*

Various other actors are mentioned in the discussion, although none of them participates in it. They include, of course, the mourners who attend funerals; accompanying staff like chauffeurs, funeral directors, clergymen and priests; administrative staff and municipality officers responsible for issues relating to the site such as highways and environmental qualities. Anna represents the needs and views of those who work in the facility, as well as others such as mourners. She reports their concerns and wishes, and is anxious to involve them to the point of holding a meeting between them and the architects. Other absent clients are represented mostly by Charles.

3 DESIGN TEAM ROLES

9.
Shani, A.B. and Lau, J.B. (2000) *Behaviour in Organizations: An Experiential Approach (seventh edition)*, McGraw-Hill.

Shani and Lau[9] assert that role differentiation occurs in every group that works together. They distinguish between what they call task roles and group-building and relationship maintenance roles (a third type, individual roles, is not relevant to the current chapter). We weave this distinction into our discussion, which is cast in terms of formal and informal roles. Role playing is, of course, a major reason for our theatrical association in construing what the actors do in these design meetings. In this analysis we group the actors together into 'architects' and 'clients', but we still maintain relevant distinctions within each group.

3.1 *Formal roles*

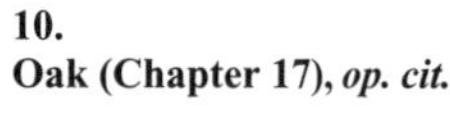

10.
Oak (Chapter 17), ***op. cit.***

The core design team includes architects and clients, both of whom arrive at the task aware of their roles, including social (group-building and relationship maintenance) roles[10] which guide their interactional behaviour. The formal task role of the architects is to design the crematorium and to present their design to the clients. That of the clients is to provide information and feedback. We shall inspect how they fulfil these roles.

3.1.1 *The architect as form giver*

The design sessions we analysed took place after the building concept had been agreed upon. Adam, in the interview, relates the history of his concept which is strongly influenced by the work of the famous American architect Louis Kahn (1901–1974). Brief references to the precedent he uses can be found in the data as well. Apart from the Kimbell precedent, the architects make very few references to other well-known precedents (an exception is Le Corbusier's Ronchamp chapel, which Adam uses to exemplify small coloured windows in a thick wall).

A considerable portion of each of the two meetings is dedicated to a detailed presentation of updated drawings of the project, during which the architects explain the design features and intentions for the design and the clients react by asking questions and offering commentary, approval or disapproval. Of the utterances in the discourse, 26% have been coded as DESIGN DESCRIPTION; of these the architects are responsible for 345 utterances, or 15% of the total number of utterances (which are 59% of the utterances under this code). On the face of it this is a low percentage which may reflect the state of the design at this phase: major decisions have already been taken, presumably following detailed descriptions and discussions. At this stage only details and small revisions are being discussed and the architects do not consider it relevant to present and discuss anything but relatively minor topics which have not yet been resolved. DESIGN DESCRIPTION is the only code under which utterances by the architects are the majority; under all other code headings the clients have the upper hand in terms of the number of utterances they make (an exception is the code DESIGN ATTRIBUTES where the count is almost equal, 135 versus 133 in favour of the architects). It is not surprising that under the code BEHAVIOUR the architects are responsible for significantly fewer

		OVERVIEW		ENVIRONMENT		DESIGNED ELEMENTS		MATERIALS		EQUIPMENT		REPRESENTATION		Total	
Category		No.	%	No.	%	No.	%	No.	%	No.	%	No.	%	No.	%
DESIGN DESCRIPTION	Ar.	23	22.5	31	14.1	172	16.9	23	9.5	89	14.6	7	11.3	345	15.3
	Cl.	10	9.8	27	12.3	102	10.0	16	6.6	85	13.9	1	1.6	241	10.7
	Ar.	2	2.0	19	8.6	44	4.3	33	13.7	35	5.7	2	3.2	135	6.0
ATTRIBUTES	Cl.	3	2.9	27	12.3	63	6.2	19	7.9	19	3.1	2	3.2	133	5.9
	Ar.	2	2.0	1	0.5	24	2.4	0	0	5	0.8	0	0	32	1.4
BEHAVIOUR	Cl.	1	1.0	3	1.4	92	9.0	14	5.8	36	5.9	0	0	146	6.5
	Ar.	1	1.0	2	0.9	3	0.3	4	1.7	2	0.3	0	0	12	0.5
PRECEDENTS	Cl.	2	2.0	3	1.4	10	1.0	3	1.2	5	0.8	0	0	23	1.0
	Ar.	5	4.9	2	0.9	42	4.1	3	1.2	12	2.0	2	3.2	66	2.9
QUESTIONS	Cl.	4	3.9	9	4.1	61	6.0	19	7.9	42	6.9	2	3.2	137	6.1
	Ar.	10	9.8	17	7.7	64	6.3	11	4.6	55	9.0	2	3.2	159	7.1
CONFIRMATION	Cl.	17	16.7	36	16.4	241	23.6	34	14.1	156	25.6	12	19.4	496	22.0
	Ar.	11	10.8	20	9.1	44	4.3	26	10.8	29	4.8	15	24.2	145	6.4
OTHER	Cl.	11	10.8	23	10.5	58	5.7	36	14.9	40	6.6	17	27.4	185	8.2
	Ar.	54	52.9	92	41.8	393	38.5	100	41.5	227	37.2	28	45.2	894	39.6
Total	Cl.	48	47.1	128	58.2	627	61.5	141	58.5	383	62.8	34	54.8	1361	60.4

Table 2. Architects' and clients' utterances, by code and category (A1 and A2).

utterances than the clients, as BEHAVIOUR describes people's actions in the various spaces, and this is indeed the clients' turf. However, it is surprising that under PRECEDENTS the architects' utterances are only half as many as those of the clients, as architects frequently refer to precedents both among themselves and in deliberations with clients. Table 2 summarises the contribution of utterances by both parties in both meetings, according to utterance codes in the different design categories.

The breakdown into categories in Table 2 reveals a somewhat more complex picture. Whereas DESIGN DESCRIPTIONS are dominated by the architects in all categories (note that for EQUIPMENT the difference is very small), they also have a slight advantage in the OVERVIEW category under the BEHAVIOUR and QUESTIONS codes, and for MATERIALS they rate higher under the ATTRIBUTES and PRECEDENTS codes. They are also higher on ATTRIBUTES in the EQUIPMENT category. There are a few 'ties', especially in the REPRESENTATION category. If we leave out the less design focused codes (QUESTION, CONFIRMATION and OTHER), which we comment on later, we see that in terms of utterances the architects are dominant only under DESIGN DESCRIPTION and in terms of categories they have the upper hand under REPRESENTATION, and mainly so under MATERIALS. OVERVIEW and EQUIPMENT are balanced categories.

3.1.2 *The client as information and feedback provider*

In her capacity as the manager of the crematorium, Anna is expected to contribute her knowledge about the various needs of its users, based on her rich experience. She does so with gusto. Charles is also a very interested client, although he is not as closely involved in the daily life of the crematorium and therefore he does not address the direct needs of users as frequently as Anna. We expect the clients to have a lot to say about users' behaviour in and around the designed building and therefore it is not surprising that under the code of BEHAVIOUR, their utterances outnumber those of the architects by more than four to one (146 versus 32 utterances, or 6.5% against 1.4%).

When we look at the breakdown in Table 2, we notice that the clients are also the dominant speakers in the categories of ENVIRONMENT and, surprisingly, also DESIGNED ELEMENTS. Their contribution is very significant under the codes of PRECEDENTS and ATTRIBUTES (disregarding the last three codes – QUESTIONS, CONFIRMATION and OTHER). This is not a trivial finding[11] and we shall discuss it in the following section.

11. We must bear in mind that the clients' participation in the meetings was more intense than that of the architects, because Tony's participation, in the second meeting only, was very partial. This resulted in the clients being responsible for a higher total number of coded utterances: 1361 (60.4%) versus 894 (39.6%) by the architects.

3.2 *Informal roles*

The literature points out that the division of labour in teams according to formal (mostly disciplinary) roles is often accompanied, and sometimes even replaced, by informal roles that team members play[12]. The latter come into being when team members feel they have something to contribute to the process that is not part of their formal, or task role. Informal roles are sometimes taken in an impromptu manner.

12. Hare, A.P. (1992) *Groups, Teams, and Social Interaction*, Praeger, New York.

On the face of it the architects in the data have not taken on any roles beyond their formal ones; in fact they fall somewhat short of what one may expect of them in this kind of discourse, namely more reference to ATTRIBUTES of designed spaces rather than their mere DESCRIPTION. Spaces are designed with specific attributes that are meant to serve a purpose, but in the data the number of utterances by the architects coded as ATTRIBUTES is not very high. Maybe they felt that the design was self-explanatory and attributes are to be taken for granted once the design has been described. However, the relatively large number of clients' utterances under this code suggests that stating ATTRIBUTES is necessary and helpful, and if the architects do not come forth with a satisfactory statement, the clients do so instead. Sometimes the clients (Anna in particular) couple utterances coded BEHAVIOUR, made in their formal capacity as information and feedback providers, with utterances coded ATTRIBUTES as Extract 1 shows.

Extract 1, A1, Clients coupling of BEHAVIOUR utterances with ATTRIBUTE utterances.

Adam	OK my only point is that it would be relatively simple to widen that [*points*] a wee bit if that was helpful to you
Anna	yes
Charles	I think it possibly would be
Anna	yes [*points*] in a sense to just these bits here what is- is that-
Adam	this is a planting bed at the moment so it would be quite simple to widen it a wee bit [*sketches*] if you thought it was helpful but
Anna	so what we have is the [*points*] doors open here so they're opened on to
	particularly attractive so it's got to be wide enough for people to have the doors open and also wheelchair access sometimes that's what we're looking at people obviously arriving in wheelchair vehicles as well and sometimes you get limousines and then you get perhaps a disabled bus as well for people being delivered funerals

Adam's utterance at A1, 307–08 was coded as DESIGN DESCRIPTION; Anna's at A1, 318–322 are both ATTRIBUTES and BEHAVIOUR. The architect describes the change he can make in the width of a paved area, but it is the client who states the reason – being able to have the doors open – and the behaviour that gave rise to that need – the arrival of people and vehicles that require a greater width in order to move in. This type of exchange, with the client reacting to a DESCRIPTION given by the architect with a BEHAVIOURAL scenario and with design ATTRIBUTES, is very typical.

As already pointed out earlier, also quite unusual is the fact that the client, and again primarily Anna, brings up more PRECEDENTS than does the architect Adam. He refers only to two well known architectural masterpieces (the Kimbell and Ronchamp) and three less well-known buildings (Robertson College in Cambridge, the Church at Charles the Cornerstone, and the theatre of the city in which the crematorium is being designed). Anna, in addition to also referring to the Kimbell (which she has studied thoroughly after learning about it from Adam), also talks of other buildings. She mentions Coventry Cathedral and Edinburgh (probably also a cathedral – building not specified by Anna), she alludes to the Alhambra in Spain and to the Albert Hall in London, and brings up the Turringham Clinic. She compares the open space around the building, and its audio systems for the recording of funeral services, with Stevenage crematorium.

We would like to supplement this account with three further informal roles the actors in these meetings took upon themselves, as part of possible divergences from 'prescribed' roles in a design team.

3.2.1 *Initiating and bringing up topics: who moves the cogs of the process?*

The purposes of the meetings are to obtain approval of plans, clarifications regarding issues that are not yet sufficiently clear, and to sort out points of disagreement that require revisions in the design. The meetings are not formally chaired by anybody but, although they take place in the clients' offices, it is in fact Adam who acts as the informal chair. This is evident from the fact that most introductions of new topics are made by him. McDonnell[13] has also referred to topic change as a significant factor. In our analysis, of 37 segments in the transcriptions, Adam is responsible for the initiation of over half. Table 3 shows the breakdown of segment initiators.

13.
McDonnell, J. (Chapter 14), ***op. cit.***

This breakdown indicates that Adam came to the meetings well prepared, with a list of questions and points he wanted to clarify. When he felt a topic had been resolved, he moved on to the next item on his list. The clients, on the other hand, were more spontaneous in their initiation of new topics for discussion, following the flow of the conversation. They clearly did not feel entirely responsible

Actor	A1	A2	Total No.	%
Adam	10	11	21	56.8
Anna	4	7	11	29.7
Charles	3	2	5	13.5
Total	17	20	37	100.0

Table 3. Segment initiators.

for the resolution of problems, and felt that their job was to help do so whenever asked for information or for their opinion.

3.2.2 *Raising questions and expressing confirmation*

Although the 'agenda' for the two meetings was set by Adam, the clients ask twice as many QUESTIONS as the architects during the meetings (137 versus 66, respectively). Their many QUESTIONS show that the clients have a genuine interest in the design features and they make absolutely sure they understand them in depth. The more modest number of QUESTIONS by the architects may indicate a high level of confidence in their design, and possibly a smaller concern with its precise adaptation to user needs (although this would be in contradiction of their stated goals). Alternatively, the architects expected the clients to be attentive to every detail and therefore thought it their business to ask questions only where they lacked information or needed a decision on the part of the clients.

Another interesting finding is the very large number of utterances coded CONFIRMATION by the clients. The difference between architects and clients is sharper here than under any other code: 159 versus 496. CONFIRMATION takes the shape of either a repetition of something stated by the previous speaker, most frequently, simply by a word like 'yes' or 'yeah', or a short chain of words that expresses consent, like: "yes, I'm sure there is space for that" (A1, 292). Often CONFIRMATION within a longer verbalisation is coupled with at least one other code.

The need for so many expressions of confirmation is interesting. Whereas it may be simply indicative of a style of conversation, it may also point to the social structure of design discourse wherein team members find it necessary to build trust and collaboration on mutual agreement and its frequent expression. We shall return to this point in Section 4 below.

3.2.3 *Easing the atmosphere – the use of humour in deliberations*

Another social construct that we find in the discourse is the use of humour which, without failure, helps create a pleasant team atmosphere. We see this social construct as part of Shani and Lau's[14] group-building and relationship maintenance activity, and also as closely related to what Dong, Kleinsmann and Valkenburg[15] call 'affect-in-cognition', i.e. the creation of positive emotions that are conducive to attaining successful results; in this case, design outcomes. Contributing along such lines is not really something that any formal role could possibly prescribe, but rather depends on the actors' personalities. In these meetings it is Anna who has the lead. Adam is responsive to her sense of humour to some degree. Consider for example Extract 2.

14.
Shani and Lau (2000), *op. cit.*

15.
Dong, A., Kleinsmann, M. and Valkenburg, R. (Chapter 7) Affect-in-Cognition through the Language of Appraisals.

Extract 2, A1, Anna and Adam's use of humour.

1068	Anna	the men can put their make up on
1069	Adam	absolutely very important I never go out without my lipstick

In Extract 3 Anna makes fun of herself.

Extract 3, A1, Self-deprecation through humour.

2195	Anna	and ladies my size too I'm building up to getting stuck in the cremator
2196		doors that's what I'm building up to [*all laugh*]
2197	Anna	they won't forget me when they cremate me that's what I'm looking
2198		for yes

Humorous expressions are not very frequent in the transcriptions and rather than giving them a special code they have been coded as OTHER.

We shall now look at the balance between two groups of encoded utterances, the 'Net' and 'Gross' discourse groups, and try to understand the relationship between them.

4 DESIGN TALK: 'NET' VERSUS 'GROSS' DISCOURSE

As evident from Section 3 and particularly from Table 2, half of the coded utterances pertain to 'hard' design issues (i.e. DESIGN DESCRIPTION, ATTRIBUTES, BEHAVIOUR and PRECEDENTS), which we call 'Net' design narrative. The other half, which we call the 'non-Net' design narrative, consists of utterances that complete the discussion and turn it into an understandable and acceptable exchange that allows continuity and the taking of decisions (QUESTIONS, expression of CONFIRMATION, and OTHER – various subject matters including jokes). Together, 'Net' and 'non-Net' utterances make up the 'Gross' discourse. Figure 1 presents the breakdown of 'Net' and 'non-Net' coded utterances in the discourse.

The architects contribute 39.6% of the utterances, the clients contribute 60.4%. The proportion of 'non-Net' narrative is surprisingly high. We do expect questions to be asked, of course, and expressions of confirmation are inevitable if decisions are to be taken. However, the very high percentage of CONFIRMATION is responsible for the fact that the 'non-Net' codes, slightly surpass 50% of the entire discourse. We have already speculated that the particular personalities of the actors may contribute to a mode of speech that is characterised by many repetitions and frequent confirmations. Even if the proportion of client confirmation may be somewhat higher than typical in this case, we can still regard the tendency as

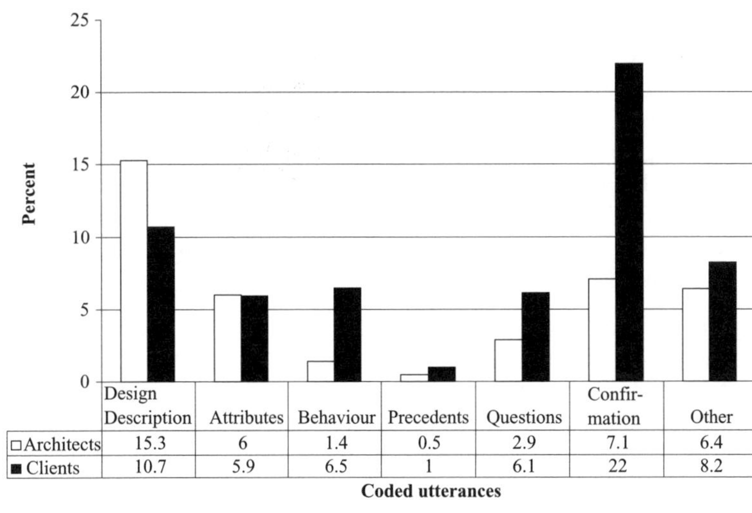

	Design Description	Attributes	Behaviour	Precedents	Questions	Confir-mation	Other
□ Architects	15.3	6	1.4	0.5	2.9	7.1	6.4
■ Clients	10.7	5.9	6.5	1	6.1	22	8.2

'Net' = 47.3% "non-Net" = 52.7%

Figure 1. 'Net' and 'non-Net' coded utterances (%).

normal, because we interpret it as a social strategy for smooth team interactions. By asking a large number of questions, expressing approval by way of confirming statements made by others and by sidestepping to make personal statements, tell little stories or joke about something, the team members ensure a collaborative and cooperative atmosphere that makes it easy to reach decisions and move the design forward. It would not have been possible for the team to converse for some two hours at a time, if they exchanged utterances concerning 'Net' subject matters only.

Oak's analysis[16] of these design meetings attribute to Anna a certain vagueness in her responses to Adam's requests for information. In our view, this mild elusiveness decreased the 'efficiency' of the meetings somewhat (lack of information is responsible for inefficiency in other types of work-related performance as well)[17]. It seems that in design as well as other types of work it is not possible to aspire to 'Net' activity alone.

16.
Oak (Chapter 17), ***op. cit.***

17.
See Lee and Akin for a recent example:
Lee, S. and Akin, Ö. (under review) Shadowing Tradespersons: Inefficiency in Maintenance Fieldwork, *Journal of Automation in Construction*.

5 CONCLUSIONS

The DTRS7 architectural data is typical of run-of-the-mill sessions in which the architectural design is being refined, collaboratively, by a team of architects and clients. There is nothing special about these sessions, nor is there anything out of the ordinary about the design itself. The subject matter that is discussed is therefore quite normal. What we find interesting in the data is mainly the nature of the collaboration among the participating actors, and the interaction strategies that facilitate this collaboration. All actors are experienced professionals in their fields, and they bring their expertise to bear on the design, thereby performing their formal roles. In the case of the architects this means presenting the current state of the design and asking for additional information in order to finalise 'open' issues. In the case of the clients the role asks for the supply of information about needs, mostly related to the likely behaviour of users of the crematorium being designed. The detailed information they offer is given in the form of vivid scenarios.

Informally, actors also assume additional roles. The clients, in particular, 'invade' the architects' territory by bringing up PRECEDENTS and ascribing ATTRIBUTES to the design descriptions by the architects. The latter, focused on the wish to finalise the design, take it on themselves to move the process forward by sticking to a pre-determined agenda. Adam seems to regard the design process as he regards the functioning of the designed building itself: "like a production line" as he mentions in the interview accompanying the data. It happens to be the client Anna who, by virtue of the use of humour, makes the largest contribution towards an amicable atmosphere that, as we know, is most helpful in building a positive team spirit through social constructs.

The most prominent finding is that utterances under codes that do not pertain directly to a design category are very numerous – one half of the total codes. This finding supports the view that collaboration cannot be built and maintained without a massive investment in social constructs. Apart from humour the actors must express frequent approval of what is being proposed; they must ask many questions in order to be totally involved, and they must side-track to foster a friendly and supportive atmosphere.

ACKNOWLEDGEMENTS

The writing of the chapter was partially supported by a grant to the first author from the fund for the promotion of research at the Technion to Research Proposal #1007141; it is hereby gratefully acknowledged.

19

Exploring the Boundaries: Language, Roles and Structures in Cross-Disciplinary Design Teams

Robin Adams, Llewellyn Mann, Shawn Jordan & Shanna Daly

In this chapter we characterise cross-disciplinary boundary work in an authentic engineering design situation in terms of language, roles, and structures. The analysis emerged inductively from participants' words and actions. Five examples are provided to describe, firstly, different boundary work practices including the nature of boundaries, how boundaries emerged, and how they were navigated; and, secondly, different cross-disciplinary and disciplinary practices. Implications are discussed regarding social processes in design, the relationship between considerations of use and different cross-disciplinary practices, co-evolutionary and transformative processes in design, and facilitating cross-disciplinary practices and work environments.

Many complex problems facing society today require cross-disciplinary approaches. For example, the core of the engineering profession lies in integrating broad knowledge toward some purpose[1]. Engineering involves working between technical and non-technical considerations, negotiating among different social worlds,[2] and managing trade-offs where solutions are judged by interdisciplinary criteria[3]. Reports on the future of engineering education emphasise the importance of preparing engineers to become 'emerging professionals' who can deal with complexity, innovate on demand, and bridge disciplinary boundaries[4].

Cross-disciplinary practice is quickly gaining momentum as an important topic for empirical investigation[5,6,7]. Where disciplinary approaches to design are situated in specific bodies of knowledge, cross-disciplinary approaches focus on the nature of the problem, integrating several perspectives to synthesise a collective whole. In this way, cross-disciplinary interactions can enable innovation and amplify creative potential[8].

The two engineering meetings provided good examples of cross-disciplinary practice and boundary work in design teams based on definitions from scholarly literature and our own experiences. This is a cross-disciplinary boundary work situation because: 1) multiple disciplines or perspectives are represented; 2) the nature of the problem is a joint disciplinary problem and not simply confined to a single perspective; 3) there are efforts to broaden and

1.
Bordogna, J., Fromm, E. and Ernst, E.W. (1993) Engineering Education: Innovation through Integration, *Journal of Engineering Education*, 82 (1), pp. 3–8.

2.
Bucciarelli, L.L. (1996) *Designing Engineers*, MIT Press.

3.
Jonassen, D.H., Strobel, J. and Lee, C.B. (2006) Everyday Problem Solving in Engineering: Lessons for Engineering Educators, *Journal of Engineering Education*, 95, pp. 139–151.

4.
NAE (2004) *The Engineer of 2020: Visions of Engineering in the New Century*, National Academy Press, Washington DC.

5.
Bromme, R. (2000) Beyond One's Own Perspective: The Psychology of Cognitive Interdisciplinarity, in Weingart, P. and Stehr, N. (eds) *Practising Interdisciplinarity*, University of Toronto Press, pp. 115–133.

6.
Klein, J.T. (1996) *Crossing Boundaries: Knowledge, Disciplinarities, and Interdisciplinarities*, University Press of Virginia.

limit 'boundaries' around the problem and the design process; and 4) there appears to be a focus on innovation by the inclusion of team members with diverse disciplinary backgrounds. We identified distinct ways of characterising cross-disciplinary boundary work in engineering design teams in terms of language, roles, and structures. These characterisations enabled us to frame our research by exploring the following questions:

- What are characteristics of cross-disciplinary boundary work in engineering design teams?
- What is the nature of these boundaries and how are they navigated?
- What are factors that affect or support cross-disciplinary practice?

7.
Lattuca, L.R. (2001) *Creating Interdisciplinarity: Interdisciplinary Research and Teaching Among College and University Faculty*, Vanderbilt University Press, Nashville.

8.
Petre, M. (2004) How Expert Engineering Teams Use Disciplines of Innovation, *Design Studies* 25, pp. 477–493.

1 LENSES FOR INVESTIGATION

Cross-disciplinary practice involves transgressing into and across other disciplines. Where the term *disciplinary* signifies a particular set of tools, methods, exempla, concepts and theories, the term *cross-disciplinary* characterises a collection of practices associated with thinking and working across disciplinary perspectives. These include multidisciplinary, interdisciplinary, and transdisciplinary practices. Movements between and across disciplinary boundaries are marked by clashing paradigms, or ways of seeing the world, as well as challenges in borrowing and integrating concepts in new contexts. A literature review (see Table 1) reveals important differences in cross-disciplinary practices regarding an orientation to the problem, mode and outcome of knowledge production, and social interaction structures and discourse practices[9,10,11]. Although some of the ideas presented in Table 1 have not been empirically grounded, they provide useful lenses for investigating cross-disciplinary practice.

In the context of cross-disciplinary practice, being inside, outside, or somewhere in between disciplinary zones suggests the existence of 'boundaries' between different knowledge practices. For this study we draw on Gieryn's[12] definition of boundary work as the cooperative pursuit of tasks in spite of boundaries that could prevent separate social worlds from achieving goals. Here, boundary work involves interaction among multiple worlds and competing world views (some of which may be disciplinary). In this way, the concept of boundary work offers a new perspective for understanding interaction structures and social processes in heterogeneous cross-disciplinary design teams[13]. One way of investigating boundary work involves attending to the nature of boundaries and how they are maintained, navigated, reformulated, policed, or negotiated. For example, boundaries could be areas of expertise (i.e., disciplinary knowledge), epistemological differences (i.e., views on the nature of knowledge and knowing), cultural

9.
Aligica, P.D. (2004) The Challenge of the Future and the Institutionalization of Interdisciplinarity: Notes on Herman Kahn's Legacy, *Futures*, 36, pp. 67–83.

10.
Balsiger, P.W. (2004) Supradisciplinary Research Practices: History, Objectives and Rationale, *Futures*, 36, pp. 407–421.

11.
Klein, J.T. (2004) Prospects for Transdisciplinarity, *Futures* 36, pp. 515–526.

12.
Gieryn, T. (1983) Boundary-Work and the Demarcation of Science from Non-Science: Strains and Interests in Professional Ideologies of Scientists, *American Sociological Review* 48, pp. 781–795.

13.
Bucciarelli (1996) *op. cit.*

	Multidisciplinary	Interdisciplinary	Transdisciplinary
Definition	Joining together of disciplines to work on common problems; split apart when work is done	Joining together of disciplines to work or identify common problems; interaction may form new knowledge	Beyond interdisciplinary combinations to new understanding of relationships between science and society
Problem orientation	Not a problem solving orientation but rather thematically oriented projects where several disciplines contribute to a theme	Problem solving orientation in which solution focus is either instrumental (pragmatic problem solving) or conceptual (philosophical enterprise)	Problem solving orientation in which solution focus explicitly includes experiences of affected persons
Mode of knowledge production	Additive, juxtaposition of perspective as separate voices	Integrative synthesis, holistic mixing of perspectives	Integrative and action-oriented transformation that transcends disciplinary views
Outcome of knowledge production	No new cross-disciplinary knowledge	New interdisciplinary knowledge	Knowledge fusion characterised by critical reflection
Interaction and discourse structures	Divide and conquer approaches	Beyond academic disciplinary structures	Participatory – science and society
	Collaborate as disciplinarians with different perspectives; no shared home	Close collaboration; development of common ground	Close and continuous collaboration; elaboration of new language, logic, and concepts

Table 1. Synthesis of cross-disciplinary practices.

differences (i.e., language, values, norms), and organisational structures (i.e., how work environments are structured). The process of navigating boundaries highlights the role of intermediaries: people and objects that can mediate social processes, translate the unfamiliar into the familiar, and fluidly take on different roles in relation to different situations[14,15,16]. In these situations groups create new languages, professional roles, and forms of knowledge[17,18]. Similarly, Newstetter and Kurz-Milcke[19] describe successful cross-disciplinary work environments in which typical boundaries seem to no longer be active. These 'agentive environments' are non-hierarchical such that no one person is expert, involve distributed expertise, and cultivate a culture of failure and impasse to foster resiliency.

The concepts of cross-disciplinary practices and boundary work extend and allow deeper investigation into current theories in design research, in particular those on social processes in design. For example, cognitive artefacts such as representations and prototypes[20] may serve as boundary objects that mediate social interaction. A boundary work lens may provide a way of understanding co-evolutionary[21] and transformative design practices[22] in cross-disciplinary design teams. A focus on cross-disciplinary practice may reveal how interdisciplinary interactions influence innovation in engineering design teams,[23] how storytelling promotes common ground,[24] and how the nature of design tasks influences design processes and thinking[25,26]. Points of synergy in this volume

14.
Bowker, G.C. and Star, S.L. (1999) *Sorting Things Out: Classification and Its Consequences*, MIT Press.

15.
Nersessian, N.J. (2006) The Cognitive-Cultural Systems of the Research Laboratory, *Organizational Studies*, 27, pp. 125–145.

16.
Star, S.L. and Griesemer, J.R. (1999) Institutional Ecology, 'Translations' and Boundary Objects: Amateurs and Professionals in Berkeley's Museum of Vertebrate Zoology, in Biagioli, M. (ed) *The Science Studies Reader*, Routledge.

17.
Galison, P. (1997) *Image and Logic*, The University of Chicago Press.

18.
Klein (1996), *op. cit.*

19.
Newstetter, W. and Kurz-Milcke, E. (2004) Agentive Learning in Engineering

Research Labs in *Frontiers in Education*, Savanah, Georgia.

20.
Visser, W. (2006) *The Cognitive Artifacts of Designing*, Lawrence Erlbaum.

21.
Dorst, K. and Cross, N. (2001) Creativity in the Design Process: Co-Evolution of Problem-Solution, *Design Studies*, 22, pp. 425–437.

22.
Adams, R.S., Turns, J. and Atman, C.J. (2003) Educating Effective Engineering Designers: The Role of Reflective Practice, *Design Studies*, 24, pp. 275–294.

23.
Petre (2004), *op. cit.*

24.
Lloyd, P. (2000) Storytelling and the Development of Discourse in the Engineering Design Process, *Design Studies* 21, pp. 357–373.

25.
Goel, V. and Pirolli, P. (1992) The Structure of Design Problem Spaces, *Cognitive Science* 16, pp. 395–429.

26.
Jonassen, D.H. (2000) Toward a Design Theory of Problem Solving, *Educational Technology Research and Development*, 48, pp. 63–85.

27.
McDonnell, J. (Chapter 14) Collaborative Negotiation in Design: A Study of Design Conversations between Architect and Building Users.

28.
Matthews, B. (Chapter 2) Intersections of Brainstorming Rules and Social Order.

include practices of negotiation among territories of expertise,[27] the intersection between social order and brainstorming[28], understanding problem 'scope',[29] how intermediary design objects may regulate and legitimate design moves,[30] and the role of analogies and metaphors as language tools that elaborate mental representations[31] and reduce uncertainty[32].

2 METHODS

We analysed the two engineering meetings (E1 and E2), focusing on the interactions among people with different perspectives and disciplinary knowledge. This includes what participants did, what they said and how they said it, as well as the context of their actions. The analysis followed a Grounded Theory approach[33] where characterisations of the data emerged inductively from the participants' words and actions. In the first pass through the data, we watched the videos of the meetings and used the meeting layout diagrams to immerse ourselves in the workings of the meetings. We did not assume that disciplinary identifiers provided in the meeting layouts represented how participants might identify themselves (see Table 2). In the first pass through the data, we identified language, roles, and structure as possible descriptive dimensions of cross-disciplinary boundary work. In the second pass, we created personas of the participants to further distinguish aspects of these themes. The personas included a description of the language that the participant used and their actions over the course of the meeting.

In the final pass, we created a narrative timeline of E1 and E2 meeting processes by 'living' the data – reading the transcripts as if they were the script of a play with characters, episodes, and climaxes. This resulted in the identification of narrative episodes or passages in the transcript that marked different stages and topics in the meetings. This process of creating episodes is similar to the work of Reyman, Dorst and Smulders[34] who defined episodes in terms of problem and solution shifts, and McDonnell[35] who defined episodes in terms of feature, function, and detail phases.

Meeting E1	Meeting E2
Alan	*Jack (Mechanical Engineering)*
Chad (Mechanical Engineering)	Patrick (Electrical Engineering, Software)
Jack (Mechanical Engineering)	*Rodney (Industrial Design)*
Rodney (Industrial Design)	Roman (Electrical Engineering, Software)
Sandra (Ergonomics)	*Sandra (Ergonomics)*
Todd (Mechanical Engineering)	Stuart (Electrical Engineering)
Tommy (Electrical Engineering, Business)	*Tommy (Electrical Engineering, Business)*

Table 2. Participant descriptions (*italics* denote presence in both meetings).

The outcome of this process was robust characterisations around *language, roles* and *structures* and how they interact to illuminate aspects of cross-disciplinary boundary work.

Language classifications describe what participants talked about from a disciplinary perspective, and how it was communicated. This is similar to Luck's[36] focus on language as a linguistic marker and boundary object. Six language types were identified (see Table 3): COMPUTER SCIENCE, ELECTRICAL ENGINEERING, MECHANICAL ENGINEERING, BUSINESS, TECHNOLOGY, and PRINTER COMPANY. PRINTER COMPANY language is environmentally situated and appeared to have a unique meaning within this professional setting (i.e., a printer company) and may have served as a common language in the group. For example, "firing the dots" was a way of talking about how the printer head worked in relation to the media and would likely be unfamiliar when used outside this situation. We did not include a specific language code for discussions related to the user, which might be an ergonomic disciplinary language, because it was too difficult to delineate from everyday conversational aspects of talking about the use of the product or its users.

29.
Atman, C., Borgford-Parnell, J., Deibel, K., Kang, A., Ng, W.H., Kilgore, D., and Turns, J. (Chapter 22) Matters of Context in Design.

30.
Luck, R. (Chapter 13) 'Does this compromise your design?' Socially Producing a Design Concept in Talk-in-Interaction.

31.
Stacey, M., Eckert, C. and Earl, C (Chapter 20) From Ronchamp by Sledge: On the Pragmatics of Object References.

32.
Ball, L. and Christensen, B. (Chapter 8) Analogical Reasoning and Mental Simulation in Design: Two Strategies Linked to Uncertainty Resolution.

33.
Glaser, B.G. and Strauss, A. (1967) *The Discovery of Grounded Theory: Strategies for Qualitative Research*, Aldine, New York.

34.
Reymen, I., Dorst, K. and Smulders, F. (Chapter 4) Co-evolution in Design Practice.

35.
McDonnell (Chapter 14), *op. cit.*

36.
Luck (Chapter 13), *op. cit.*

LANGUAGE	DESCRIPTION
COMPUTER SCIENCE	Language associated with the computer science profession and/or ideas associated with programming, writing software and protocols, but not at a hardware level (e.g., digital format, digital signatures, binary, and prestore).
ELECTRICAL ENGINEERING	Language associated with the electrical engineering profession and/or ideas that are electromechanical, related to power, design architecture and interface, involve electronic technologies that are not specifically computer-related (e.g., architecture, sensor, energy per dot, CCD, just a peak, shifthead register, and sinusoidal pattern).
MECHANICAL ENGINEERING	Language associated with the mechanical engineering profession and/or refers to ideas related to forces, angles, temperature, mass, friction, etc. (e.g., controlling the forces, thermal mass, grammage, compressed, and angle control).
TECHNOLOGY	Language associated with using computers, but not designing or programming them (e.g. upload to laptop, download, Wifi, and USB).
BUSINESS	Language associated with the business profession and/or ideas associated with market issues (e.g., risk adverse, demonstrator stage, engineer the cost, profit from the media, on the cheap, and market for it).
MANAGEMENT	Language associated with managing the meeting (e.g., first thing to do, what we already know, moving it to the side, keep brainstorming going, and it's going well).
PRINTER COMPANY	Language associated with the printing profession (e.g., heat it up, dpi, fire the dots, and laid down in stripes).

Table 3. Language classification scheme.

We also characterised how the participants communicated with each other: their use of metaphors and analogies, humour,[37] gestures, representations, imprecision related to technical details, and mixing of languages. Where others in this volume provide a more detailed account of the nature and use of metaphors,[38,39] we focused on how metaphors and analogies triggered different kinds of cross-disciplinary practices. We also noted the role of gestures in which participants used their bodies to communicate a particular idea (e.g., various participants pounding the table to evaluate the maximum force that could be applied to the pen under design). Visser[40] provides a richer characterisation of gestures in this volume. Representations (e.g., flip charts, prototypes, sketches, presentation slides) served as tools for communicating design ideas and providing common reference points[41]. 'Imprecise' language distinguished a type of everyday language that lacked technical precision or specificity (e.g., "wobbly bits" when referring to the part of the pen design that holds the print head) or hedging words about project goals[42]. Finally, mixing of languages refers to linking together different kinds of language (see Table 3) in uninterrupted talk and may signify efforts to translate, bridge, or integrate multiple perspectives.

Role classifications describe participants' actions in the engineering meetings in terms of the project, processes, and experiences. They do not represent participants' areas of expertise. We observed eight role types (see Table 4). Two characterised how the project was managed (FACILITATOR) and how information was brought into the meeting (INFORMER). Four characterised design process activities: EVALUATOR, IDEA GENERATOR, INTERPRETER and QUESTIONER. Two characterised users or experiences of use: STORYTELLER and USER CONTEXTUALISER. In terms of cross-disciplinary boundary work, we might anticipate how acting as a FACILITATOR involves managing the boundaries of the project and the group process. Similarly, the other roles provide different ways for bringing in multiple perspectives such as translating ideas (e.g., INTERPRETER), advocating the needs of users (e.g., USER CONTEXTUALISER), challenging perspectives (e.g., QUESTIONER and EVALUATOR), or providing information to the group that may not be common across the group (e.g., INFORMER, IDEA GENERATOR and STORYTELLER).

While other classifications exist,[43,44,45,46] these classifications emerged inductively as a way to characterise the *combination* of design and social process issues that collectively characterise cross-disciplinary boundary work. These role classifications share similarities with other analyses of the engineering meetings in this volume in terms of design processes[47] and ways of prescribing use[48]. They also parallel analyses of the architecture meetings. For example, Goldschmidt and Eshel[49] observed a role of someone who initiates and brings up topics, asks questions, expresses

37.
See also **Dong, Kleinsmann and Valkenburg (Chapter 7)** and Lloyd (2000), *op. cit.*

38.
Stacey, Eckert and Earl (Chapter 20), *op. cit.*

39.
Ball and Christensen (Chapter 8), *op. cit.*

40.
Visser, W. (Chapter 15) The Function of Gesture in an Architectural Design Meeting.

41.
See also McDonnell (Chapter 14), *op. cit.*, and Luck (Chapter 13), *op. cit.*

42.
See also Ball and Christensen (Chapter 8), *op. cit.*

43.
Stempfle, J. and P. Badke-Schaub (2002) Thinking in Design Teams – An Analysis of Team Communication, *Design Studies* 23, pp. 473–496.

44.
Redström, J. (2006) Towards User Design? On the Shift from Object to User as the Subject of Design, *Design Studies* 27, pp. 123–139.

45.
Kelley, T. and Littman, J. (2005) *The Ten Faces of Innovation: IDEO's Strategies for Defeating the Devil's Advocate and Driving Creativity Throughout Your Organization*, Doubleday.

46.
Lloyd (2000), *op. cit.*

47.
Atman, Borgford-Parnell, Deibel, Ng, Kilgore and Turns (Chapter 22), *op. cit.*

48.
Lloyd, P. (Chapter 5) Ethical Imagination and Design.

Role	Description
FACILITATOR	Directs the meeting by (1) *managing* – directing the flow of the meeting, providing structure, and keeping people on task (e.g., Alan's activities in E1); (2) *encouraging* – encouraging others to be involved and affirming inputs (e.g., Tommy positively affirmed people's brainstorming ideas in E2); and (3) being a *historian* – recording the meeting outcomes (e.g., Alan and Tommy captured ideas on the flip chart, and Rodney gathered sketches in E1).
INFORMER	Brings outside information into the meeting such as (1) *project information* that already exists (e.g., Tommy provided information regarding the client (E2, 7)); (2) *company information* gained through experience within the printer company (e.g., Patrick provided information about drivers he had used in the past (E2, 1611)); and (3) *disciplinary information* that is unique to a particular discipline (e.g., Patrick provided disciplinary expertise regarding PCs and power supplies (E2, 1334)).
EVALUATOR	Makes judgments regarding the ideas discussed in the meeting and identifies a need to conduct an evaluation (e.g., Patrick compared the pen to a simple pencil (E2, 981) or the idea of having a PC connection for the pen was evaluated (E2, 1325)).
IDEA GENERATOR	Presents new ideas about the topic being discussed. Ideas can be technical or non technical, and range from very specific to quite abstract (e.g., Chad proposed a new idea on how to keep the print head on the paper using a rolling ball (E1, 224)).
INTERPRETER	Involves translating information such as (1) *clarifying* a concept or idea presented to the group (e.g., Rodney suggested an application for the thermal pen and Tommy giving an analogy (E2, 645)); (2) *building* on information or ideas on the table by providing additional details (e.g., Patrick suggesting having all of the user interface for the thermal pen on a PC (E2, 1007) and Sandra building onto this by suggesting a library of patterns (E2, 1024)); and (3) *manipulating* an idea or concept or connecting it to another idea (e.g., Stuart manipulated a previous idea of lottery tickets to one of pin codes (E2, 193) and Jack manipulated that by linking it to private letters (E2, 212–213)).
QUESTIONER	Asks for information that has not already been brought into the conversation (e.g., Patrick solicited technical information about the product (E2, 113;169; 354–355; 507) and Sandra asked about the plan for the product (E2, 371–372)).
STORYTELLER	Contextualises an idea or clarifies an idea by telling a story of a personal experience (e.g., Todd presented his idea about controlling the placement of the thermal pen on the paper as a story about his son (E1, 397–416)).
USER CONTEXTUALISER	Provides a 'voice of the user' to the discussions both (1) *internally* – by picturing themselves as users (e.g., Todd described how hard it is to draw a picture with a mouse (E1, 633)) and (2) *externally* – by projecting or advocating the view of the user (e.g., Sandra raised the idea that older children may want to use the pen to create something to keep (E2, 758)).

Table 4. Role classification scheme.

confirmation, and plays the role of absent clients. Their observations map to our classifications of IDEA GENERATOR, QUESTIONER, EVALUATOR and USER CONTEXTUALISER. McDonnell[50] observed the role of asking questions to defer and assert expertise (often disciplinary) which triggered conversations. It is interesting to note that McDonnell[51] did not observe tactics to control the conversation in the architecture meetings.

Structure classifications describe the action context: the structure of the design space and the organisational structure of the meetings. The design space classifications are illustrated in Figure 1.

49. **Goldschmidt, G. and Eshel, D. (Chapter 18) Behind the Scenes of the Design Theatre: Actors, Roles and the Dynamics of Communication.**

50. **McDonnell (this volume), *op. cit.***

51. ***ibid.***

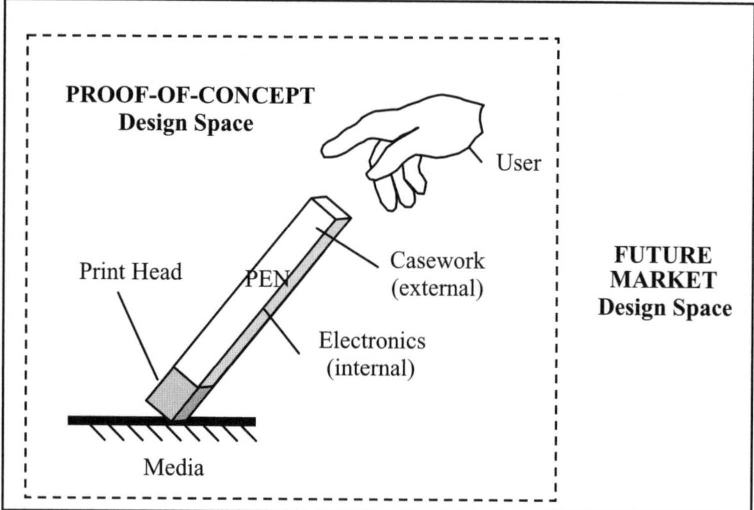

Figure 1. Engineering meeting design space.

Here, the problem-solution space boundaries include the PROOF-OF-CONCEPT design space (dashed box) and the FUTURE MARKET design space, which includes the PROOF-OF-CONCEPT space. Each of these classifications relate to the goals of the two meetings: to produce a proof-of-concept demonstration of a thermal pen for young children, and to develop new product ideas that utilise the novel print head and media technologies. These PROOF-OF-CONCEPT and FUTURE MARKET design space classifications are similar to the brainstorming topic codes developed by Atman et al.[52] (i.e., aspects of the proof-of-concept pen (features, interface, and architecture) and applications (FUTURE MARKET)). The interface boundaries in Figure 1 include the media, thermal print head, electronics, casework, and user interface. Many of these interface boundaries relate to the disciplinary backgrounds of the engineering team meeting participants (e.g., mechanical engineering, electrical engineering, industrial design, business, and ergonomics). These map to the conversation focus codes developed by Atman et al.[53] (i.e. the pen, paper, users, and interactions between them).

52.
Atman, Borgford-Parnell, Deibel, Ng, Kilgore and Turns (Chapter 22), ***op. cit.***

53.
ibid.

The organisational structure includes both the meeting goals and the presence of different participants in different meetings (see Table 2, italics). The goals of both engineering meetings were clearly stated in an agenda. This differed from what was observed in the architecture meeting where the client and architect came with their own agenda[54]. Participants in the meetings were different for potentially intentional reasons. For example, a goal of the E1 meeting was to have a 'mechanical brainstorm' on the casework of the proof-of-concept pen, and as such many of the participants had a mechanical engineering background. Similarly, goals of the E2 meeting were to have an 'electrical brainstorm' for

54.
McDonnell (Chapter 14), ***op. cit.***

the proof-of-concept pen and to brainstorm future applications for the print head and media technologies. Most of the participants in the E2 meeting had an electrical or electronics background. Perhaps to facilitate cross-talk, some participants were present in both meetings (Tommy, Jack, Rodney and Sandra).

3 OBSERVATIONS

We observed two consecutive engineering meetings occurring in the early stages of the design of a new thermal pen. As such there was a major effort to brainstorm new and novel ideas for both how the design would function and how it would be used.

Almost everyone in both meetings used every type of language identified, regardless of how frequently they talked. PRINTER COMPANY language was predominant throughout both meetings, suggesting that it was a common language for the group. There was substantial language mixing, with some participants such as Alan and Tommy mixing four or more language types in a single utterance. This suggests some level of multilingual capability amongst the participants (an ability to link or integrate multiple perspectives) and of people borrowing concepts from outside of their expertise. The mixing of language is often seen linking technical issues to user or business issues, which is similar to McDonnell's observation[55] of switching 'territories of expertise' in the architecture meeting. While meetings were marked by considerable mixing of language, there were times when only a few disciplinary languages were evident. We also observed the evolution of new PRINTER COMPANY language: phrases around "ducks" (E2, 146–157) (e.g., fat ducks, skinny ducks, a line of ducks) that appeared to help participants create a common understanding of how the printer head interacted with the thermal media. Also, language (via gestures, representations, metaphors and analogies) was often used as a boundary object that helped create conversational spaces or relay user experiences around the pen. As Ball and Christensen[56] observed, most of the analogies and metaphors were non-disciplinary in nature, referencing everyday objects and lived experiences.

55.
ibid.

56.
Ball and Christensen (Chapter 8), *op. cit.*

Participants switched among a variety of roles, with some engaging in particular roles more often than others. Everyone engaged in an INTERPRETER role, which often involved efforts to understand the context of the problem such as user and client issues[57]. The prevalence of the INTERPRETER, USER CONTEXTUALISER and STORYTELLER roles suggests that understanding the user and the new technology became places for team members to build a shared understanding of the problem. Disciplinary expertise was most often relayed through the INFORMER role (e.g., providing details about the

57.
See also McDonnell (Chapter 14) *op. cit.*

technology or the existing proof-of-concept prototype). The FACILITATOR role often involved managing the meeting flow, including determining which aspects of the design space could be discussed. This is similar to an observation by Matthews[58] on how the meeting facilitators (Alan and Tommy) mediated the social dynamics and brainstorming rules. Overall, participants fluidly enacted different roles which suggest that roles were flexible, impromptu, and not assigned to any one person based on disciplinary expertise. A similar finding was observed for the architect meetings[59].

58.
Matthews (Chapter 2), ***op. cit.***

59.
Goldschmidt and Eshel (Chapter 18), ***op. cit.***

An analysis of structure classifications revealed boundaries that limited or enabled opportunities to interact across disciplines and aspects of the design space, including potentially co-evolutionary interactions between the proof-of-concept and future market applications. The meeting structure decomposed the PROOF-OF-CONCEPT space into areas of disciplinary expertise – providing opportunities for some disciplinary interactions (i.e., electronic, industrial design, and ergonomics) and potentially limiting others (i.e., electronic and mechanical). In some situations participants transgressed boundaries (this was particularly the case when goals were ambiguous); in other situations boundaries were strongly policed (e.g., what could be talked about or negotiated in the different meetings). Examples of these observations are provided later in this section.

While language, roles, and structures characterisations were insightful as separate analyses, a more comprehensive picture emerged when they were viewed in combination. This allowed characterisations of cross-disciplinary boundary work as revealed through conversational and social interactions in reference to different work environment situations (i.e., the design space and the structure of the meeting). Five examples of cross-disciplinary boundary work are presented below in relative chronological order. Collectively these examples characterise social processes in cross-disciplinary teams as signified by shifts: 1) in topic or language associated with aspects of the design space (see Figure 1), 2) in focus between PROOF-OF-CONCEPT and FUTURE MARKET design spaces, 3) in phases of the design process, and 4) in disciplinary, multidisciplinary, and interdisciplinary practices. In presenting these examples, we will highlight the influence of language, roles, and structures in triggering and navigating these 'shifts' as well as describing the situational context in which they occur.

3.1 *Towards interdisciplinarity: Todd's ghost story (E1)*

During engineering meeting E1, Alan facilitated a brainstorming session to have participants take turns presenting their ideas. This conversation was often characteristic of a multidisciplinary 'divide and conquer' approach since the design space was limited to brainstorming only mechanical engineering solutions and efforts to consider other features of the design space in Figure 1 were often curtailed. However, these organisational and design space boundaries were transgressed as illustrated in the episode of 'Todd's Ghost Story', which resulted in a push towards interdisciplinary practices and the emergence of a novel concept.

After Rodney suggested a mechanism analogous to the way a razor works (E1, 307) and Sandra initiated a short conversation by asking about different options for the media (E1, 378), Alan asked for additional thoughts (E1, 391). Todd began his idea by suggesting that he: "only brought a magic marker … cos I didn't think there's a problem" (E1, 392–394) which prompted laughter in the group. He then put forward his idea by means of a story about a toy ghost his son found in a cereal box. By adopting a STORYTELLER role, Todd helped the group develop a common understanding of the product and situated his example in a familiar context others might understand. Sketches and gestures served as a means of communication between participants, and supported the push toward interdisciplinarity. Todd explained how the toy worked by using a gesture of moving his pen around the paper in front of him, and Rodney asked him to draw it (E1, 400). Everyone engaged the role of INTERPRETER, by clarifying and building upon Todd's idea, using analogies and metaphors that served as bridges between the idea and a group understanding: "pastry brush" (E1, 419), "white line machines" (E1, 422), and "wide like a caravan" (E1, 432). While the focus was on brainstorming mechanical solutions to the pen casework problem MECHANICAL ENGINEERING language did not dominate the conversation. Instead, there was a considerable amount of language mixing, particularly among MECHANICAL ENGINEERING, user, and PRINTER COMPANY languages.

The earlier controlled turn-taking structure of previous episodes was relaxed during Todd's episode, allowing for all participants to freely contribute without Alan's interruption as facilitator. This provided opportunity for a common understanding to emerge, which is characteristic of interdisciplinary practices. Participants actively contributed to the discussion, sought clarification verbally and in the forms of drawings or gestures, and built on each others' ideas. Humour appeared to soften the managerial and social structure of this episode. Todd began his idea with a humourous comment, and others added humour throughout the meeting. For example, Tommy clarified the idea Todd was explaining by saying: "blimey does that go all over the walls and everything" (E1, 413). Alan ended this episode by returning to a facilitated turn-based process and asked the group for additional ideas (E1, 465).

Our observations around 'Todd's Ghost Story' parallel others in this volume. For example, Ball and Christensen[60] described Todd's

60.
Ball and Christensen (Chapter 8), ***op. cit.***

61.
Matthews (Chapter 2)
op. cit.

ghost story as a non-printer based analogy that helped create a common space. Matthews[61] observed that this brainstorming episode did not clearly fit the rules of social order because the idea lacked relevance and went 'off topic'; however, it triggered a conversation that resulted in a novel idea. While it may be true that the idea had limited technical relevance, a different kind of relevance was established through the introduction of a personal story *as a user*.

3.2 *Towards interdisciplinarity: Sandra's choice (E2)*

In the second engineering meeting, E2, Sandra posed a question that triggered a move towards interdisciplinary practices. Prior to the question, the previous episode was marked predominantly by ELECTRICAL ENGINEERING language and revolved around the printing resolution of the print head and the paper (E2, 343–365). Sandra initiated this move by asking if patterns would be stored in the pen (E2, 371), shifting the conversation from detailing the proof-of-concept solution to opening up a new market consideration regarding using external technologies. Patrick and Tommy then initiated a predominantly disciplinary conversation by suggesting ways of interfacing the pen with external technologies (e.g., incorporating a USB port (E2, 378–384)), in the process shifting the conversation back to detailing solutions. This was followed by Roman asking about the business issue of the price of paper (E2, 385), which again triggered a disciplinarily-oriented conversation about the proof-of-concept solution: "you may energise the dots for about two milliseconds … and during that period the centre of the dot might go to more than two hundred degrees" (E2, 408–411).

Sandra redirected this conversation back to the FUTURE MARKET design space by asking: "there's actually quite different applications aren't they the one where you're printing a pattern and the one where you're just uncovering things" (E2, 424–426). Here, Sandra clarified two forms of future market opportunities, writing with the pen and revealing with the pen (uncovering hidden text in the print media). This option of 'revealing' text was a novel idea for how the pen could be used. Then the group shifted to a more general conversation about possible uses of the pen based on the options introduced by Sandra.

Overall, the structure of 'Sandra's choice' resembled a ping pong game where the discussion bounced back and forth between the problem and solution spaces, as well as the PROOF-OF-CONCEPT and FUTURE MARKET design spaces. This had a co-evolutionary quality in which shifts were triggered by asking questions about use of the product or its users. These shifts (which were each of the order of several minutes) marked moves between disciplinary and interdisciplinary practices, where disciplinary practices were often associated with detailing a proof-of-concept solution and interdisciplinary practices were associated with understanding user and media issues. As such, the participants often enacted a USER CONTEXTUALISER role to understand issues about how users might interact with the pen (E2, 457–472). This was signified by language that had an everyday, imprecise character to it in which everyone was involved in the conversation: "you could have a story about I don't know a teddy bear or a snowman" (E2, 448). In contrast, the pushes towards disciplinarity involved predominantly one or two types of language, most often being MECHANICAL ENGINEERING or ELECTRICAL ENGINEERING.

3.3 *Towards multidisciplinarity: hitting the boundaries (E1 and E2)*

After the group from the first meeting had identified a number of 'mechanical brainstorm' ideas, Tommy stated that they had at least three ideas for the pen casework and reminded the group about the importance of controlling the movement of the pen (E1, 550–555). This sparked a nearly half-hour conversation in which efforts (often by Tommy and Alan) to direct the group toward generating ideas about details of the mechanical aspects of the proof-of-concept solution were sidetracked by efforts to evaluate how children might actually use the pen. This evolved into an iterative process of concurrent development of ideas for the final solution and future market opportunities, while generating greater clarity into the requirements for the proof-of-concept. Prominent roles throughout this episode include INTERPRETER, IDEA GENERATOR, EVALUATOR and USER CONTEXTUALISER.

During this conversation, the group continually hit boundaries related to the goals of the meeting. The meeting facilitator pushed the group to focus on the mechanical aspects of the pen while redirecting discussion of future market scenarios and user issues to the second meeting. Alan, Tommy, and sometimes Jamie took on the role of FACILITATOR in order to manage these boundaries. The first contact with an E1 meeting boundary (see Extract 1) began with Sandra:

Extract 1, E1, First meeting boundary.

622	Sandra	what do we sorry what do we actually envisage kids doing with this pen I mean
623		are they drawing pictures or making invitations and Christmas cards or-
624	Tommy	erm ( ) we're going to try to deal with that a fair bit on Monday
625	Sandra	oh alright
626	Tommy	as well as electronics and some of the control and user interface features
627		and things whereas erm this is the kind of functional end of the mechanics and
628		keeping the head on
629	Sandra	it does affect a little bit-
630	Tommy	it does yeah you're right

This sparked a brief interdisciplinary conversation about user issues which was then redirected back to the multidisciplinary space of the 'mechanical brainstorm' where MECHANICAL ENGINEERING and ELECTRICAL ENGINEERING language was predominant. User issues returned to the foreground when Rodney contextualised issues by talking about his own preferences regarding pens: "I think if I was in their shoes using this I'd prefer there'd be something where I decide whether it's in the right position or not..." (E1, 685). Participants responded to Rodney's comments by considering how to train people to use the pen properly through the use of various feedback mechanisms (E1, 714). This sparked another interdisciplinary conversation where participants connected their own perspectives on how they use pens (E1, 716) with ways to design the shape of the pen so that it would be held in a particular orientation. This shift was marked by the mixing of language types and analogies (e.g., graphic art, calligraphy, paintbrush, and roller).

Sandra continued to take a role in critiquing and clarifying user issues, including asking questions about whether the pen was to be a training aid or a toy (E1, 893). These moves towards interdisciplinary practices were cut short by managerial redirections to table the discussion for another time (e.g., Tommy's response that the purpose of the pen is 'for fun' (E1, 894) which effectively ended this initiated switch). This switching back and forth between interdisciplinary phases of activity around user issues and multidisciplinary phases of activity around designing the casework continued until Tommy redirected the group to consider issues about protecting the print head from user abuse (E1, 1070). This unveiled a crucial design issue at the intersection of the user-casework interface that could have seriously impacted the functionality of the pen (E1, 830). However, the group was directed away from spending too much time on this issue that would be discussed during the second meeting. This resulted in a push back toward multidisciplinary 'divide and conquer' practices.

Overall, this example illustrates a continuous push-pull of multidisciplinary-interdisciplinary practices marked by shifts between the PROOF-OF-CONCEPT and FUTURE MARKET design spaces and triggered by issues surrounding the user. Boundaries regarding which meeting would discuss issues surrounding the user were policed by FACILITATORS (primarily Tommy and Alan) and were challenged by participants asking questions about the use of the pen. These observations parallel other analyses in this volume. For example Matthews[62] observed how social action was facilitated by Tommy and Alan and how this created boundaries that limited activity. Similarly, Lloyd[63] observed boundaries associated with use and issues related to the user. Also, the process of 'hitting a boundary' was not unique to the first meeting. An example of a similar scenario

62.
ibid.

63.
Lloyd (Chapter 5), ***op. cit.***

in E2 was when Patrick said: "we probably ought to go through this list and look at which ones you can actually do with a pencil" (E2, 981), to which Tommy replied: "that's the screening stage yea ... let's move on" (E2, 985) cutting off further discussions on this topic.

3.4 *Towards disciplinarity: Jack's sketches (E1)*

Near the end of the first meeting Jack tried to focus the conversation on detailing the design of the casework. This initiated a move towards disciplinary practices.

After a short discussion on whether or not ways to clean the print head should be a design requirement (E1, 1361–1393), Jack took on a FACILITATOR role by requesting that the group move towards detailing the design of the casework: "OK want to do some more detail a bit more the mechanism" (E1, 1395). However, Alan first checked to see if the group had covered everything in the agenda (E1, 1396) and then started a brief interdisciplinary discussion on the issue of children using the thermal pen on something other than the provided media (e.g., their skin or clothes). Tommy tabled this user-safety discussion to the second meeting (E1, 1416), creating a boundary around the allowed topics for discussion. However, questions about the heat generated and how quickly the heat dissipates continued until Alan directed attention back towards Jack (E1, 1478). At this point, Jack asked the group to identify details of the casework and produce sketches (E1, 1479), marking a move to a disciplinary discussion. This also marked a shift from brainstorming to detailing a solution, as well as a shift from discussing user issues to discussing technical functionality issues.

Jack started this disciplinary conversation by sketching the proof-of-concept, which focused the group's attention on the sketch. Unlike the sketches earlier in the meeting, which helped develop a common understanding of the functional issues within the group, these sketches were used to develop and record design details. Alan sat down for the first time in the meeting (E1, 1498), which potentially signalled a switch of the FACILITATOR role between Alan and Jack. The language used from this point on was predominantly MECHANICAL ENGINEERING, with the concepts discussed revolving around the sketches Jack drew: "you'd need an additional part of the mechanism ... you could have contacts on top and bottom of that piston" (E1, 1504), "this stays at a fixed angle in that way it can move up and down and it can move side to side it's fixed that way" (E1, 1521). This marked a change from a potpourri of language types to the predominant use of one type of language (MECHANICAL ENGINEERING). During this time, Alan, Todd, and Tommy took on the role of INTERPRETER by clarifying, manipulating and building on the design ideas on the table. This focus on disciplinary issues continued until the end of the meeting.

3.5 *Towards disciplinarity: Tommy's technical time (E2)*

Late in the second engineering meeting, E2, the group discussed possible features and uses for the proof-of-concept pen. Tommy changed the direction of the meeting by asking for detailed clarification on one of the ideas presented, which moved the discussion towards a disciplinary discussion about electrical issues.

The prior conversation was marked by mixing multiple kinds of language: ELECTRICAL ENGINEERING (e.g., watts, charging, power supply, electrics), MECHANICAL ENGINEERING (e.g., spacing, drag, longish), and PRINTER COMPANY (e.g.: "you could have one character per line per line or something" (E2, 1266–1267), "pre-printed or just a grid that you colour in" (E2, 1294)). After a discussion of potential features of the pen, Tommy interrupted the conversation flow with a key question: "you can probably just get electrics for that stuff can't you" (E2, 1354). Tommy's question marked the beginning of a shift to detailed design, since he asked for clarification on how the ideas previously discussed could be implemented. Jack then reinforced the move Tommy initiated by asking: "how would you want to select the different features on the pen then" (E2, 1360). The level of technical ELECTRICAL ENGINEERING language is maintained for the duration of the disciplinary episode. They talk about LCD panels, buttons, scrolling, and ideas including: "a lot of controllers microcontrollers have got LCD drivers wired in" (E2, 1398). The role of disciplinary INFORMER is prevalent in this episode, such as when Patrick said: "just say that on the paper built in LCD driver" (E2, 1400). However, some mixing of language occurred, such as discussions about business issues and cost as a way to evaluate technical electrical ideas (e.g., a discussion about the cost versus benefits of adding wireless capabilities to the pen design (E2, 1409–1455)). These divergences reinforce the disciplinarity of the primary conversation by allowing them to evaluate ideas and generate or modify ideas based on these evaluations.

This disciplinary conversation ended when Tommy facilitated a change of discussion: "right erm OK we've done the other ones erm I'd like to talk a bit about erm the choice of architecture" (E2, 1479–1480). This signalled the beginning of a new disciplinary conversation about a different technical issue.

3.6 *Comparison of examples*

These five examples of cross-disciplinary boundary work all provide lenses for revealing social processes in cross-disciplinary design teams. They illuminate different cross-disciplinary practices and how movements among these practices are initiated.

They also suggest that cross-disciplinary boundary work is less about crossing disciplines and more about crossing perspectives (where disciplines are one of many perspectives).

The examples are also different in important ways. Sandra's choice (Section 3.2) differs from Todd's ghost story (Section 3.1) in the kinds of interdisciplinary shifts; the first is situated in the problem space, the second in the solution space. This suggests that the context of what is being discussed doesn't guarantee an interdisciplinary shift since interdisciplinary practices were observed in both problem formulation and detail design phases of activity. The multidisciplinary example (Section 3.3) emphasised how boundaries, often imposed by a facilitator, can trigger multidisciplinary-interdisciplinary shifts. The two disciplinary examples were both situated in the detailed design space; however Jack's (Section 3.4) was a facilitated shift, whereas Tommy's (Section 3.5) shift was in response to a general question.

4 INTERPRETATION AND DISCUSSION

An analysis of language, roles, and structures revealed:

- characteristics of cross-disciplinary boundary work in engineering design teams,
- the kinds of boundaries encountered and how they were navigated at an individual and group level, and
- actions that limited or enabled different cross-disciplinary and disciplinary practices.

Each of these addresses our original research questions.

As summarised in Table 5, language classifications revealed what perspectives were included (and in some cases excluded) and marked cross-disciplinary and disciplinary conversational shifts. Role classifications revealed actions to police, maintain, reformulate, or modify boundaries associated with the design space, the way the meetings were organised, and the design process followed in the two meetings. In this way, roles triggered shifts among cross-disciplinary and disciplinary conversations. Structure classifications provided a means for describing participatory boundaries, conversational topic boundaries, and problem-solution boundaries as well as their evolution in relation to participants' actions. These boundaries impacted the kinds of cross-disciplinary and disciplinary interactions possible within the two engineering meetings. When boundaries were relaxed, important conversations occurred that resulted in pushes towards interdisciplinary practices and the generation of novel ideas. For example, Sandra's push on user issues opened up new avenues for future markets (Section 3.2); Tom's humorous story of a ghost

Lens	Nature of Boundaries and Characteristics of cross-disciplinary boundary work
Language	• Language marked cross-disciplinary and disciplinary conversational shifts • Predominance of PRINTER COMPANY language suggests a history of cross-disciplinary collaboration in this group • Language served as boundary object to enable common ground and synthesis via (1) language mixing and people using disciplinary language outside of their training, (2) use of analogies and metaphors, (3) sketches, (4) imprecision and hedging words around project goals, (5) using gestures to communicate issue about using the pen, and (6) generating new language (e.g., "ducks" as new PRINTER COMPANY language)
Roles	• Roles triggered shifts among cross-disciplinary and disciplinary conversations or practices, and therefore triggered different modes of knowledge production and social interactions • High level of role switching suggests the meeting environment was non-hierarchical where access to roles was unlimited • Roles illuminated how people and actions mediated and facilitated cross-disciplinary practices by (1) bridging and synthesising multiple perspectives (particularly issues of use and users), (2) encouraging discussion, (3) stretching and stimulating imaginations, and (4) negotiating ideas. For example: – FACILITATOR enabled or limited participation in a conversation, policed and reformulated what could be discussed – INFORMER enabled bringing knowledge into the conversation and was the only specific disciplinary role observed – USER CONTEXTUALISER and STORYTELLER enabled including knowledge about use and users – QUESTIONER challenged problem-solution ideas, what could be discussed (or not), when ideas could be discussed, and how ideas could be discussed. Often these actions were associated with questions about user issues (e.g., Sections 3.1, 3.2, 3.3)
Structures	• Structures impacted social interactions by creating participatory boundaries, conversational topic boundaries, and problem-solution boundaries • Structural boundaries created an exclusion-inclusion dynamic that prompted participants to enact roles that pushed on boundaries or brought outside information into the design space • Participation and process structures revealed multidisciplinary practices (e.g., divide and conquer approaches), interdisciplinary practices (e.g., creating common ground), and disciplinary practices (e.g., focusing on technical specificity)

Table 5. Language, roles and structure in cross-disciplinary boundary work.

provided a new way of configuring the proof-of-concept pen design (Section 3.1).

Through a combined lens of language, roles, and structures we characterised different outcomes of boundary work in this setting. These outcomes describe different cross-disciplinary and disciplinary practices in terms of modes of knowledge production, and social interaction and discourse structures (see Table 1). These are summarised below:

- In *disciplinary boundary work* practices, a single discipline dominates the language, roles, and representations within the group. Disciplinary practices were often strongly mediated by a facilitator who encouraged the group to focus on a specific subset of issues. It is important to note that in this situation this occurred even though people from multiple disciplines were present and participating in a group environment. We observed how disciplinary boundaries limited the group from considering ideas or practices outside of the dominant disciplinary space.
- In *multidisciplinary boundary work* practices, individuals retain disciplinary identities while working with others with different disciplinary backgrounds. In this situation the group had a history and a common language around a shared topic (i.e., electronic printer technologies). Participants engaged in the group conversation through roles where they brought their disciplinary expertise into the design space. Managerial structures and the FACILITATOR role played a crucial role in defining multidisciplinary environments in three ways: divide and conquer approaches, juxtaposition rather than integrating two or more perspectives, and creating artificial disciplinary boundaries.
- In *interdisciplinary boundary work* practices, individuals from multiple disciplines work together in a way that is not specific to a single discipline. Part of the process includes creating common ground to overcome challenges due to differences in disciplinary languages and world views. In this situation, group processes and culture involved active participation, collaborative building around ideas, language mixing and other efforts such as using everyday and imprecise language to enable conversation and the creation of common ground. Creating opportunities to question issues at the user-product interface relaxed work boundaries and triggered the development of novel ideas.

This study illustrates how a lens focused on cross-disciplinary boundary work can help interpret events, actions, and social processes in design settings. Here, language, roles, and structures classifications illuminated, firstly, different boundary work practices regarding the nature of boundaries, how boundaries emerged, and how boundaries were navigated, and, secondly different cross-disciplinary and disciplinary practices regarding modes of knowledge production and social interactions.

We feel this research has important implications for engineering design research and education. First, it provides another way to understand design as a social process[64]. Here, social processes are revealed by exploring how people interact within a design work environment: the boundaries they experience, the way individuals and groups navigate these boundaries, and the outcomes of this boundary work. In this way, this study complements other studies in this volume such as McDonnell's work[65] that illustrate the subtleties

64. Bucciarelli (1996), *op. cit.*

65. McDonnell (Chapter 14), *op. cit.*

of collaboration and negotiation of progress with a design. Second, this study emphasises the importance of considering use and users and how these considerations can trigger innovative outcomes and pushes towards interdisciplinary practices. This supports Lloyd's[66] observation about how considering 'unethical' user behaviours sparked creative imagination. It also provides a complementary framework for the observation of Reyman, Dorst and Smulders[67] about how 'use' was a bridging concept in the engineering meetings that linked problems and solutions in a functional way[68]. This is similar to McDonnell's[69] observation in the architecture meeting on how appealing to use or user created conversational openings.

66. **Lloyd (Chapter 5), *op. cit.***

67. **Reymen, Dorst and Smulders (Chapter 4), *op. cit.***

68. **See also Atman et al. (Chapter 22), *op. cit.***

69. **McDonnell (Chapter 14), *op. cit.***

Third, this study provided another framework for considering co-evolutionary and transformative processes in design[70,71]. Efforts that challenged 'boundaries' often triggered abstract-concrete cycles around problem requirements and solution alternatives, which is one way of describing co-evolution in design. These often involved shifts toward interdisciplinary practices. However, problem-solution cycles were not the only kinds of co-evolution observed. There were also co-evolutionary cycles between the PROOF-OF-CONCEPT space and the FUTURE MARKET space. In other words, conversations around future market applications sparked a dialogue by linking to an issue regarding the proof-of-concept design, and vice versa. This is an observation of co-evolution in design that may be novel. Fourth, a cross-disciplinary boundary work lens illuminated features of innovation environments and innovation social practices[72]. These were often associated with shifts toward interdisciplinary practices triggered by considerations of use and users.

70. Dorst and Cross (2001), *op. cit.*

71. Adams et al. (2003), *op. cit.*

72. See also **Lloyd (Chapter 5) *op. cit.***, Petre (2004), *op. cit.*, and Newstetter and Kurz-Milcke (2004), *op. cit.*

This study also provides insights for supporting cross-disciplinary work environments and facilitating cross-disciplinary practices. Because we observed many pushes and pulls towards different disciplinary and cross-disciplinary practices, facilitating cross-disciplinarity may be less about enabling a particular practice and more about enabling multiple practices and switching among these practices. This makes sense since different practices may be appropriate for different phases of the design process or facilitating different kinds of outcomes (i.e., novel solutions or applications). Similarly, this study illustrates that boundaries (participatory, conversational, and structural) can be transgressed as well as policed. Individuals can trigger a shift through their actions (e.g., USER CONTEXTUALISER and FACILITATOR roles) as well as their use of language. This suggests that enabling cross-disciplinary environments involves fluid access to roles and promoting multilingual capability as a precursor to developing common ground. Finally, structures appeared to have the greatest impact on facilitating cross-disciplinary shifts. Cultures that support cross-disciplinary work should provide opportunities for individuals or groups to challenge and reformulate existing structures or boundaries.

Part 7

Objects, References, Context

20
From Ronchamp by Sledge: On the Pragmatics of Object References

Martin Stacey, Claudia Eckert & Chris Earl

References to previous designs and other objects play an important role in the synthesis of new design ideas, but object references are used for a wide variety of other purposes in design thinking. This chapter reports on the roles that object references played in design meetings in projects developing two very different products: a crematorium, and a hand-held device with a thermal print head for drawing on heat-sensitive paper. These roles depended on the moment-to-moment needs of the participants in the meetings, which varied rapidly within meetings, and which were determined largely by the type of product and the state of the project. Almost all the references used were concise identifiers of concepts or features, or exemplars of categories.

Designing does not take place in a vacuum: designs for new artefacts are powerfully influenced by other designed artefacts, especially similar products. But the public rhetoric of the creative industries, as well as views of the design process widely held by both laypeople and designers, radically underestimates the variety of roles that references to other objects play in design processes. For instance Leclerq and Heylighen[1] contrast their observation of 5.8 analogies an hour in episodes of architectural design with a prevailing myth of one analogy per architectural masterpiece.

This study examines some of the ways object references are employed in the development of design ideas in meetings. Idea creation by analogical transfer and the fusion of different concepts played a crucial role in the DTRS7 engineering design meetings exemplified in lines 137 and 160–161 of Extract 1 from the first meeting, E1. But the uses of object references in the meetings went well beyond the invention of new design ideas. Our aim was to understand how references to other objects function in moment-to-moment design thinking and design discussion, starting from our own observations of object references serving different functions in coarser-grained studies of design processes,[2,3] as well as other studies suggesting different ways to look at how designers use analogies and references to objects[4,5,6]. Beyond looking for occurrences of different types of object references, we did no specific hypothesis testing.

1.
Leclerq, P. and Heylighen, A. (2002) 5.8 analogies per hour in Gero, J.S. (ed), *Artificial Intelligence in Design '02*, Kluwer Academic Publishers, Dordrecht, pp. 285–303.

2.
Eckert, C.M. and Stacey, M.K. (2003b) Sources of Inspiration in Industrial Practice. The Case of Knitwear Design, *Journal of Design Research*, 3.

3.
Eckert, C.M., Stacey, M.K. and Earl, C.F. (2005) References to past designs in Gero, J. Bonhardel, N. (eds) *Proceedings of Studying Designers '05*, Aix-en-Provence, France, pp. 3–19.

4.
Leclerq and Heylighen (2002), *op. cit.*

5.
Strickfaden, M. (2006) *(In)tangibles: Sociocultural References in the Design Process Milieu*, PhD Thesis, Napier University, Edinburgh.

6.
Christensen, B.T. and Schunn, C.D. (2007) The Relationship of Analogical Distance to Analogical Function and Preinventive Structure: The Case of Engineering Design, *Memory & Cognition*, 35, pp. 29–38.

Extract 1, E1, Prepackaged analogies; concept fusion; explicit reflection.

Alan	… what did you come up with Jack
Jack	I ended up with a + hold on + sledge
Alan	a sledge excellent
All	[*laugh*]
Alan	so what did that generate then
Jack	well a sledge manages to keep level by having quite a wide base and then a main force in the middle so unlike a set of skis where quite narrow and you go up on an edge-
Alan	yeah
Jack	when you're turning
Alan	yeah
Jack	a sledge is er quite broad and then you have the weight right in the middle so they manage to keep both runners on the snow-
Alan	yeah
Jack	more often than say a sledge or a snowboar- a skis or snowboard
Alan	so so would you potential see some some some guiders almost down the side of this
Jack	well I guess the easiest way to keep the pen at a right angle would be to have a set of stabilisers on it
Alan	yeah
Jack	like a bicycle or like a sledge
Alan	yeah no problem ++ stabilisers +++ like a bicycle yeah that's a good idea any other things that that sort of generated either for you or for anybody else
Chad	I was thinking that sort of maybe like a flat base with a sort of universal joint like a windsurf mast
Alan	yeah
Chad	if the face is quite big but it stays flat but the bit you hold onto can be at different angles

In Section 1 we place this study in context by contrasting the different types of studies of the roles of object references. We coded the references for specificity as individuals or categories and in Section 2 we look at the range of objects which are used as reference points and sources of analogies. In Section 3 we consider how closely the referred-to objects are related to the aspects of the design they are used to develop. We found that classifying the purpose of the reference as idea generation, explanation or problem finding[7] was insufficient to capture the richness of the purposes of design discourse; so we coded the role the reference served in the discourse separately from what kinds of ideas were being developed. In Section 4 we discuss the types of communication through object references we found in the meetings; and in Section 5 we relate the communicative and ideational purposes of object references to the types of mappings between mental spaces they involve. In Section 6 we examine the types of design information the object references are used to generate.

7. cf. Christensen and Schunn (2007), *ibid.*

1 METHODOLOGIES FOR STUDYING OBJECT REFERENCES IN DESIGN

Different methodologies for studying the use of object references provide different kinds of information about design processes and design thinking. Three main research methodologies have been used: experiments on design behaviour in artificial scenarios, close analysis of records of meetings in real design projects (as here), and larger-scale observations and case studies of design processes.

Artificial scenarios remove knowledge of background information and context, especially specialist knowledge of solutions to similar problems affording close analogies. These methods are generally limited to making inferences about universal cognitive mechanisms and processes. Experiments that fix the source objects[8] bypass source-selection processes, and can push designers into more difficult or unnatural transfer processes than they would use in real life. Other research on the role of analogies in idea generation in design has employed experiments allowing free choice of references and analogies, or none, in solving artificial problems[9]. In a typical example of this approach, Dahl and Moreau[10] manipulated the pressure on engineering students to identify and use analogies, and found that the use of distant analogies had a significant positive effect on the perceived originality of their designs.

Ethnographic and observational case studies[11] reveal how design processes are structured and organised, and how the participants themselves see their activities and environments. Such studies can identify types of information and types of thinking that are crucial for important design activities but skate over the complexity and variety of moment-to-moment thinking. In many areas of engineering design, reasoned comparisons to previous designs and previous projects play a central role beyond the synthesis of new designs, in process planning, costing, change assessment, the selection of starting points for design modification, and corroborating design proposals. These reasoned comparisons, systematically selected from a small and known set of recent similar designs, function very differently from the spontaneous analogies and comparisons that occur in design discussions.

The close examination of episodes of real life designing embeds the study of individual object references in the rich context of a real project. The strength of this method is that it exposes the variety and subtlety of object references employed in improvising solutions to a wide range of moment-to-moment problems in designing and communicating. An inevitable limitation is that any one situation covers a small subset of the range of problems confronted by different kinds of designers: there is more to be seen in other situations. The problems faced by the participants in design meetings,

8. Eckert, C.M. and Stacey, M.K. (2003a) Adaptation of Sources of Inspiration in Knitwear Design, *Creativity Research Journal*, 15, pp. 355–384.

9. Such as Leclerq and Heylighen (2002), *op. cit.*

10. Dahl, D.W. and Moreau, P. (2002) The Influence and Value of Analogical Thinking During New Product Ideation, *Journal of Marketing Research*, 29, pp. 47–60.

11. For instance Eckert and Stacey (2003b), *op. cit.*

in both designing and communicating, govern the uses they make of object references. These are very different in the crematorium and thermal pen projects.

Our method was to identify object references in the transcripts by identifying words and phrases referring to 'things' outside the design, classify them according to several coding schemes addressing different aspects of the object references, revise the coding schemes as and when they proved inadequate, and use the classifications primarily as an attention-focusing device in examining how the object references function, rather than for quantitative analysis. The first author did all the coding, with discussion of tricky issues. The coding schemes are examined further when discussing the observations we used them to generate.

For this study we use the term *object reference* to encompass all explicit uses of earlier designs and objects in design processes, without prejudging what these uses are or what exactly the objects comprise. We considered all references to physical objects outside the design itself, unless the surrounding discourse was unconnected to the design, but excluded components of the design, such as power sources for the thermal pen and types of stone for the crematorium. We also collected and coded *implicit references*, where the creation of a new concept combines aspects of the current design with aspects of another conceptual space whose retrieval is not mentioned independently of the new concept. Almost all these implicit references were new potential uses for the thermal pen.

The uses we observe in the DTRS7 transcripts include some that are absent, or at least not remarked on, in other studies of designers who were concerned with different problems. For instance, Christensen and Schunn[12] observed 102 explicit uses of analogies over seven meetings held by a design team in the field of medical plastics, and were able to classify the purposes of all of them as: generating solutions, finding problems with potential solutions, or explanation. We found this classification inadequate for the DTRS7 data. This is because our participants had different purposes[13]. Although any individual study of object references is not 'typical', it is possible to pick out some consistent patterns.

12.
Christensen and Schunn (2007), *op. cit.*

13.
According to Ball and Christensen the use of analogies to generate new uses for a design, which both they and we observe in the engineering meetings, has not been reported previously.
Ball, L.T. and Christensen, B.T. (Chapter 8) Analogical Reasoning and Mental Simulation in Design: Two Strategies Linked to Uncertainty Resolution.

2 WHAT? THE RANGE OF OBJECT REFERENCES

What objects were referred to, and what was their relationship to the design? The vast majority of object references were used to elaborate the designers' understanding of the design problem by creating mappings between the object and some aspect of the design, though the thermal pen and crematorium discussions involved very different kinds of references. A few object references were exemplars of categories, serving to elaborate the categories.

2.1 *Abstraction*

Other authors have noted that the analogies used by designers differ in their level of abstraction; notably Ball, Ormerod and Morley[14] found that in an experiment, expert engineers used more schema-based analogies (activating directly applicable solution principles) and fewer case-based analogies (requiring the construction of a mapping from a solution to a concrete problem to the new situation) than novices. In the present study we looked at a separate issue: how abstract the 'things' were that participants in the meetings referred to.

14.
Ball, L.T., Ormerod, T.C. and Morley, N.J. (2004) Spontaneous Analogising in Engineering Design: A Comparative Analysis of Experts and Novices, *Design Studies*, 25, pp. 495–508.

The overwhelming majority of the 70 coded object references in the first engineering meeting, E1, were to categories of designed objects (55, of which 43 mapped to some aspect of the design), rather than individual designs (8, all mapped to the design) or other things, people or situations. In the second engineering meeting, E2, as many of the 119 coded references related to information or images the pen might produce (34) or to other elements of the product-user-environment-activity system (37) as to the thermal pen itself (34), but only 6 were to individuals rather than categories.

The category labels that the designers used were consistently concise ways to name an exemplar concept embodying the feature sets the designers intended to compare or transfer to the design, for instance the windsurfer mast in Extract 1, 160–161. In the cases where more general or precise categories could be described, they would be difficult or verbose to describe. The individual machine designs mentioned are either singleton categories, because unique embodiments of a solution principle (the customised Avro Lancasters referred to in Extract 2, which were at the correct altitude to release their bouncing bombs when two searchlights pointed at the same spot), or unique in the speaker's experience.

Extract 2, E1, Solution principle with unique exemplar; confirmation and explanation.

652	Rodney	you could project something onto the paper in front of the pen
653	Tommy	like DAMBUSTERS ( )
654	Rodney	yeah
668	Rodney	I was going to say an optical mouse if you lift it off the page you can
669		actually see it's got the pattern it creates separates into two areas so it's when
670		it's actually on the surface two points meet

However references to exemplars were used to convey category concepts for applications and contexts when they could communicate object categories more concisely and powerfully than category descriptions. Thus the object reference in: "you're never gonna draw a whole pic – you're never gonna draw – paint the SISTINE CHAPEL" (E1, 914–915) indicates the space of artworks beyond the capability of the thermal pen. Similarly: "things like PAC-MAN" (E2, 897) communicates the concept of a computer game producing information needing to be recorded in an unforgettable way.

A few object references served as concrete exemplars of already-understood categories. These were suggestions of images the thermal pen might draw (there were 5, for instance: "holly leaves" (E2, 776) or text it could reveal (there were 2: "teddy" and "snowman" as characters in stories, E2, 448), which served to elaborate understanding of the space of images the thermal pen *could* draw. The extensive discussion of possible applications in the second engineering meeting included several very specific applications, but the only other examples of concrete cases serving as exemplars for larger categories in generative thinking are 'go to jail … MONOPOLY' and 'SNAKES AND LADDERS' in Extract 3 (where the context is discussing possible applications of writing not controlled or predicted by the user).

Extract 3, E2, Generative exemplars, concept fusion, confirmation.

614	Stuart	could it give you instructions go to jail or something like in MONOPOLY
615	Sandra	yeah
616	Tommy	erm you could build it into a game
617	Stuart	SNAKES AND LADDERS or something you've got to have go up a snake go
618		up [*laughs*]
619	Tommy	er kind of like CHANCE cards yeah

The architectural design meetings for the crematorium, A1 and A2, differed in that a majority (31) of the 57 design-relevant object references were to individual buildings, with 18 to groups or categories of buildings. But these individual buildings were either unique embodiments of particular features, or a succinct way to refer to them, apart from two exemplars of buildings with stained glass designed by a professional artist (A1, 1423–1424). Most of the references to broad classes of buildings – "chicken huts" (A1, 1268), "TESCOS" (A1, 54), "NISSEN HUTS" (A2, 1518) – were intended to convey experiential properties that are the consequences of their overall appearance. Anna, the crematorium registrar, made several references to the shared characteristics of categories of other crematoria, mainly to point out aesthetic or practical problems with design possibilities.

2.2 *The range of targets: Whatever needs elaboration*

Most of the object references in the DTRS7 data were related to the product being designed, but the participants in the meetings used object references to elaborate whatever aspect of the design problem was of interest (see Table 1).

	A1	A2	E1	E2
Product itself	21	21	53	34
Other part of system	1	3	4	37
Other object reference	9	3	9	14
Output of product	n/a	n/a	4	34
Total	30	27	70	119

Table 1. Targets of the object references in each meeting.

Of 57 object references involved in design thinking in the architectural design meetings, 42 mapped to some aspect of the design, 3 mapped to the current crematorium, and 3 were comparisons to previous object references. Three others are shown in Extract 4, where modern crematoria generally are compared to McDonalds and Tescos buildings; then the Nottingham crematorium (shown in a photograph) is compared to McDonalds and Tescos as an exemplar of modern crematoria, to buttress an assertion by providing an illustration.

Extract 4, A1, Emergent properties: categories and category exemplar.

Anna no the new ones yeah well that – no I said to you they all look like TESCOS but ANDY said to me you know what they look like are MCDONALDS drive through and they do they do look like that they look awful look that's the new one up in NOTTINGHAM no inventiveness

Two more references (shown below in Extract 6) elaborate understanding of the context of the design problem, by articulating Anna's understanding of other people's understanding of it. The other three design-relevant object references in the crematorium meetings mapped to other parts of the product-user-environment-activity socio-technical system; two of these were exemplars of the range of music mourners might ask for. The remaining case is shown in Extract 5. Here the mapping is to users of the crematorium, with difficult interpersonal relationships, interacting with each other in the physical environment provided by the crematorium.

Extract 5, A1, Human behaviour as part of the system being designed

Anna yeah and they sit separately in the chapel as well it's all to do with money and you know they've left someone something wonderful that's most of the time what it is or they – the other family are cross because one family has arranged it and they used they never visited her while she was alive and how dare they get involved with this and it all escalates
Peter it's like east EASTENDERS [*all laugh*]
Anna indeed I mean yes it can escalate to sort of violence at times not here so far but threats of it at times so the idea that number one we have people for other services arriving perhaps at the same time they want to keep separate families like to keep separate

The engineering meetings contained a wider variety of references: the discussions of how the device might work or what shape it would be featured object references serving as analogies to aspects of the design itself, as in Extracts 1 and 2. The extensive discussion of what the thing might be used for included references suggesting alternative components of the product-user-environment-activity system: for instance 'MONOPOLY' in Extract 3 indicates both context and purpose, as do: "library" (E2, 918) and: "post office" (E2, 922). Many of the references were to objects that suggest possible uses for heat sensitive paper, such as Sudoku games with hidden answers (E2, 644); images the thermal pen might draw: "clock" (E2, 539); or alternative heat sensitive media, such as 'the cat' (see Extract 11), or 'mugs' and 'fabric' (see Extract 10), which we discuss further in Section 5.

When the object reference was related to an aspect of the design itself, the context made clear which aspect of the design was intended. The retrieval of referents from memory appeared to be guided by the direction of attention to particular aspects of the design or the product-user-environment-activity system. The designers appeared to use object references to elaborate any aspect of the representation of the design situation that needed it.

3 HOW FAR? SOURCES OF OBJECT REFERENCES IN AND BEYOND THE DESIGN CONTEXT

Studies of the use of analogies and sources of inspiration have examined the reach of designers' reference-finding using cognitive interpretations of closeness, as conceptual distance from the source of an analogy to the target; and cultural interpretations of closeness, as contextual closeness to the designing situation.

3.1 *Analogical distance*

In experimental research on analogical mappings in design, attention has focused primarily on two issues, what the analogy is for (which we will come back to in Sections 4 and 6), and how similar or different the source object is from the target object[15,16]. In their *in vivo* study, Christensen and Schunn[17] divided their 102 explicit analogies into within-domain and between-domain. Our observational work on large-scale engineering design[18] leads us to the conclusion that there is another distinction that is crucial for how other designs are used to create new ones: between "within-domain" objects, and objects that do the same job in the same way as the new design, so they are candidates for adaptation (other crematoria for a new one; there are none for the thermal pen).

All the 57 object references related to design thinking in the architectural design meetings were 'within-domain' in the sense that

15.
Dahl and Moreau (2002), *op. cit.*

16.
Casakin, H. (2004) Visual Analogy as a Cognitive Strategy in the Design Process: Expert Versus Novice Performance, *Journal of Design Research*, 4.

17.
Christensen and Schunn (2007), *op. cit.*

18.
Eckert, Stacey and Earl (2005), *op. cit.*

they belonged to the same category of entity as the part of the design they were being mapped to. These were mostly buildings (49, of which 22 were crematoria). Close mappings to elements of the product-user-environment-activity system included artworks in the extensive discussion of stained glass windows, the groups of people in Extract 5, and stump cameras in cricket (A2, 1317) to corroborate the possibility of unobtrusive static cameras to film funerals. The references, in Extract 6 line 1359, to possible styles for the stained glass window cite relatively vague category concepts formed from experience of within-domain objects.

Extract 6, A1, Abstract references for styles.

1354	Adam	well what I'd like to do is something contemporary perhaps even
1355		cubist or whatever something that is contemporary and architectural
1356		that has lots of colours in it and maybe I could have a go at that having
1357		just taken on board what you've said and then have a go at that
1358	Anna	yes I mean most people when they think of stained glass they think of
1359		sort of gothic patterns or actual church stained glass

The peculiarity of the thermal pen design problem, where no product performing the same task exists, makes deciding what qualifies as 'within-domain' rather difficult. The obvious domain is thermal printing technology. The first engineering meeting included one reference to a previous design members of the team had worked on (see Extract 7); the second meeting contained eight references to thermal printing technology. Almost all the object references in the first engineering meeting were to types of machines (a couple of them fictional), which were the closest or most obvious embodiments of the particular solution principles the references convey as were the outside-engineering references to gloves and finger puppets. The participants in the engineering meetings were adept at retrieving 'distant' analogies, such as bicycles with stabilizers (in Extract 1) and aircraft (in Extract 2), that matched aspects of the thermal pen problem at the level of the system of relationships between elements of the situations, when they had the system of relationships in mind to cue the retrieval. These systems of relationships can be quite subtle: consider the discussion of sledge versus skis in Extract 1. This corroborates the structure-mapping view of analogy[19], and Dunbar's[20] observation that people perform more impressively in natural analogizing, where they start with the abstract relationship structure, than they do in laboratory experiments, where they start with superficial features of a situation and need to find the right set of relationships between them.

19.
Gentner, D. (1983) Structure-mapping: A Theoretical Framework for Analogy, *Cognitive Science*, 7, pp. 155–170.

20.
Dunbar, K. (2001) The Analogical Paradox: Why Analogy is so Easy in Naturalistic Settings, Yet so Difficult in the Psychological Laboratory in Gentner, D., Holyoak, K.J. and Kokinov, B.N. (eds) *The Analogical Mind*, MIT Press.

Meeting E1 included 2 references to other heat producers, including the statement: "it could be a soldering iron" (E1, 85), and 16 references to different kinds of pens and marking devices, plus 5 to computer mice and 3 to writable surfaces (see Extract 10).

Meeting E2 contained 3 references to heat producing devices, 2 to other kinds of printers, 7 to handheld marking devices, 8 to computer pointer technology, 7 to writable or information-revealing surfaces (including “wet t-shirt” (E2, 247) but not counting imagined applications), as well as many that were outside engineering but within the domain of children’s activities.

Extract 7, E1, A reference to a previous design.

Tommy yeah the thing that we did a few years ago which had a kind of sort of – we termed it a forced balanced print head we tended to do it on fairly wide print heads to try and keep them in contact with the medium it appears a bit different less controlled

3.2 *Object references as the shared culture of several communities*

In an investigation of the ‘cultural capital’ product designers bring to their work, Strickfaden[21,22] coded the references design students made in discussions of their projects as ‘local’ or ‘universal’ (that is, shared globally by a community of expertise); and as ‘inside’ or ‘outside’ their design environment, that is, inside or outside the knowledge or experience acquired through training and working as a designer, as opposed to the experiences shared by designers and non-designers. Boundaries are hard to draw, but these two dichotomies define four quadrants to categorize the specificity of an object reference to a designer’s local culture. The engineers make references to objects in all four quadrants, though only two references to the company’s thermal printers (in Extract 7 and at E2, 518) and the reference to an alternative proposal for the current design at E2, 548 are LOCAL-INSIDE. The type of project determines how many LOCAL-INSIDE references are actually possible; the thermal pen project is extreme in how different the product is from any precedents in the designers’ own experience and thus how little they have to draw on beyond general knowledge. Beyond Anna’s (the crematorium registrar) arcane knowledge of Britain’s crematoria, applying Strickfaden’s quadrants to the object references in the crematorium project becomes too arbitrary. Our view is that the categories make more sense as dimensions. How local? Which community of expertise? ‘Universal’ isn’t as universal as all that: the cultural capital used by the participants of both the thermal pen and crematorium projects was distinctively British.

21. Strickfaden, M (2006), *op. cit.*

22. Strickfaden, M. and Rodgers, P.A. (2007) References in the Design Process Milieu in *Proceedings of the 16th International Conference on Engineering Design (ICED’07)*, Paris.

Shared understanding of object references depends on life experiences, and is often unpredictable. Tommy’s ‘Dambusters’ reference in Extract 2 made perfect sense to Rodney but confused others in the meeting. The engineering design and architectural design meetings illustrate the active development of shared cultures within both projects: for instance, the Kimbell Art Museum (A1, 15, 1481; A2, 1364) is a reference point all the participants in the crematorium project are already familiar with; and the discussion of characteristics of another crematorium in Extract 4 is only indirectly related to design thinking, but creates shared awareness of possibilities and requirements. The participants in the thermal pen brainstorming session were encouraged to look for sources of analogies and bring these to the meeting. Not sharing the necessary reference points can cause serious problems in design processes. We,[23,24] found that knitwear designers work in a culture that gives them a remarkably uniform set of shared reference points, and that this gives them a means to communicate ideas to other designers by object references that they could not easily communicate otherwise, but their technicians often do not share enough of their cultural knowledge to interpret their designs correctly[25].

23.
Eckert, C.M. and Stacey, M.K. (2001) Designing in the Context of Fashion – Designing the Fashion Context in Lloyd, P. and Christiaans, H. (eds) *Design in Context: Proceedings of the Fifth Design Thinking Research Symposium*, Delft University Press, The Netherlands, pp. 113–129.

24.
Eckert and Stacey (2003b), *op. cit.*

25.
Eckert, C.M. and Stacey, M.K. (2000) Sources of Inspiration: A Language of Design, *Design Studies*, 21, pp. 523–538.

4 FOR WHOM? GENERATION, COMMUNICATION AND CONFIRMATION

Designers use references to other objects, whether as analogies, sources of inspiration, precedents or reference points, both in developing ideas on their own and as tools for communicating ideas to others. We coded the object references according *to whom* they conveyed new information. We found that 'inventing' and 'communicating' did not cover the ways object references were used in the DTRS7 transcripts, so we coded the expressive purpose of the references in the discourse into five categories: GENERATION (of new ideas), EXPLANATION (of existing ideas), CONFIRMATION, PREPACKAGED, and OTHER (comprising just three: two references to the current crematorium building, that helped establish context (A1, 1445; A2, 593) and one audiovisual system mentioned as having been recommended by consultants as an example of their work (A2, 1105)).

Where ideas are developed communally, generative and explanatory references are intermixed, and not always easy to distinguish, as in Extract 8.

Extract 8, A1, GENERATION versus EXPLANATION

Adam if you want to we could puncture the wall with some – some more ++ holes if
you like what I'm thinking of is like LE CORBUSIER's chapel at
RONCHAMP I'm not sure if you're familiar with that but this chapel has some
[*begins to sketch*] holes that might have a tiny bit of stained glass in them
but might in three dimensions look something like that so that if that's the
outside and this is the inside you got small amount of covered light
reflecting itself off the walls

A feature of both the crematorium review meetings and the thermal pen brainstorming meetings was the appearance of several PREPACKAGED analogies: object references that appear to have been generative analogies for the designers working on their own before the meeting, then used as tools for explanations. As Bo Christensen points out,[26] a PREPACKAGED analogy is a combination of GENERATION (earlier) and EXPLANATION (current); however explicit reporting of a PREPACKAGED analogy conveys information about when and how the idea was generated, plus information about the speaker's activities. In the thermal pen case, the generation of PREPACKAGED analogies was explicitly requested before the meeting; in the crematorium case, they reflect a lot of earlier thinking about issues and the development of a shared project culture. Extract 1 illustrates both explicit reflection on the use of analogies and a more subtle purpose: offering the prepackaged sledge and windsurfer analogies to the group for further exploration.

26.
Ball and Christensen (Chapter 8), ***op. cit.***

In the crematorium meetings, EXPLANATION and PREPACKAGED object references predominated, while the large majority of object references in the second brainstorming session in meeting E2 were generative (see Table 2). But another type of reference played an important role in the thermal pen meetings: CONFIRMATION – an analogy from the design to an external object, used to confirm that an explanation has been correctly understood. Extract 2 includes both a CONFIRMATION reference: 'DAMBUSTERS' (showing correct understanding), and an explanatory reference: 'optical mouse'; as does Extract 9: 'paintbrush' and 'roller'. Extract 3 includes a GENERATION reference: 'MONOPOLY', and a CONFIRMATION reference: 'CHANCE cards'.

The architectural design meetings, dealing with much more concrete plans and well-understood concepts than the engineering meetings, contained four CONFIRMATION references, one of which is discussed in Section 2.2 and shown in Extract 5; two checked understanding of design features through references to: "dovecote" (A1, 2020) and: "COVENTRY CATHEDRAL" (A2, 825); and one checked understanding of an engineering technique through mention of: "ALBERT HALL" (A2, 1200).

	A1	A2	E1	E2
GENERATION	4	4	37	81
EXPLANATION	19	10	13	19
PREPACKAGED	4	9	8	6
CONFIRMATION	2	2	12	13
OTHER	1	2	0	0
Total	30	25	70	119

Table 2. Expressive purposes of the object references.

5 HOW? MAPPINGS BETWEEN MENTAL SPACES

Many aspects of thinking involve the construction of coherent combinations of mental entities,[27] and many of these, not just the use of similarities and analogies in reasoning, involve making mappings between different mental spaces comprising different sets of conceptual entities and relationships between them[28]. Holyoak and Thagard[29,30] argue that analogical mappings are constructed through a process of constraint satisfaction that optimises similarity, structure (similarity in the system of relationships between features), and appropriateness to purpose; the construction of coherent mappings is extremely fast.

27.
Thagard, P. (2000) *Coherence in Thought and Action*, MIT Press.

28.
Fauconnier, G. (1997) *Mappings in Thought and Language*, Cambridge University Press.

29.
Holyoak, K.J. and Thagard, P. (1989) Analogical Mapping by Constraint Satisfaction, *Cognitive Science*, 13, pp. 295–355.

30.
Holyoak, K.J. and Thagard, P. (1995) *Mental Leaps*, MIT Press.

5.1 *Mappings and expressive purpose*

Observations of the ways object references are used in the meetings reveal a variety of different types of mapping. Classic GENERATION analogies like the windsurfer in Extract 1 involve creating a mapping between corresponding features and relationships in the two situations, then extending the mapping to include matches between elements of one situation to elements that are newly conjectured to be in the other situation (a universal joint to link the print head to the handle of the thermal pen). CONFIRMATION analogies like the 'DAMBUSTERS' reference quoted in Extract 2 employ mappings that are complete in the sense that the initial mapping between corresponding known elements of the two situations contains all the information needed for the task – confirming that the mapping is possible – so that no new elements need to be conjectured by analogical transfer. Whether EXPLANATION analogies like 'optical mouse' in Extract 2 involve transfer as well as alignment depends on the knowledge of the hearer. A large fraction of the object references in the meetings are *contrasts* (for instance sledge and skis in Extract 1): mappings between very similar elements force mappings between alignable differences and direct attention to them – often to communicate the point that the way that the design is different from the contrasting situation is important for the success of the design, as in Extract 9.

Many of the object references were simply mapped to some aspect of the design or product-user-environment-activity system. However, some mapping were not binary, but created a more complex structure comprising mappings between three or more objects, as in Extract 9.

Extract 9, E1, Three-way mapping; CONFIRMATION and EXPLANATION; contrast.

819	Alan	the other thing to to think about is in almost all cases when I look at pens
820		the apart from re-wired sort of micropens the the tip is actually the
821		narrowest part of the product whereas in what we're looking at it could
822		actually be as wide or wider
823	Tommy	mmmm
824	Alan	than
825	Chad	so it's more like a paintbrush isn't it like a DIY paintbrush
826	Alan	yeah
827	Tommy	it's more like a roller like a roller yeah

5.2 *Implicit references*

We included in our analysis implicit references where an object is named but only as part of a new possible product-user-environment-activity system: the explicit reference depends on a conceptual combination including elements mapped from an external object. For instance, 'thermal wallpaper' in Extract 11 depends on the idea of wallpaper. Six of the 70 references identified in the first thermal pen meeting and 30 out of 119 in the second were coded as IMPLICIT. Almost all were proposals for new applications of the thermal pen (see Extract 11) or alternative components for the product-user-environment-activity system, as in Extract 10. There was only one in the architectural design meetings; it was: "I see it as a spiritual modern art gallery flavour sort of space" (A1, 1389).

Extract 10, E1, The substrate as part of the system being designed.

129	Sandra	is it is it only paper you're thinking about or could it be other things like
130		mugs or fabric or pottery
131	Jack	it could be anything

5.3 *Novel concepts*

A large fraction of the object references in the engineering meetings were used to generate new concepts by combining features of the thermal pen with features of other objects or situations, especially new applications for the thermal pen in the second meeting. Most of the IMPLICIT references reflected concept fusion.

For instance: "lottery tickets" (E2, 183) creates a new concept fusing the structural features of non-forgeable marks on heat-sensitive paper with the functional features of a lottery ticket. 'MONOPOLY' in Extract 3 matches a proposed behavioural feature of the thermal pen, that it should generate writing not controlled by the user, with the feature of Monopoly that it involves unpredictable instructions (involving an alignment of elements and relationships to do with performing an action to reveal hidden information), to create a merged concept combining board game with the heat-pen and heat-sensitive paper.

In Extract 10, an analogy is constructed from a component of the product-user-environment-activity system (the heat sensitive paper) to potential alternative components (mugs or fabric) sharing the essential abstract property of having a writeable surface. Substituting the analogous objects creates new concepts: mugs and fabric with heat-sensitive coatings. Several times one object reference creating a fused concept triggered an analogy to another object with similar features and the same relationship to the thermal pen: for instance: "Advent calendar" and: "branching stories" (E2, 438–443).

Extract 11 shows the fusion of heat-sensitive paper, wallpaper and writing on walls, to create the concept of thermal wallpaper. This involves an analogy between using a thermal pen inappropriately (on the cat) and using a pen inappropriately, to draw on walls – a pen or pencil is part of a mental space contributing to the blend but is not explicitly mentioned.

Extract 11, E2, Envisionment of misuse; concept fusion.

1470	Patrick	using it on the cat or something
1471	Tommy	[*laughs*]
1472		yes at least you can't draw on walls and things with this can you + it's just on wire
1473		[*laughs*]
1474	Jack	that's another point
1475	Tommy	thermal wall paper for kids bedrooms

There is no *a priori* reason why concept fusion of the sort discussed here should involve analogy, in that each mental space might only provide elements to fill roles that the other leaves open, rather than the alignment of features and relationships in both. Nonetheless the cases in the engineering meetings all appear to involve some element of analogical mapping between corresponding features. However the: "spiritual modern art gallery sort of space" (A1, 1389) appears to create a conceptual combination in the same way as a metaphor.

6 WHY? THE PURPOSES OF OBJECT REFERENCES

Most research on the use of analogies in designing has focused on its role in the synthesis of new design elements. But our observations of the uses of other designs as reference points in complex engineering projects shows that object references can play an important part in analytical activities as well[31]. Christensen and Schunn[32] restricted their analysis of their medical plastics

31. Eckert, Stacey and Earl (2005), *op. cit.*

32. Christensen and Schunn (2007), *op. cit.*

project to clear-cut cases of analogical transfer, and found that all the analogies that weren't used for idea generation or explanation were employed to find problems with designs. In analysing the DTRS7 transcripts we chose to be more inclusive, and found that matters weren't quite so simple. We coded the ideation purpose of the object references – what sort of information they were used to produce – as SYNTHESIS or ANALYSIS. Drawing on the set of analytical activities identified by Eckert, Stacey and Earl[33] in which object references can be used, we divided ANALYSIS purposes into REQUIREMENTS, CONSTRAINTS, FUNCTIONS, CORROBORATION, PROBLEMS, and USE. These purposes are often closely linked, as envisionment of USE directs attention to CONSTRAINTS and PROBLEMS, making it difficult to accurately identify a single purpose in every case; for this reason we are reporting broad patterns rather than counts.

33.
op. cit.

The first engineering meeting was dominated by SYNTHESIS activities directed to meeting a small number of central REQUIREMENTS and CONSTRAINTS so the large majority of the object references were used to generate design elements, as illustrated in Extracts 1 and 2. Extract 1 also illustrates SYNTHESIS by exclusion: the reference to skis narrows the imagined solution space by indicating what is not intended.

Much of the second engineering meeting was devoted to finding possible applications for the thermal pen. We coded object references for this purpose as ANALYSIS of FUNCTION, but this activity might be viewed as SYNTHESIS by composition of new product-user-environment-activity systems. Unsurprisingly, none of the object references in the architectural design meetings were concerned with ANALYSIS of FUNCTION.

The engineering meetings included object references being used in a range of ANALYSIS activities, including envisioning the use of the product – Extract 11 shows envisionment of misuse to identify potential problems[34]. Object references were also used in the ANALYSIS of REQUIREMENTS (for example: "like a pen" (E2, 988), an assumption about shape framing a discussion of what controls the thermal pen should have), and for CORROBORATION, that is, using an example of a comparable situation to show that a proposal is feasible (for example, cost of heat sensitive paper compared to fax paper and till receipts, E2, 385–395). There was only one reference, in any of the meetings, used in the ANALYSIS of CONSTRAINTS separately from a discussion of a problem with a particular proposal. (This was a light emitting diode used previously, in a discussion of the limitations of LCDs (E2, 1750).) Several suggestions of things the thermal pen might draw appeared in envisionments of use ("fat ducks" and "skinny ducks" (E2, 147–9) depending on how quickly the pen was moved).

34.
See Lloyd for a consideration of how the participants in the meetings discussed potential misuse.
Lloyd, P. (Chapter 5) Ethical Imagination and Design.

As the architectural design meetings were review meetings discussing modifications to fairly detailed proposals, uses of references for SYNTHESIS purposes, as in Extract 8, were infrequent. Contrasts were used for the ANALYSIS purposes of identifying a PROBLEM with a design proposal, as in Extract 12; and elaborating a REQUIREMENT, as in Extract 13 – an amorphous aesthetic requirement clarified by exclusion.

Extract 12, A1, Comparison for analysis: identifying a problem.

1154	Adam	… I've always wanted to do a stepping stone
1155		type bridge across a pond but maybe this isn't
1156	Anna	perhaps we can put glass inserts in you know they can it looks like
1157	Charles	it would be very slippy +++
1158	Anna	oh he's poo pooing the idea I don't know but yes it would be nice to
1159		have that I don't see that that's a problem but I can see that in the end
1160		what will happen it will have to be probably sort of taken away as a – as just a design and /( )\
1161	Charles	/there is one a very\ small one of these there was in erm the winter
1162		garden wasn't there
1163	Adam	was there
1164	Charles	wasn't there a winter garden in central NEWTOWN
1165	Adam	I know the big (common) you mean I didn't know there was a
1166		stepping stone running across it
1167	Charles	there was a stream
1168	Adam	oh right so you could walk through it
1169	Charles	there was a stream running across and you came down into the winter
1170		garden down the steps and you had to go across stepping stones across
1171		a stream and this was to a health club remember and they had so
1172		many accidents people twisting their ankles or ( ) that they've actually
1173		filled in the gaps between them

Extract 13, A1, Comparison for analysis: elaborating a requirement.

1607	Adam	so you know we could reduce this if you like to make it work but what
1608		I was trying to do was to really open up this end of the site to make it
1609		more useful
1610	Anna	yeah well it it needn't also there it's nice 'cos it balances the
1611		building doesn't it there's got to be a certain amount of balance
1612		between the building and the design features otherwise it looks sort of a
1613		bit lost in nothingness around it which is what STEVENAGE has got just
1614		two little bays in front of it really they've got this lovely big building
1615		and then they've got these two little boxed bays in front of the building
1616		which are awful

7 CONCLUSIONS

The most striking feature of how the participants in the meetings used references to other objects is how diverse the references were, both in the range of what was referred to, and in how the references were used to alter the participants' understanding of the design situation. The uses of object references go well beyond idea generation and well beyond classic analogical transfer.

The participants in the meetings covered a wide range of issues in a short period of time, mixing synthetic and analytical thinking. The types of analytical and synthetic thinking they performed were dictated by the moment-to-moment needs of the situation, and their memories for other objects and situations served as a resource for whatever reasoning they were doing. This resource was used both in planned and calculated ways and opportunistically. Object references were used to develop the designers' understanding of the design situation. Depending on need object references were used in constructing the design itself, the requirements it must meet, envisionments of how it will be used, the environment it is used in, and so on. But references to external objects are not ubiquitous: they are absent from the DTRS7 data for quite long periods.

Object references were used both as sources of features introduced into mental representations of the design through analogical transfer, and as comparison points. When object references serve as contrasts, the purpose-relevant features of the target are not constructed but already known, they are highlighted as similar to or different from the corresponding features of the referred-to object. In some cases this sharpens understanding of the target by narrowing the range of possibilities for what it might be by saying explicitly what it is not.

In the architecture project, two other types of object reference appeared: exemplars of an understood category, used to elaborate and sharpen understanding of its features; and examples of objects with particular features, used to support envisionment of the feature in the new design in analytical thinking.

The participants in the meetings used the most succinct descriptions they could of objects embodying particular features; in the engineering project these were usually fairly general classes of objects, such as 'optical mouse' or 'windsurfer', and references to individual designs were frequently to unique exemplars of particular features – and thus the most succinct category descriptions. The object references were seldom more specific than they needed to be. This corresponds to our observations of knitwear designers' use and naming of garment categories, which is frequently by reference to individual designs[35]. Which features were intended to be mapped was either clear from context or explicitly stated, so the hearer's attention was directed to the features essential to the comparison. While many of the object references in the engineering meetings were 'distant' from thermal printing technology, they were close and obvious matches to the sets of often relatively abstract features and relationships that they shared with the target. What the use of object references in all four of the meetings indicates is that given a system of relationships between the elements of a situation, people are adept at retrieving matches to it[36].

35. Eckert and Stacey (2000), *op. cit.*

36. cf. Dunbar (2001), *op. cit.*

ACKNOWLEDGEMENTS

Claudia Eckert's contribution to this research was supported by the EPSRC block grant to the Cambridge University Engineering Design Centre. We thank Bo Christensen for his very helpful comments on an earlier version of this paper.

21
Keeping Traces of Design Meetings through Intermediary Objects

Emine Serap Arıkoğlu, Eric Blanco & Franck Pourroy

This chapter presents an approach for analysing the use of intermediary objects involved in a design meeting. The aim of this research is to provide engineers with some indicators in order to help them to identify the critical objects around which relevant design meeting minutes should be constructed. Examples of such indicators are proposed from the analysis of an actual engineering meeting.

Collaboration over product design within companies often takes place between designers located at distant geographical sites. Our goal is to understand the ways of working in these settings, and to support them by proposing new methods and tools that efficiently assist designers during design activity.

A number of engineering and research tools have been proposed and developed in recent years in order to capture and support systematic decision making and design rationale in particular. For example, DRed,[1] DREW[2] and Compendium[3] have been proposed as means to formalise design rationale as graphs of arguments for and against ideas relevant to a design task. It has been suggested that these graphical representations can be useful tools for engineers to formalise the minutes of design meetings and that the tracking of argumentation during design meetings is of much importance for working logically through a design, for improving decision making and communication, and for reducing the effort associated with design documentation[4].

At the same time, the role played by sketches and intermediary objects in the design processes has been highlighted by many research studies. Sketches play an important role in the first stages of the design process[5,6] but their inherent ambiguity presents opportunities as well as certain difficulties for design teams. Although they are useful for supporting thinking and negotiation during design meetings, they are less useful for recalling the process afterwards[7,8].

Many designers include sketches in their minutes of design reviews, however, not all of the sketches produced during design review are relevant for keeping in design minutes. The main idea of our work is to develop an activity indicator that allows designers to define which sketches and intermediary objects are most important tc a design meeting. In the case of virtual team shared spaces or Internet meetings where designers use computers to interact with

1.
Bracewell, R.H. and Wallace, K. (2003) A Tool for Capturing Design Rationale, in *Proceedings of the 14th International Conference on Engineering Design*, Stockholm, Sweden, pp. 185–186.

2.
Corbel, A., Girardot, J.J. and Jaillon, P. (2002) DREW: A Dialogical Reasoning Web Tool in Proceedings International Conference on ICT's in Education (ICTE2002), Badajoz, Spain.

3.
Compendium Institute http://compendium.open.ac.uk/institute

4.
Bracewell and Wallace (2003) *op. cit.*

5.
Goel, V. (1995) *Sketches of Thought*, MIT Press.

6.
Purcell, A.T. and Gero, J.S. (1998) Drawings and the Design Process: A Review of Protocol Studies in Design and other Disciplines and Related Research in Cognitive Psychology, *Design Studies*, 19, pp. 389–430.

7.
Gardoni, M., Blanco, E., and Rueger, S. (2005) MICA-Graph: A Tool for Managing Text and Sketches During Design Processes, *Journal of Intelligent Manufacturing*, 16, pp. 397–407.

8.
Stacey, M. and Eckert, C. (2003) Against Ambiguity, *Computer Supported Cooperative Work*, 12, pp. 153–183.

these objects, we assume that such a measurable indicator might enable an automatic evaluation of the relevance of these objects. A specific tool might then be developed to help designers select objects to be integrated into the minutes of design meetings.

In this chapter we focus on the first engineering meeting, E1, and we propose a method for codifying and quantifying the activity of an intermediary object. The interaction between stakeholders and objects are modelled and quantified. From this codification, some basic quantitative indicators are proposed to represent intermediary object relevance. In the first section we discuss related work on the role of sketches in design meetings. In the second section we describe our methodology. Tools and methods are proposed to define the relevant sketches. Finally, in the third section we discuss the results of this analysis.

1 THE ROLE OF SKETCHES IN DESIGN MEETINGS

We are especially interested in the status and role of intermediary objects[9] such as scribbled drawings or annotations on drawings made by several designers. Our investigations into innovation and design activities showed that intermediary objects are always present during these processes. We have already underlined their number and diversity, as well as the importance and multiplicity of interactions channelled by them[10,11]. Depending on the case, their status varies. For example, they are used to back up the action, to introduce information, to make intentions or compromises come true, to impose constraints on actors, and to mediate between diverse logic-based actions. Studying intermediary objects reveals actors, interactions, organisations and effective processes. As revealing agents, these graphical objects help to describe the process and its timing: work periods of varying intensity, breaks in time. As mediators, they refer to changes in the nature of the job at hand and to exchanges between the actors. They also translate and constitute representations of the product during design. They are the product of this activity, of which they leave some traces. They are also resources mobilised by the actors to convince, explain, remember, revise, imagine, agree, and so on. Hence, they mark the activity, the state of relations between the actors and the current representation of the future product[12]. As a result of these findings we assume that keeping trace of intermediary objects in design meetings can be useful to capture and retain the design rationale. Clearly, not all the objects produced during a design meeting are of importance. Thus, the question is to decide which have to be kept.

In previous work, we showed that however many steps are invisible, sketches are traces of the emerging process of the design

9. Vinck, D. and Jeantet, A. (1995) Mediating and Commissioning Objects in the Sociotechnical Process of Product Design: A Conceptual Approach, in MacLean, D., Saviotti, P. and Vinck, D. (eds.), *Designs, Networks and Strategies*, European Community, pp. 111–129.

10. Blanco, E. (2003) Rough Drafts Revealing and Mediating Design, *Everyday engineering. An ethnography of design and innovation*, in Vinck, D. (ed.), MIT press, pp. 181–201.

11. Boujut, J. and Blanco, E. (2003) Intermediary Objects as a Means to Foster Cooperation in Engineering Design, *Journal of CSCW*, 12, pp. 205–219.

12. Finger, S., Konda, S., and Subrahmanian, E. (1995) Concurrent Design Happens at the Interfaces, *Artificial Intelligence for Engineering Design Analysis and Manufacturing*, 9, pp. 89–99.

solution[13]. Analysis of sketching activity shows that knowledge of their construction process is necessary to be able to understand the sketches themselves. During the process of sketching, designers create symbolic devices which are connected to semantic representations in a local convention. This allows them to let some parts of the device function at a low definition level: fuzzy and partial. Sketches also play a mediating role in a collective socio-cognitive process. Principally, they act as communicative devices and support collective sharing. Finally, they act as an individual and a collective memory support. They take the strain off the memory, and prevent actors from repeating what has already been said or demonstrated. They are therefore cognitive artefacts.

13.
Blanco (2003), *op. cit.*

We want to highlight the fact that the sketches act as pragmatic conventional supports. These conventional supports are negotiated in the interactions between actors, even if they are also using shared, higher level cultural conventions, such as the rules of technical drawing. In this sense, sketches are built into the action process and conventions and allow their interpretation to be quite local.

However, actors making use of intermediary objects need to have a level of detail in the representation that is consistent with the current problem and solution they are dealing with. The sketches become a synthesis of the work performed by the group. They act in the process of creating a new solution (and in a way in the decision making process) as the 'crystallisation' of some product features. We can say that sketches create some irreversibility in the design process, because once the sketch is on the paper, it becomes a reference for the participants which implicitly orients them towards future choices.

Thus, we assume that linking sketches with a representation capable of capturing design rationale can be useful. In the analysis presented here we used a tool based on the IBIS model (Issue-Based Information System). IBIS was developed by Horst Rittel and colleagues during the early 1970's[14]. The heart of IBIS is the matrix of QUESTIONS, IDEAS, and ARGUMENTS that combine together to create a conversation. QUESTION states a question; IDEA proposes a possible resolution for the question; and ARGUMENT states an opinion or judgment that either supports or objects to one or more ideas.

14.
Rittel, H.W.J. and Kunz, W., (1979) Issues as Elements of Information Systems, *Working paper 131*, Centre of Planning and Development Research, Berkeley, USA.

Sketches can relate to any of these elements of design rationale. But of course sketches vary in their importance to decision making and to the construction of the design solution. Some are based on very local conventions and just help the common understanding at that instant (see Figure 1, sketch 3 from E1), while some are more complex and represent technical solutions and use technical drawing conventions. But even in this type of object, some sketches are related to general principles (Figure 1, idea 1 in sketch 6) while others represent details of solutions or address a specific problem

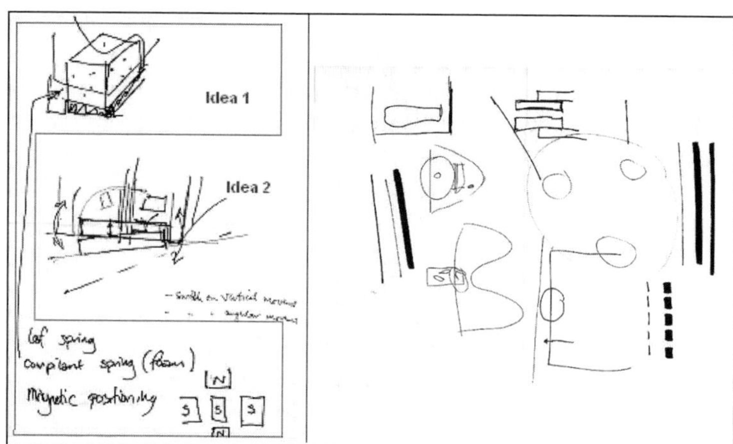

Figure 1. Two sketches from E1 sketch 6 (left) and sketch 3 (right).

(Figure 1, idea 2 in sketch 6). These different levels can co-exist on the same worksheet with no logical connection to link them.

Thus we need to qualify sketches as intermediary objects. Fergusson[15] identifies three kinds of sketches: THINKING, TALKING, and PRESCRIPTIVE ones. In the meeting studied here most of the intermediary objects are of the second type; most TALKING sketches involve collaborative activities. As an example, sketch 3 in Figure 1 is a TALKING sketch used to support the exchange of ideas and design arguments between the actors. In contrast, studies which have focused on a single designer's activity mainly point to the THINKING role of sketches. PRESCRIPTIVE sketches are those that are intended to convey design solutions to individuals outside of the design interaction, and therefore rely more heavily on formal conventions for representing design features. Observations of industrial design activity show that the three categories frequently co-occur. We make the assumption that TALKING and PRESCRIPTIVE sketches are more important to keep as a record of a design interaction.

15.
Fergusson, E.S. (1992) *Engineering and the Mind's Eye*, MIT press.

In the next section we characterize and rank intermediary objects by using indicators of activity, based on a model of their usage by the group. To achieve this, we develop and present a method to identify a measure of this activity.

2 METHOD

We focus here on the first engineering meeting, E1 which has two main advantages for our objectives. Firstly, this meeting relates to mechanical engineering issues, which match with our own disciplinary backgrounds. This is a key aspect as we need to have a good understanding of the design rationale emerging during the

	Stage	Objectives	Analysis tools
Global	Overview of the data	• Becoming familiar with the data • Noting the sequence of the events	• Compendium • Videograph
↓	Identification of the objects	• Listing the objects • Construction of a coding scheme • Identification of the active objects	• Videograph
Detail	Detailed analysis	• Identification of the critical phase • Characterisation of the objects	• Compendium • Videograph • Excel

Table 1. Three stages of analysis summarised.

meeting. Secondly, this meeting seems to be richer than the others in the DTRS7 dataset in terms of the number of intermediary objects involved in the interactions. This is another key aspect as we choose to focus on these objects for analysing the data.

The method that we developed to analyse the E1 meeting data covers three main stages. As a first step we undertook a global analysis in order to become familiar with the data and to develop an overview of the meeting. Then we carried out a deeper, systematic analysis to identify the intermediary objects and their activity levels. Finally, a more detailed analysis of a particularly relevant sample from the meeting was carried out. Table 1 itemises the objectives of these three stages and the corresponding analysis tools: Compendium and Videograph.

Compendium[16] is a software tool providing a flexible visual interface for managing the connections between information and ideas. This tool is intended to "break down the boundaries between dialogue, artefact, knowledge and data". Compendium extends the use of IBIS from modelling a discussion, to a more systematic modelling of a problem. IDEAS, QUESTIONS and ARGUMENTS are expressed as icons (forming nodes). They can be connected to other IDEAS via links to form maps, networks of linked nodes where the nodes can be placed anywhere in a 2D space[17,18].

Videograph[19] is a video-analysis tool. It is used to play and assess digitised videos simultaneously. The software enables the construction of observation categories and rating scales that can be used by the viewer as a 'measuring instrument' to analyse and code the contents of the video. The coding can take place synchronously while the video is playing; coding can be compartmentalized into time intervals (time sampling) or refer to what is taking place in the video (event sampling). After their creation the data sets are graphically presented on the screen and can be transferred to spreadsheet software such as Microsoft Excel.

The use of these tools in the three main stages of our analysis are now described in more detail.

16. Compendium Institute website *op. cit.*

17. Concklin, E.J. and Yekemovic, K.C. (1991) A Process Oriented Approach to Design Rationale, *Human-Computer Interaction*, 6, pp. 357–391.

18. Conklin, J., Selvin, A., Buckingham Shum, S. and Sierhuis, M. (2001) Facilitated Hypertext for Collective Sense Making: 15 Years on from gIBIS, in *Proceedings 12th ACM Conference on Hypertext and Hypermedia*, Århus, Denmark.

19. Information on Videograph is available at: http://www.qurl.com/jf1rg, (accessed 25 August 2008).

2.1 *Overview of the data*

Our first task was to analyse and understand the transcripts of the meeting dialogues. To fulfil this task we needed a global view of the meeting, including the drawings and gestures observable on the video. Therefore, we used Videograph to review selected sections of the video. Our aim was to become familiar with the data and to create a workable representation of it. We looked at the discussion topics, objects, organisation of participants (formal authority hierarchy) and the way they expressed their ideas, and so on. We defined the sequence of events via Videograph for a closer examination.

As a first step, we produced a Compendium map of this event sequence, and subsequently attempted to define the main phases of the meeting by analysing the interaction between designers and the topics they were discussing. We defined transitions between the phases formed in terms of topic changes or changes in the structure of communication between designers.

2.2 *Identification of the objects*

We use the term 'object' for all sorts of physical and electronic artefacts which can be pointed out, talked about or sketched on. These include, for example, sketches, technical drawings, pictures, and existing products. We developed a coding scheme for all objects that we observed during the meeting with Videograph and analysed time periods when these objects were active. We define an object as 'active' as long as it is being used or produced by any of the meeting participants. The aim was to identify those object types which were more active than others, at different phases of the meeting. As a result, we observed whether transitions between the phases resulted in any change in the nature of the objects. The results of the initial two stages of analysis provided the basis for moving onto the third stage. The three stages of our analysis and the sequence of objectives for each is summarised in Table 1.

2.3 *Detailed analysis*

The first step of this final stage used the results of the initial two stages of the analysis to identify the 'critical phase'. We defined this critical phase as a time period during which key decisions occurred (sometimes implicitly). We hypothesise that in this critical phase, the designers complete the definition of the design problem and focus on solutions. This is the synthesis part of idea creation in which designers try to reduce the number of ideas generated to a workable subset. Thus, to define the critical phase we analysed the Compendium map to identify the phase in which designers focused on detailing and evaluating proposed solutions. Previous work has shown that the appearance of technical sketches indicates a key stage in decision making and solution emergence[20]. Frequency of occurrence of sketches was therefore used as an additional indicator to identify the critical phase.

20. Blanco (2003), *op. cit.*

The second step of this third stage was to scrutinise the critical phase in order to analyse the interactions between the participants and the different objects which are active in this phase. To do this, we performed an interaction-based analysis of the objects. The objective was to see if we could find any indicator that could be a reference for selecting critical objects. In our research, critical objects are those that play an important role in the group communication process. They provide a trace of the solutions – or important arguments regarding the solutions – which must be captured for the future understanding of the design process. Such indicators may be useful in the design of tools to assist in selecting critical objects for an effective capturing of the design rationale.

Initially, to qualify the interactions on the objects, we proposed a model of interaction. We identified from the video all kinds of visible interactions that engineers could have with an object. These interactions are listed in Table 2, which also defines the descriptors for each category defining the engagement of the actors in the interaction.

It is important to note that the idea behind these categories is to identify different levels of interactions which could be used to identify objects that cannot be ignored for the design record. The hypothesis is that the importance of the object to the design rationale is related to the interaction level associated with that object.

The analysis described was carried out for each of the objects active during the critical phase. To do that, we developed a coding

Interaction with object	Indicators, criteria
INTERACTING PHYSICALLY	The actor holds the object in his/her hand and/or places it on the table.
SKETCHING	The actor sketches. This action is defined as 'pen on paper'. As it is difficult to distinguish writing from sketching, the former is incorporated in drawing.
TALKING ABOUT	The actor explains an idea with reference to an object. The transcription provides the verbal record of this activity.
LISTENING ABOUT	The actor listens whilst looking towards the individual talking about the object.
LOOKING	The actor is looking at the sketch that he is producing or at objects that other participants are sketching or interacting physically with. If the actor who is sketching or interacting physically with the object talks about the object at the same time, the actor looking at the object is considered to be engaged in listening (not looking).

Table 2. The interaction model proposed for coding the object activity periods.

scheme within Videograph to figure out for each actor the time periods when he or she was engaged in each of the interactions.

3 RESULTS OF THE ANALYSIS

The following sub-sections describe the results that we obtained from the three stages of our analysis.

3.1 *Overview of the data*

The overall aim of the design meeting was not only to find a good solution, but also to generate a range of different ideas. According to brainstorming rules, participants are expected to respect the ideas of others and to avoid being critical. The document the participants received before meeting together already contained relevant information and gave them the opportunity, prior to meeting, to think about the answers for some questions to be discussed during the meeting itself. They also had the possibility of preparing some materials in order to illustrate their proposals. Alan, one of the participants, chaired the meeting and took notes of discussion points on a flipchart. The other participants were peers, and not subject to any formal hierarchy. At the outset of the meeting Alan listed the discussion topics and, as they proceeded, he introduced new ones.

As mentioned before, we started to analyse the meeting by structuring the overall discussion with Compendium, which led us to breakdown the meeting into four main time periods:

- The start of the meeting covers the global presentation and brainstorming rules statement (E1, 0–112). This time period corresponding to phase 0 will be discounted;
- The first phase covers the discussion about two main problems (E1, 112–1166);
- The second phase covers the discussion about two further problems (E1, 1166–1487);
- The third phase covers the organisation of ideas on the basis of sketches (E1, 1487–1960).

As an illustration, Figure 2 shows the Compendium map of the second phase of the meeting. The particular fan-shaped structure of the map and the existence of a restricted number of arguments, are typical of a brainstorming session.

The third phase of the meeting shows some differences even though the structure of the map is quite similar. This phase of the meeting seemed to be more convergent as no new topic was discussed. Instead, participants tried to organize the ideas proposed in the first and second parts of the meeting using the support of

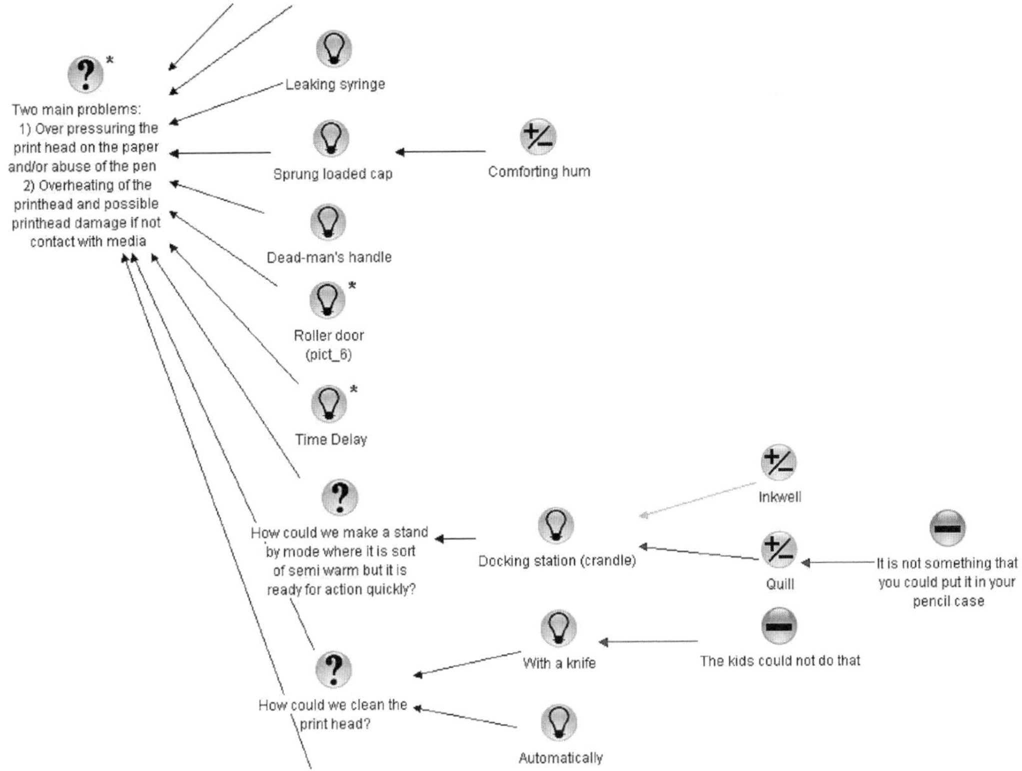
Two main problems:
1) Over pressuring the print head on the paper and/or abuse of the pen
2) Overheating of the printhead and possible printhead damage if not contact with media
Leaking syringe
Sprung loaded cap
Comforting hum
Dead-man's handle
Roller door (pict_6)
Time Delay
Inkwell
How could we make a stand by mode where it is sort of semi warm but it is ready for action quickly?
Docking station (crandle)
Quill
It is not something that you could put it in your pencil case
With a knife
The kids could not do that
How could we clean the print head?
Automatically

Figure 2. Compendium schema of the second phase of the meeting.

sketches (see Figure 4). This result corresponds to the observations of Mabogunge et al.[21] and Adams et al.[22] who both identify similar features of this last period in meeting E1.

21.
Mabogunge, A., Eris, O., Sonalkar, N., Jung, M. and Leifer, L. (Chapter 3) Spider Webbing: A Paradigm for Engineering Design Conversations during Concept Generation.

22.
Adams, R., Mann, L., Jordan, S. and Daly, S. (Chapter 19) Exploring the Boundaries: Language, Roles and Structures in Cross-Disciplinary Design Teams.

3.2 *Identification of the objects*

At this stage of the analysis, we introduced our coding scheme. Data analysis led us to define two object categories: those brought to the meeting by the participants and those produced during the meeting. As a first task we identified all objects for both categories and divided these into subgroups depending on their nature. We assigned a code and a number to each object within these subgroups. To give an example, the pictures of physical objects that were brought to the meeting by the participants were given codes such as 'pict_n' (where n represents the number of the object). A total of 37 objects were identified by this means. Table 3 shows the distribution of these objects among the different categories.

This coding scheme was used in Videograph to identify the time periods when mentioned objects were active. Figure 3 gives an overview of the results. Each horizontal line refers to a particular object. The horizontal axis represents time (it covers the duration of the

Objects brought to the meeting			
Type	Description	No. (n)	Code
Picture	Pictures of real objects	6	pict_n
Notebook	Notebooks of participants	4*	notebk_n
Presentation	PowerPoint presentation explaining the topics of discussion	5**	pre_n
Object	Real objects used by participants to explain ideas	6	obj_n
Information page	Briefing document sent to inform the participants about the meeting	1	inf_page
Prototype	Prototype of the thermal pen	1	proto_n
Objects produced during the meeting			
Type	Description	No. (n)	Code
Sketch	Sketches designed by the participants during the meeting	9	sketch_n
Flipchart	Flipchart containing the notes of the meeting	5	flip_n

* Some participants did not use a notebook
** We assigned a number to each page

Table 3. Coding scheme for the objects.

entire meeting); blocks in the time lines record the active periods for each of the different objects. The three vertical dotted lines identify the limits of the four phases. It can be observed from this figure that many objects are active even in the early stages of the meeting, although most of these early objects were those brought into to the meeting. Indeed, Alan brought in the PowerPoint presentation and the briefing sheet (the information page). The other participants came to the meeting with pictures and physical objects to illustrate their ideas related to the design issues the meeting was about.

We notice that obj_6 (a certain type of pen) was active for a long time during the first and second phases of the meeting. In fact, after it was analysed the participants used it as a regular pen, from this point therefore it was no longer considered to be an intermediary object. However we believe that objects are active as long as they are used by actors. This object gave us cause to consider the direction of our on-going studies and forced us to question for ourselves when an object becomes an intermediary object, for instance, designers often use parts of their immediate environment to simulate or mock-up product behaviour.

Figure 3 highlights the change of activity at the beginning of the third period. Jack proposes organising the mechanisms in some sketches (E1, 1487). Adams et al.[23] also observed this change stating that this moment 'marked a shift from brainstorming to detailing a solution, as well as a shift from discussing user issues to discussing

23.
ibid.

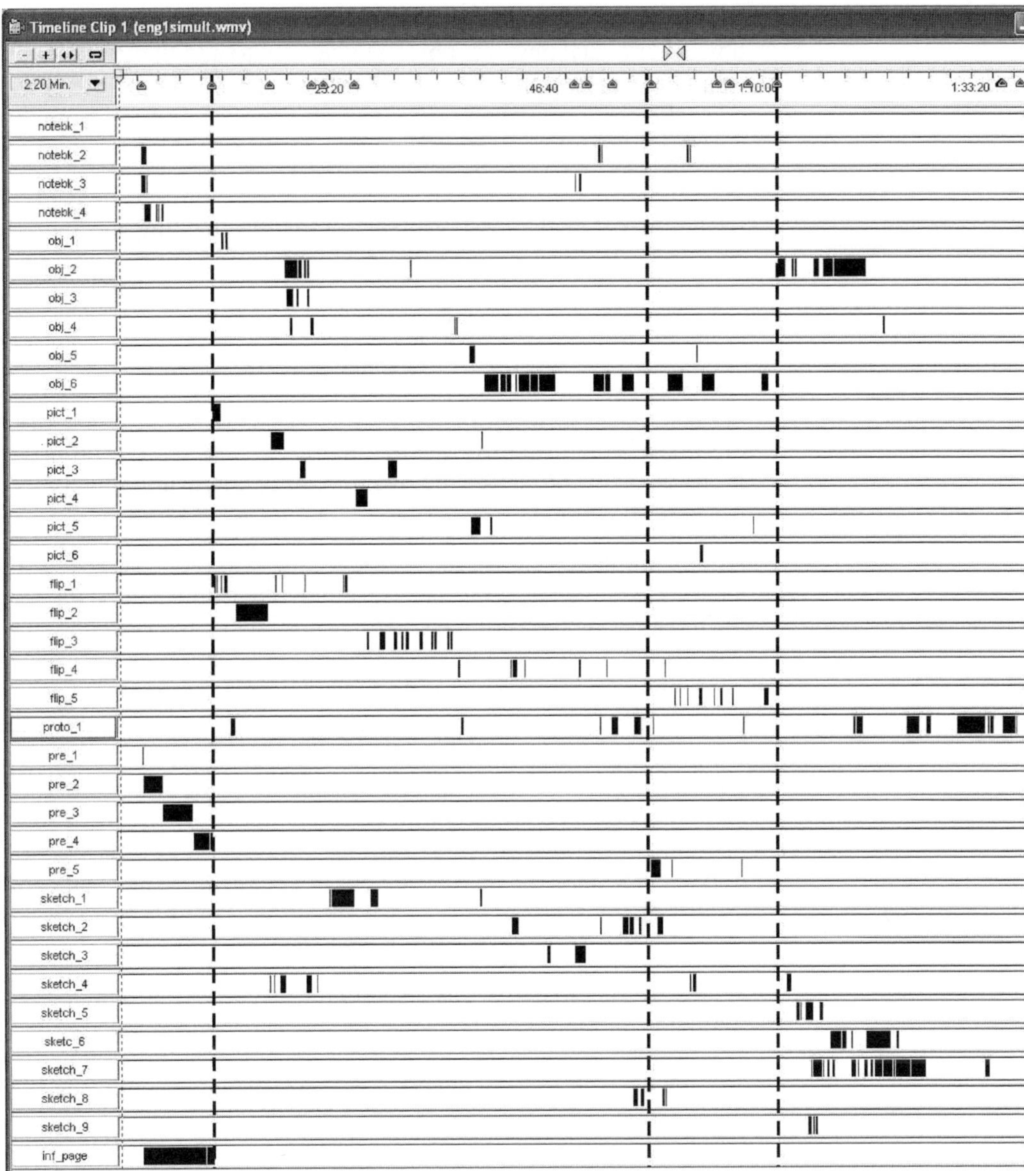

Figure 3. Timeline showing periods when objects were active during meeting E1.

technical functionality issues'. This was the starting point of the last phase of the meeting. In this final phase we do not see any more pictures or objects (except obj_2 and the prototype) but instead mostly sketches. This change in the character of intermediary objects is consistent with the analysis of Dahl, Chattopadhyay and Gorn[24]. They describe two different kinds of visualisation; memory visualisation (which refers to events or occasions that one has personally experienced or observed) and imaginary visualisation (instead of recalling a past experience from the image, the creation of a new, never before experienced event). Similarly, we

24.
Dahl, D.W., Chattopadhyay, A. and Gorn, G.J. (2001) The Importance of Visualisation in Concept Design, *Design Studies*, 22, pp. 5–26.

have seen in this meeting, memory visualisation (pictures of real objects or actual physical objects) give way to imaginary visualisation (sketches) as the meeting progresses. For example the group takes the electronic razor (obj_2) as a reference point and from this generates hypotheses about the mechanism that the pen can have by way of drawing sketches (sketch_5).

In summary, in the first two phases of the meeting the participants mainly discuss the problem statement and connect this with existing products that can become part of a design solution. This discussion is supported by representations of real products, either their photos or the products themselves. Simultaneously, the group was listing brief descriptions of their ideas on the flipchart. But in the last phase of the meeting they focused on the mechanism of the razor. They tried to organize the ideas which were proposed in the first two phases of the meeting, generally using much more freehand sketching.

At this point we can use the results of the initial two stages of the analysis to define the critical phase. Firstly, the Compendium maps of the three main phases of the meeting showed that the last phase was more convergent than the preceding phases. Designers focused on generated ideas, engaging in 'deep reasoning'[25]. It is in this phase of the meeting that we can see the greatest creativity. Secondly, we observed that the nature of the active objects changed from physical objects, or pictures of existing objects, to the production of sketches and drawings. The last phase of the meeting was richer in terms of the number of sketches produced. So, according to our criteria (as defined in Section 2.3), we define this last phase of the meeting as the critical phase. Clearly, much more attention should be paid to this phase if we are seeking to make a record of the design process.

25.
See Mabogunje et al. (Chapter 3), ***op. cit.***

We also observed that in this phase of the meeting, Alan stopped taking notes on the flipchart. However, participants continued to take notes on the sketches. Figure 4 shows the Compendium map of the critical phase, including the related intermediary objects. This map illustrates that with existing tools it is possible to record digitally intermediary objects when capturing design rationale. In this figure what we initially identified as sketch_6 exists physically on a single sheet of paper. However, the sketch presents two different ideas (see Figure 1) so we separated it into 'sub-sketches' (sketch_6bis1 and sketch_6bis2 in Figure 4) to capture the two roles the sketch(es) play in the developing design rationale. This observation raises new questions about defining the extents of intermediary objects.

3.3 *Detailed analysis – towards the identification of critical objects*

The final step of our approach consists of identifying the critical objects. Using Videograph, we analyzed in detail the first 10 minutes

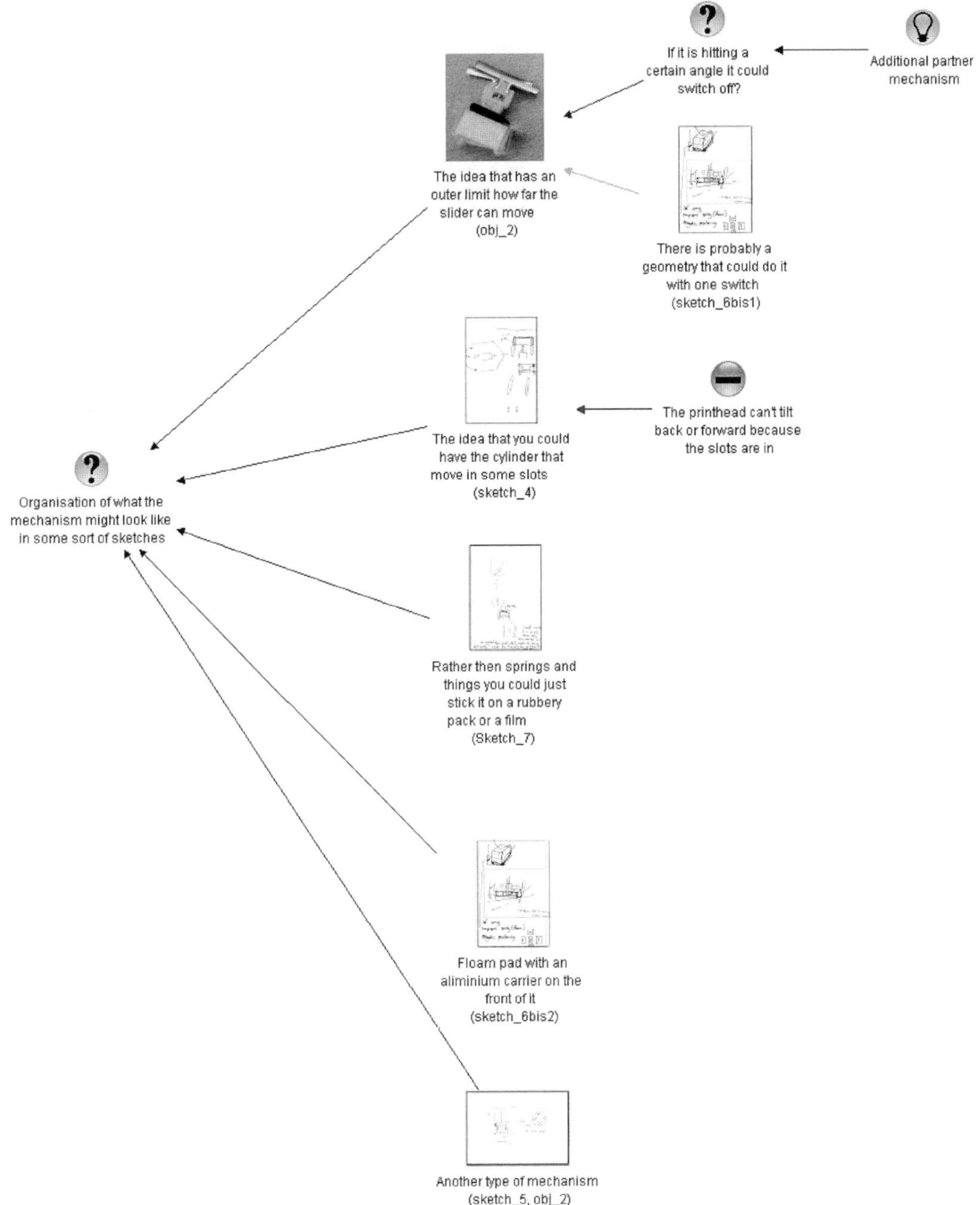

Figure 4. Part of the Compendium schema for the third phase of the meeting.

of the critical phase (E1, 1487–1661). As mentioned in Section 2.3, we carried out an interaction based analysis for each object used in this time period (obj_2, proto_1, sketch_4, sketch_5, sketch_6, sketch_7 and sketch_9). In this period, only six participants were present, as one, Sandra, had left the meeting (E1, 1383).

The analysis consisted in identifying, for each participant, the time periods of interaction (from the interaction model in Table 2) associated with each of the active objects. For example, we coded that Jack is SKETCHING 'sketch_n' between 'starting time' and 'finishing time'. The results of this analysis were exported from Videograph in tab-delimited text format to recover them in a calculation worksheet. As an example, Table 4 shows an extract of results of the interaction-based analysis of the objects active in the critical phase. The table depicts the total length of time each object was active and the amount of time that this activity was a 'collective activity' involving three or more members of the group.

These two indicators are represented in Figure 5 as percentages of the total time period. Looking at this figure we can observe that sketches 4 and 6 involve greater collective activity than obj_2, even though the latter has a longer period of activity. Considering the duration of activity, sketches 5, 6, and 7 seem roughly equivalent. However sketches 4 and 9 have shorter lives. Sketches use pragmatic conventions and they are often ambiguous and incomplete as we have already noted, so common understanding needs to involve interaction around the object. We make the assumption that collective activity is an important element of critical sketches. If we combine the two criteria sketch 6 appears to be the more critical object. Sketches 7 and 9 clearly are non-collective objects.

These two indicators are straightforward to quantify but may be not sufficient to indicate criticality. We can go further, perhaps, by qualifying the interaction. One route is to explore *which* group members interact with the objects as shown in Figure 6. A second route is to use the interaction model to qualify the nature of the interaction as shown in Figure 7. Figure 6 depicts the total length of time each actor interacts with each object. This indicator shows that all the participants are involved in the interaction with the razor mechanism, but that Alan interacts with it the longest. Sketch 7 appears as an individual's object as does sketch 9.

In contrast, Figure 7 focuses on the type of interaction, and characterises each object in terms of the total time that different types of interaction are associated with each object. The combination of Figure 6 and Figure 7 confirm that sketches 7 and 9 are individual sketches. They can be qualified as THINKING objects following Ferguson's typology[26]. As these were not shared objects, it may be possible to exclude these from the design rationale as they may be incomprehensible to people who did not participate in the meeting.

26. *op. cit.*

Table 4. Sample results of the interaction-based analysis of actives objects in the critical phase.

Active time	obj_2	sketch_4	sketch_5	sketch_6	sketch_7	sketch_9	proto_1
Collective	4:29	0:34	1:19	1:37	0:00	0:00	0:38
Total	7:12	0:44	2:21	1:53	2:12	1:04	1:03

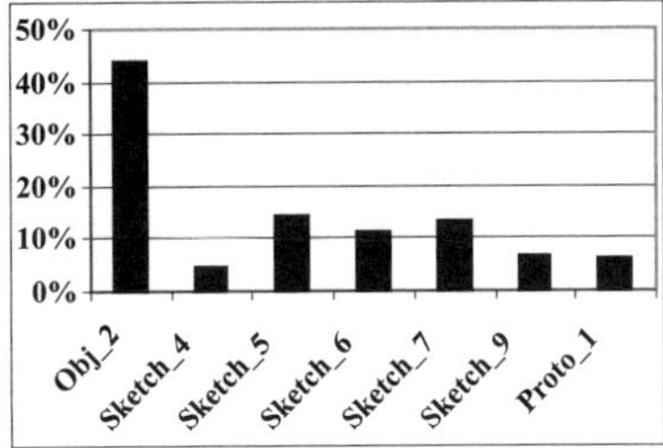

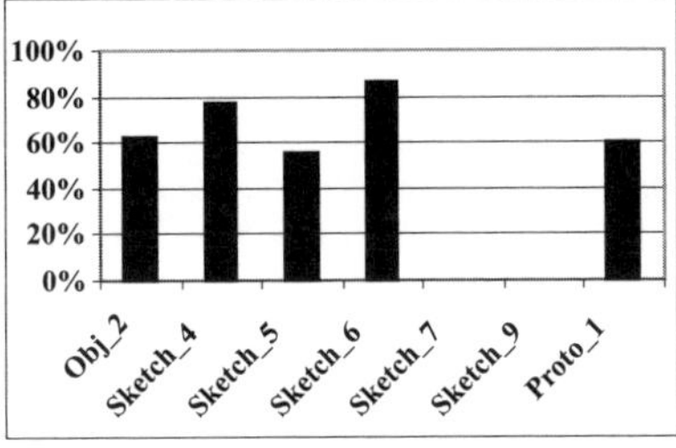

Figure 5. Object activity (left) and collective activity (right) as a percentage of total time.

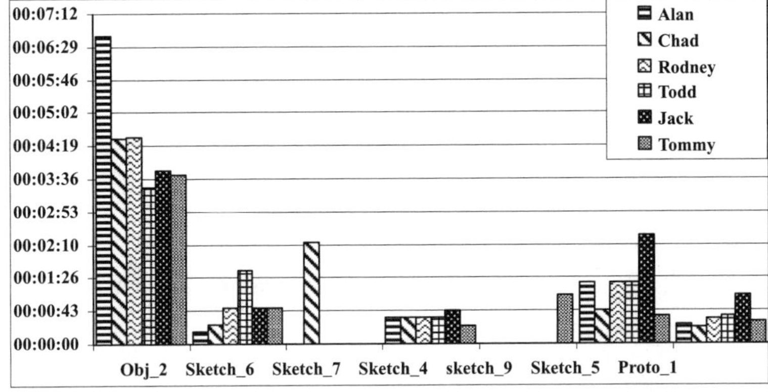

Figure 6. Total time periods that actors interact with the objects.

They could however be retained among the personal notes produced by each participant.

Figure 7 also shows that there is a higher level of verbalisation during the activity associated with obj_2. Combining Figure 6 and Figure 7, we can imagine a 'listening rate' corresponding to the ratio of listening time to number of actors involved. The high value of this ratio makes us conclude that obj_2 played a significant role in the cognitive synchronisation of the group. Thus it seems to be an important object to be retained in the record even if it is not a constructed object.

Figure 7 shows that sketch 4 was not produced during this phase but was a shared focus of some attention during this period. The duration of sketching might also be an indicator of the degree of finish of the sketch. Sketch 7 appears to be more complete than sketch 6.

Figure 8 highlights the group dynamic underlying the construction of two particular objects (sketch 6 and sketch 5). Sketches may be produced as an aside between one or two people, or more collectively between three people or more. Thus this last indicator shows the percentages of time spent in collective and aside interactions for sketch 5 and sketch 6.

This indicator shows that sketch 6 is more collectively produced than sketch 5. The implications of this result are not clear, it may

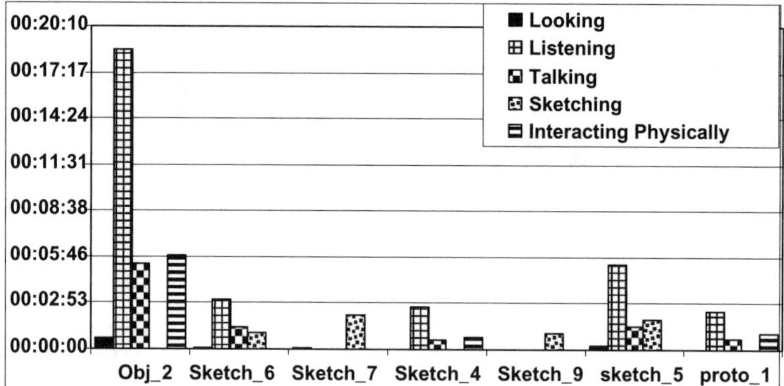

Figure 7. Total time period of the activities associated with each object.

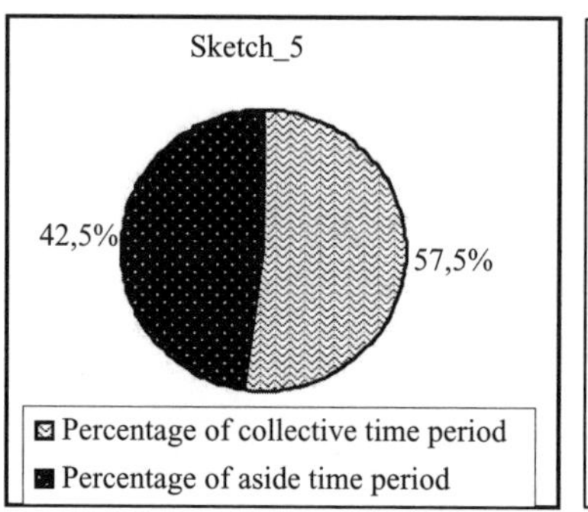

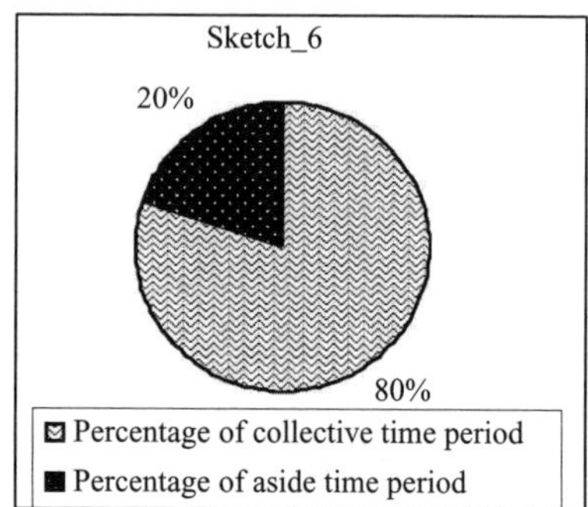

Figure 8. The percentages of collective and aside production time periods for two sketches, sketch_5 (left) and sketch_6 (right).

suggest that a collectively constructed sketch will be easier to re-interpret in the design rationale record, on the other hand, a sketch produced in an aside interaction may be adopted as a solution by the participants.

4. CONCLUDING REMARKS

Our goal is to provide engineers with tools that help to capture design rationale. To do this, we consider that it is necessary to track and retain selected intermediary objects in the minutes of design meetings. To this end, we suggest there is a need for selecting and including relevant objects and capturing some elements of context which will give sense to these objects for a future understanding of the meetings. This presupposes that we have a clear understanding of the role played by intermediary objects in the design process, and are able to identify the most relevant objects on which to concentrate engineers' formalisation efforts.

As a contribution to this clearer understanding, three main ideas were developed in this chapter. The first is the idea of focusing on what we call the critical phase of the meeting, defined as a convergence and production period, which includes and reflects the key decisions of the meeting. The second is to focus on relevant intermediary objects (referred to here as 'critical objects') which are to be identified within the critical phase. The third is to bring these critical objects closer to detailed design rationale based models of people's interactions around these objects.

We based our analysis on the first engineering meeting data in order to identify a critical phase, and to propose some indicators that reveal the critical objects. An analysis method was proposed, and some basic indicators were discussed. In this chapter, we focused on indicators that are mainly based on the collective nature of the objects. This is due to our intention to use sketches associated with collective objects in the minutes of meetings to keep a record of the design rationale.

Previous studies[27,28] have shown that collective involvement in sketch construction is a key pointer for its effective reuse and reinterpretation[29]. This is consistent with observations that cognitive synchronisation is a key issue in collective design processes[30]. These indicators allow us to exclude some objects as non-essential, and to highlight some important ones. The relevance of the indicators suggested here has to be validated in other design settings and protocols. Some lacunae in our indicators have already identified. For example, time based indicators may overlook objects that support rapid agreement and decision making of the group. There is also no guarantee that rejected ideas would be recorded, even though they may be important elements of design rational. Other metrics should certainly be developed to address these important issues in the capture of design rationale.

27.
Boujut and Blanco (2003), *op. cit.*

28.
Blanco (2003), *op. cit.*

29.
Stacey and Eckert (2003), *op. cit.*

30.
Detienne, F., Boujut, J.F., and Hohmann, B. (2004) Characterization of Collaborative Design and Interaction Management Activities in a Distant Engineering Design Situation in *Proceedings of 6th International Conference on the Design of Cooperative Systems (Coop '04)*, pp. 83–98.

Given valid indicators, it is possible to envisage a tool for recording the minutes of design team meetings. In such a situation, the creation, activation and closure of intermediary objects could be supported by IT tools and collaborative design software capable of automated identification of the active periods of objects, the levels of object sharing, and interaction levels. Such a tool might then propose candidate critical objects for the design meeting minutes. An engineer responsible for the minutes would only have to accept or reject each of the proposed objects, and to build, when appropriate, a design rationale schema of the discussion around the object.

However, there is still much to do before this can be achieved, as the research presented here raises many questions. Perhaps the most prominent of these concerns the model of interactions with the object, which needs now to be set within a theoretical framework, keeping in mind the goal to allow automatic interaction identification in the case of virtual meetings. The notion of critical phase is also open to further investigation. The identification criteria for critical phases need to be explored further alongside the more general issue of how best design meetings can be segmented into different phases. The impact of segmentation on the indicators we have introduced here also has yet to be explored.

ACKNOWLEDGEMENTS

This work had been funding by Region Rhone-Alpes as a part of project ASPIC in research cluster GOSPI.

22
Matters of Context in Design

Cynthia Atman, Jim Borgford-Parnell,
Katherine Deibel, Allison Kang, Wai Ho Ng,
Deborah Kilgore & Jennifer Turns

> "The world is changing rapidly, sometimes with intent, but too often by accident.... The one thing that makes this state of affairs tolerable is the inchoate knowledge that change – desired change – can be wrought by human intention. Human intention made visible and concrete through the instrumentality of design"[1].

1. Nelson, H.G. and Stolterman, E. (2003) *The Design Way: Intentional Change in an Unpredictable World*, Educational Technology Publications, Englewood Cliffs.

An engineer's ability to address broader contexts in design is viewed by engineering experts as a critical competency for the 21st century. This chapter examines the DTRS7 data to understand how a design team handled context issues in their design process. We use timelines to graphically illustrate when particular design and discussion phenomena occurred in the design process. We have found intriguing relationships between contextual issues and several of those phenomena, which both support contemporary understandings of the design process and point to new arenas of inquiry.

The world is increasingly becoming a designed environment wherein 'human intention' is overwhelmingly evident. Humans' interactions with the world and each other are often mediated through the objects and processes wrought by designers. Design is indeed instrumental, as Nelson and Stolterman[2] argued; it is a tool to be wielded by humans to generate intentional change. However, simply because a design is intentional does not preclude the derived object or process from being poorly designed or from affecting the world in negative and unintended ways. As Samuel Florman wrote in regard to engineering ethics, "A well intentioned, even saintly engineer may still be an inept engineer, that is, a 'bad' engineer if his [or her] work does not serve the public well.... [I]n technical work, the facts show that competence does more good than 'goodness,' that incompetence does much more harm than evil intent"[3]. This raises an important question: What constitutes competence when it comes to engineering design?

2. *ibid.*

3. Florman, S.C. (2002) Engineering Ethics: The Conversation without End, *The Bridge*, 32, *National Academy of Engineering*, http://www.qurl.com/w1ct5 (accessed 2nd August, 2008); p. 3.

Certainly an essential aspect of the answer is that a competent designer must be cognisant of and able to take into consideration appropriate contextual issues. Contextual issues extend beyond technological feasibility and local implementation concerns into the broader societal, cultural, environmental, and ethical realms. Lloyd[4] proposed that engaging in design "can function as a good way of making ethical decisions". The degree to which a designer

4. Lloyd, P. (Chapter 5) Ethical Imagination and Design.

or design team is able to competently address both the fine-grained details of a design as well as its broader contextual parameters, the greater the likelihood is of a successful outcome. Design context, from which problems stem and within which solutions are implemented, is seen as being increasingly important as the globe shrinks, and our designed environment becomes more complex.

In the engineering education community in the United States, this point has been made repeatedly and forcefully over the past few decades. In *The Green Report* in 1994, the Engineering Deans' Council and Corporate Roundtable posited that engineering "students must understand the economic, social, environmental and international context of their [future] professional activities"[5]. In *The Engineer of 2020: Visions of Engineering in the New Century*, it was argued that "successful engineers in 2020 will, as they always have, recognise the broader contexts that are intertwined in technology and its application in society"[6]. The engineering accrediting body, ABET[7]. recognised that competent engineering graduates should be able to link technologies to their broader contexts. ABET refers to this aspect of designer competence in engineering criterion 3h as understanding: "the impact of engineering solutions in a global, economic, environmental, and societal context"[8].

5. ASEE (American Society of Engineering Education) (1994) *The Green report – Engineering education for a changing world*, http://www.qurl.com/vl2kv (26th August 2008).

6. NAE (National Academy of Engineering) (2004) *The Engineer of 2020: Visions of Engineering in the New Century*, The National Academies Press.

7. ABET (2007) *Criteria for accrediting engineering programs*, http://www.qurl.com/3pyhf (accessed 26th August 2008).

8. *ibid.*

The ability of engineering designers to address the broader context in the development of new technologies and technological applications is viewed as a policy concern, a professional competency, and an educational imperative. Therefore, it is of significant importance to the authors as an area of research and was the focus for this study. Ours is one of several studies that asked different questions of a common dataset. Our primary interest in examining the design meeting data was to better understand how the team of designers handled context issues in their design process.

In the next section, we further situate the study by exploring the responses of the educational and research communities to the calls for emphasis on context in design and then identifying the opportunity the study addresses. In Section 2 where we deal with our approach, we discuss our adaptation of an analysis method to the needs of studying context in team design. The subsequent sections present and discuss the results.

1 PRIOR RESEARCH AND PRESENT OPPORTUNITIES

We identified several threads of activity that were significant to the emphasis on context: educational initiatives that emphasise learning to grapple with context, research on design processes with an eye toward the role of context, and research on student perceptions related to context. We focus this short review on the first two threads.

Several conceptual and descriptive papers focus on educational approaches addressing the need for engineers and others to understand the implications of technology in context. A sampling of these approaches includes an introductory engineering course at Smith College that incorporates socially relevant design projects[9]; Rensselaer Polytechnic Institute's Product Design and Innovation program that promotes design learning in conjunction with understanding the social and cultural context of technologies[10]; an initiative at Virginia Tech to infuse freshmen courses with experiences that provide students with: "topics essential to their preparation as globally and socially conscious engineers"[11] and the new collaborative course 'Product Realization for Global Opportunities' offered to students at the University of Pittsburgh and the University at Campinas in Brazil, in which students develop knowledge of sustainability issues in cultural contexts[12]. These educational initiatives illustrate that considerations of context in design have advanced well beyond the recognition of need and are slowly becoming important aspects of engineering design education.

Stemming from several of the reports mentioned earlier[13] and others, a policy discussion has developed regarding the *global competencies* required of contemporary engineers[14,15]. Often the context issues engendered in this discussion of competencies concern the language, culture, and disciplinary differences of engineers working with other designers, and clients in other parts of the world. In this way the larger contextual considerations encompass both the context of a designed artefact as well as the context within which a competent engineer functions. Two of the above programs[16,17] are examples of ways this discussion is manifesting in engineering education. While more educators are creating programs to help students learn to grapple with context in design, one can wonder about the knowledge base that such programs are leveraging in their design efforts. Current educational wisdom suggests that effective design of learning activities should, where possible, have the educational activity guided by a clear understanding of how the target competency is enacted in practice.

A body of work has examined the link between contextual factors and problem scoping. Problem scoping is defined as the design stage during which designers explore relevant contextual issues and set the parameters of the problem they will continue to solve[18,19]. During this process, designers gather the information they need to clarify or better define a problem, as well as identify the information necessary to formulate design solutions. Several studies illustrate the significance of problem-scoping activities as they relate to expert practice[20,21,22,23]. Studies have revealed evidence of consideration of broad contextual issues in early problem structuring activities[24,25]. In their summary of design research, Restreppo and Christiaans[26] noted that information accessed during early problem

9. Mikic, B., and Grasso, D. (2002) Socially-Relevant Design: The TOYtech Project at Smith College, *Journal of Engineering Education*, 91, p. 319.

10. Gabriele, G., Kagan, L., Bornet, F., Hess, D., and Eglash, R. (2001) Product Design and Innovation: A New Curriculum Combining the Humanities and Engineering in *Proceedings of ASEE Annual Conference*, pp. 8105–8118.

11. Terpenny, J., Goff, R., Lohani, V., Mullin, J., and Lo, J. (2007) Preparing Globally and Socially Conscious Engineers: International and Human-centered Design Projects and Activities in First Year: *Design and Engineering Education in a Flat World, Mudd Design Workshop VI Proceedings*, Harvey Mudd College, Claremont, CA, USA.

12. Mehalik, M.M., Lovell, M., and Shuman, L. (2007) Product Realization for Global Opportunities: Learning Collaborative Design in an International Setting: *Design and Engineering Education in a Flat World, Mudd Design Workshop VI Proceedings*, Harvey Mudd College, Claremont, CA, USA.

13. For example, NAE (2004), *op. cit.*

14. Chubin, D.E., and Babco E.L. (2005) Diversifying the Engineering Workforce, *Journal of Engineering Education*, 94, pp. 73–86.

15. Downey, G.L., Lucena, J.C., Moskal, B.M., Parkhurst, R., Bigley, T., Hays, C., Jesiek, B.K., Kelly, L., Miller, J., Ruff, S., Lehr, J.L., and Nichols-Belo, A. (2006) The Globally Competent Engineer: Working Effectively

structuring phases (e.g. information related to users, environments of use, etc.) is fundamentally different from the kinds of information accessed during problem solution phases (e.g. information related to materials, manufacturing conditions, etc.). Other studies have shown that broad contextual issues influence concept generation,[27] and are sometimes a cause of design iterations[28]. These and other studies illustrate that consideration of broader contextual issues is linked to key design activities and may, therefore, fundamentally influence design solutions.

While context is somewhat implicit in much of the problem scoping work, one thread of this work has explored context more explicitly[29]. This study compared the performance of first year and graduating engineering students on a lab-based engineering design problem. The researchers reported that: "on average, graduating [students] consider more and a broader array of factors than [first-year students]". This led the researchers to ponder what the problem scoping of experts might reveal.

While the above limited review suggests momentum around the issue of context, there remains *opportunity* for more research. For example, little is known about how context is introduced by multidisciplinary teams facing authentic design problems. Such research could explore questions such as: When and what type of context issues are addressed in the design process? Are context issues associated with particular phases of the design process? The analysis described here represents a response to this opportunity.

2 APPROACH

In our research, we analysed the process of an interdisciplinary team working on the design of a thermal pen. The primary question we sought to answer was: how did this team of designers integrate context into their design process? To investigate, we adapted an analytic methodology from previous research[30,31,32] to the demands of an authentic, team-based design session, with specific attention paid to tracking context information. Our analysis resulted in a systematic time-based description of the team's conversation in terms of design activities, topics underlying these activities, and whether their focus was on the context of the design or not.

We segmented and coded the second engineering meeting (E2) transcript using four different schemes of analysis – Agenda Item, Conversation Topic, Design Activity, and Context Focus – as described below. We gave the E2 meeting priority as it had a broader range of the various activities and phases of engineering design[33] and thereby provided opportunities to track the introduction of context in the design process. After coding the data, we constructed detailed timelines that provided visual representations of our four

with People who Define Problems Differently, *Journal of Engineering Education*, 95, pp. 107–122.

16.
Terpenny et al. (2007), *op. cit.*

17.
Mehalik, Lovell, and Shuman (2007), *op. cit.*

18.
Bursic, K.M. and Atman, C.J. (1997) Information Gathering: A Critical Step for Quality in the Design Process, *Quality Management Journal*, 4, pp. 60–75.

19.
Cross, N. (2001) Design cognition: Results from Protocol and other Empirical Studies of Design, in Eastman, C.M., McCracken, W.M., and Newstetter, W.C. (eds) *Design Knowing and Learning: Cognition in Design Education*, Elsevier Science.

20.
Schon, D.A. (1988) Designing: Rules, Types and Worlds, *Design Studies*, 9, pp. 181–190.

21.
Cross, N. and Clayburn Cross, A. (1998) Expertise in Engineering Design, *Research in Engineering Design*, 10, pp. 141–149.

22.
Jain, V.K. and Sobek II, D.K. (2006) Linking Design Process to Customer Satisfaction through Virtual Design of Experiments, *Research in Engineering Design*, 17, pp. 59–71.

23.
Mehalik, M.M., and Schunn, C. (2007) What Constitutes Good Design? A Review of Empirical Studies of Design Processes, *International Journal of Engineering Education (Special Issue on Learning and Engineering Design)*, 22, pp. 519–532.

coding schemes, thus allowing analyses of the design process as it unfolded over time.

2.1 *Segmenting and coding transcripts*

The coding process began by segmenting the meeting E2 transcript into units identified as distinct pause-bounded utterances, ranging from a short phrase to three sentences. This resulted in 1696 segmented units from the meeting. Start and stop times of the segments were identified with Mangold INTERACT© software.

Segments were coded using the four coding schemes. For each pass at the coding, two researchers coded the data, an inter-coder agreement was calculated, and discrepancies were resolved to reach an arbitrated code for each segment. This coding process was repeated twice for each scheme. With the first pass, a moderate reliability level was achieved for the coding (71.9% for Design Activity, 70.7% for Context Focus, 65% for Agenda Item, and 61.9% for Conversation Topic). We had variable success with the second coding attempts, with some agreement levels increasing – to a high of 81.2% for Context Focus – and others showing a decrease. This prompted us to undertake a lengthy refinement process to improve the coding definitions (refer to Table 1 for code definitions), and then to resolve all existing coding discrepancies with an additional consensus coding process. The consensus coding process involved a team of three senior researchers who collaboratively completed a final pass for each coding scheme by verifying and arbitrating each code to consensus.

We assumed from the beginning that we would encounter some difficulties in coding, since three of the coding schemes were newly developed for this study – Agenda Item, Conversation Topic, and Context Focus. The fourth scheme – Design Activity – also presented some challenges, since in previous studies it had been applied to data from individual designers, not a team process like the DTRS7 data. Additionally, the design problems used in prior studies, intentionally, did not require domain expertise. The primary difficulty we encountered was that the E2 data were highly domain dependent. This introduced several challenges for coders who were not sufficiently cognisant of the technical jargon and colloquialisms of the particular knowledge domain represented in the data. For example, terms such as media, interface, and implementation have different meanings when used in different domains. Training coders to acquire this domain knowledge before coding is a time-consuming task. Therefore, in recognition of this difficulty, the final consensus coding for each segment was completed by a team of three senior researchers who were firmly grounded in all aspects of the research process, and included one member with domain fluency. The specifics of the coding systems and the

24.
Restreppo, J. and Christiaans, H. (2004) Problem Structuring and Information Access in Design, *The Journal of Design Research*, 4, pp. 247–259.

25.
Atman, C.J., Adams, R., Mosborg, S., Cardella, M., Turns, J. and Saleem, J. (2007a) A Comparison of Students and Expert Practitioners, *Journal of Engineering Education*, 96, pp. 359–379.

26.
Restreppo and Christiaans (2004), *op. cit.*

27.
For example, Ennis, C.W. and Gyeszly S.W. (1991) Protocol Analysis of the Engineering Systems Design Process, *Research in Engineering Design*, 3, pp. 15–22.

28.
Adams, R. (2002) Understanding Design Iteration: Representations from an Empirical Study, *Common Ground: Design Research Society International Conference*, London, UK, Staffordshire University Press.

29.
Atman, C.J., Yasuhara, K., Adams, R.S., Barker, T.J., Turns, J. and Rhone, E. (2007b) Breadth in Problem-scoping: A Comparison of Freshman and Senior Engineering Students, *Design and Engineering Education in a Flat World, Mudd Design Workshop VI Proceedings*, Harvey Mudd College, Claremont, CA, USA.

30.
Atman et al. (2007a), *op. cit.*

31.
Adams, R., Punnakanta, P., Atman, C. and Lewis, C.C. (2002) Comparing Design Team Self-Reports with Actual Performance: Cross-validating Assessment Instruments, in *Proceedings of ASEE Annual Conference*, pp. 5783–5798.

Scheme	Code	Definition
Agenda Item	FFEATURES	Basic functions and additions to the base product
	INTERFACE	Interaction between the user and product
	ARCHITECTURE	Electronics and manufacturing of product
	APPLICATIONS	Potential uses for the product
Conversation Topic	PEN	Form, abilities, and engineering of the thermal pen
	MEDIA	Various media the pen interacts with
	PEN/MEDIA	How the pen and media work together
	USER/USAGE	Persons who will use the product, applications of the product, usability issues
	GROUP	Implicit: indicating understanding and support Explicit: regarding the state of the group process
Design Activity	PROB. DEFINITION	Defining the problem
	GATHERING ONFO.	Collecting information
	GENERATING IDEAS	Thinking up potential solutions
	MODELING	Detailing how to build solution or parts of a solution
	FEASIBILITY	Assessing possible or planned solutions
	EVALUATION	Comparing solutions within constraints
	DECISION	Selecting one idea or solution
	COMMUNICATION	Revealing and explaining design to others
Context Focus	BROAD CLOSE	End users, marketing, usage concerns, safety Design details, technical aspects

Table 1. Code definitions.

coding process are described below and in Table 1; further details are available from the cited technical reports.

2.2 *Agenda Item*

Agenda Item[34] covered the broad topic areas the design team was asked to discuss in the second meeting. Since the individual items were designated by the team leader, the specific codes in this scheme (FEATURES, INTERFACE, ARCHITECTURE, and APPLICATIONS) were dependent upon the agenda set for the E2 meeting. Through the coding process, each item definition was expanded and refined to account for what was actually discussed during the meeting.

2.3 *Conversation Topic*

During an early viewing of the E2 meeting, we recognised that the designers were sometimes talking about the pen, the media,

32.
Atman, C.J., Chimka, J.R., Bursic, K.M., and Nachtmann, H.L. (1999). A Comparison of Freshman and Senior Engineering Design Processes, *Design Studies*, 20, pp. 131–152.

33.
See Atman et al. (1999), *op. cit.*

34.
Ng, W.H., Deibel, K., Atman, C.J., and Borgford-Parnell, J. (2007) DTRS7 Agenda Topic Coding, *Technical Report CELT-07-07*, Center for Engineering Learning and Teaching, University of Washington, Seattle.

or the users; and sometimes about the interactions between any of the three. The Conversation Topic coding scheme[35] was designed to capture this focus: PEN, MEDIA, USER/USAGE, or PEN/MEDIA interaction. For analysis purposes, interactions between the user and any other topic were coded as USER/USAGE. We also added a code for segments that focused on the group design process itself (GROUP). GROUP segments typically consisted of utterances by one member of the design team indicating agreement with or encouragement of another member's statements.

35. Deibel, K., Atman, C.J., and Borgford-Parnell, J. (2007) DTRS7 Conversation Coding, *Technical Report CELT-07-06*, Center for Engineering Learning and Teaching, University of Washington, Seattle.

2.4 *Design Activity*

To code the design process, we used a well documented Design Activity coding scheme[36] consisting of eight activities: PROBLEM DEFINITION, GATHERING INFORMATION, GENERATING IDEA, MODELING, FEASIBILITY, EVALUATION, DECISION and COMMUNICATION. As this coding scheme has mainly been applied in studying the design processes of individuals, we expected to do some refinement for the group process in the E2 meeting. Using the coding and intercoder arbitration process described previously, refinements to the code definitions were identified[37].

36. For example, Atman et al. (2007a), *op. cit.*

37. Kang, A., Ng, W.H., Atman, C.J., and Borgford-Parnell, J. (2007) DTRS7 Design activity Coding, *Technical Report CELT-07-05*, Center for Engineering Learning and Teaching, University of Washington, Seattle.

2.5 *Context Focus*

The coding scheme to capture Context Focus was defined in prior studies[38]. However, previous code definitions did not completely suffice when applied to the group design data. Therefore (although previous general definitions still applied) we developed more specific definitions relevant to the present data set[39]. The revised Context Focus code definitions are CLOSE and BROAD. In the team design process, when CLOSE is a focus of conversation, team members are considering information and ideas about technical components, parts, or features; budgeting; or manufacturability of items; with an intent to further refine an agreed upon design concept (even if the concept is later discarded). When BROAD is a focus of conversation in the team design process, team members are considering any of the following:

- people/persons/users, either in general or specific characteristics, dimensions, or aspects of the same;
- cost and/or information regarding the availability of resources/parts to the end-users, or marketing to the users;
- health or safety; and
- other environmental, social, or political concerns.

38. Kilgore, D., Atman, C.J., Yasuhara, K., Barker, T.J., and Morozov, A.E. (2007) Considering Context: A Study of First-year Engineering Students, *Journal of Engineering Education*, 96, pp. 321–334.

39. Borgford-Parnell, J., Ng, W.H., Kilgore, D., and Atman, C.J. (2007) DTRS7 Context Focus Coding, *Technical Report CELT-07-08*, Center for Engineering Learning and Teaching, University of Washington, Seattle.

2.6 *Timelines*

To visualise the coded segments, we used a timeline representation for each coding scheme. Because our four coding schemes all use the same segmented transcript, the four timelines can be presented and viewed collectively as in Figure 1. The entire meeting is represented on each timeline with time presented from left to right. For each segment of the transcript, a mark is placed on the line corresponding to how the segment was coded and at the appropriate location given the segment's start time. The width of the mark is proportional to the duration of the segment. For a single coding scheme, a timeline shows how time is allocated among the codes within the scheme and reveals how different codes are emphasised at different points of the meeting. Juxtaposing the four timelines helped to reveal any interactions between schemes and codes. Software developed by one of the authors (Deibel) was used to generate the timelines.

2.7 *Durations, transitions, and transition rates*

Various quantitative measures were also used to give further insights about the design process as it evolved over time. Time measures such as the amount of time spent in each code in a scheme were calculated. Additionally, the number of transitions across the different codes in a scheme was measured. We also considered the transition rate (the number of transitions per minute) rather than the pure transition count. If a portion of the transcript had a low transition rate, then multiple, consecutive segments were coded with the same code. The corresponding region of the timeline will show a solid block for one code. A high transition rate area consists of consecutive segments with different codes, and the timeline will show small, thin marks for two or more of the codes.

3 FINDINGS

Visual inspection of the four juxtaposed timelines (Figure 1) suggested that meeting E2 consisted of an overall pattern of segment blocks delineating four sections of the meeting: one introductory section (referred to here as the Preamble), followed by three main sections (referred to as Episodes I, II, and III). Considering the meeting as consisting of these four sections enabled our analyses both across and along the timelines. Further analyses of the coded transcripts supported our notion that the four sections represent distinct aspects of the design process. The characteristics of each of the four sections and the analyses are described below.

3.1 *Preamble (00:00:00 to 00:14:36)*

The team leader, Tommy, dominated much of the Preamble conversation by introducing and reiterating the design problem, answering questions from other team members, and providing information

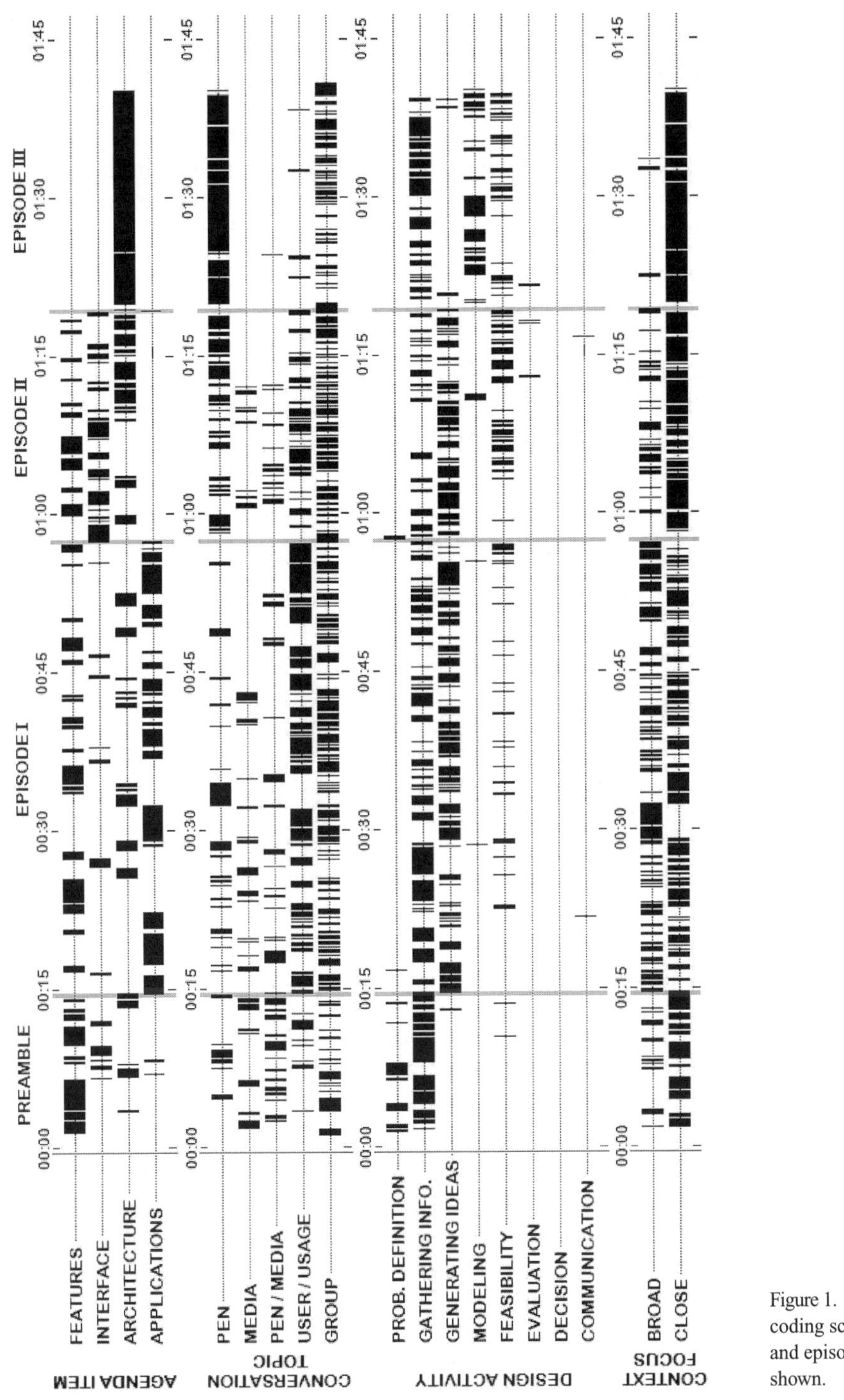

Figure 1. Timelines of the four coding schemes with preamble and episode boundaries shown.

and ideas from the previous meeting (E1). As the Design Activity timeline (Figure 1) shows, very little PROBLEM DEFINITION occurred in the overall meeting, with most of that activity taking place within the Preamble. In the first half of the Preamble, there is interplay between PROBLEM DEFINITION and GATHERING INFORMATION, but with noticeably fewer PROBLEM DEFINITION segments in the latter half. In terms of Agenda Item, FEATURES segments dominated the Preamble. GROUP segments occurred throughout the Preamble as Tommy set the scope of work for the design team.

Contextual issues discussed in the Preamble weighed heavily toward CLOSE (7.4 minutes CLOSE vs. 2.2 minutes BROAD). Since most of the Preamble was Tommy's report on decisions made in the previous meeting (E1), time was mostly spent on detail issues, such as the mechanics of thermal printing and the engineering of the current pen prototype.

3.2 *Episode I (00:14:35 to 00:57:23)*

This 43-minute episode may be characterised as a brainstorm of the potential applications and features of the thermal pen product. The episode began when Tommy explicitly opened the discussion to new ideas, although the Design Activity timeline shows that the team had already begun GENERATING IDEAS. Throughout the episode there is interplay between GATHERING INFORMATION and GENERATING IDEAS, with some instances of determining FEASIBILITY. Similar high numbers of transitions occur on the other timelines with the exception of the Agenda Item timeline. That timeline shows an emphasis on APPLICATIONS but also continues the FEATURES discussion begun in the Preamble. In contrast, there were many initial transitions among all the Conversation Topic codes, followed by an increasing focus on USER/USAGE, and a decrease in focus on the other topics (with the exception of GROUP). The initial high transition rate reflects the design team's consideration of what, how, and by whom the thermal pen might be used.

The Context Focus timeline shows a high frequency of transitions between BROAD and CLOSE, with almost equal time spent on each (16.8 minutes CLOSE to 14.5 minutes BROAD). In this episode, the discussion generally followed a BROAD-to-CLOSE sequence that begins when one person introduces a BROAD application or a BROAD issue such as safety. Subsequent conversation remains at the BROAD context level, but if the topic requires deeper insight or suggests new features for the system, an extended CLOSE, detailed conversation follows. An example was when one team member professed concern about the heating element of the pen being safe for kids, and then the conversation narrowed to a focus on how the pen's thermal head is fired. Another example was a discussion on using the pen to print Braille. After discussing the motivation and need for such an application, the group pondered how to use the pen to create the bumps/blisters needed for Braille.

3.3 *Episode II (00:57:23 to 01:19:25)*

This 22-minute episode entails the narrowing of the group's focus onto a particular design solution. This narrowing of focus was initiated by a PROBLEM DEFINITION statement by Tommy, who asked the team to turn their discussion to the pen's features and interface for the designated user-group: children aged 4–11. Gericke, Schmidt-Kretschemer, and Blessing[40] noted a similar transition point in the E2 meeting. These researchers describe the 58 minute mark as the point of a long initial analysis phase (application and requirements) changing into a very clear synthesis phase (solution). They also pointed to Tommy's intervention as the cause of the transition. There are many transitions across all of the codes in Episode II. While the previous episode had periods of focus on specific Agenda Items, this episode featured interplay between FEATURES, INTERFACE, and ARCHITECTURE with emphasis initially on FEATURES and INTERFACE and later on ARCHITECTURE. Despite its prominence in Episode I, APPLICATIONS were not discussed. This episode is also characterised by interplay between multiple Conversation Topic codes (particularly PEN, USER/USAGE, and GROUP) with increasing emphasis on PEN and a decrease in focus on USER/USAGE as the episode progressed. This change coincided with the increasing emphasis on ARCHITECTURE. For Design Activity, the emphasis continued on the interplay between GATHERING INFORMATION and GENERATING IDEAS (but no additional PROBLEM DEFINITION) and an increased discussion of FEASIBILITY and occasionally of EVALUATION.

40.
Gericke, K., Schmidt-Kretschmer, M. and Blessing, L. (Chapter 11) The Influence of the Design Task Description on the Course and Outcome of Idea Generation Meetings.

In terms of Context Focus, the episode still exhibits an interplay between BROAD and CLOSE, but the number of transitions is less pronounced (i.e., longer uninterrupted periods of CLOSE) with the attention to CLOSE (12.8 minutes) now greatly exceeding the attention to BROAD (3.1 minutes). In this episode, the sequence found in Episode I is reversed to CLOSE to BROAD. The conversation was mainly CLOSE and focused on the INTERFACE and FEATURES of the PEN and MEDIA. At times, a participant brought up a BROAD issue or idea that helped the team assess and refocus the current topic of conversation. One example was a discussion of potential FEATURES and INTERFACE details required to store multiple patterns in the pen through barcodes. As the discussion focused exclusively on technical details, Sandra brought up that the users are children. In this episode, BROAD statements typically align with USER/USAGE and CLOSE statements with PEN and MEDIA.

3.4 *Episode III (01:19:25 to 01:40:47)*

The final episode (21.5 minutes in length) marks the group's focus on the technical details of the design solution. As with Episode II, Tommy's intervention started this final episode. Tommy asked the team to stop everything else and concentrate the discussion on ARCHITECTURE. This episode has a prominent single code on each of the timelines. For Agenda Item, the emphasis is exclusively on ARCHITECTURE. For Conversation Topic, the emphasis is back on the PEN (although there is a consistent backdrop of segments devoted to GROUP process). For Design Activity, GATHERING INFORMATION is the dominant code although not as exclusively dominant as in the other categories due to the interplay between GATHERING INFORMATION, MODELING and FEASIBILITY. Most of the MODELING activities in the meeting occur in this episode.

In this episode, the Context Focus emphasised CLOSE (16.2 minutes) with a bit of BROAD (0.5 minutes). A similar sequence as in Episode II occurred: mainly CLOSE statements that occasionally triggered a BROAD statement, although there were far fewer instances of this type of sequence in Episode III. The overall CLOSE discussion focused on the pen's electronic components: capabilities, power demands, manufacturing costs, etc., whereas the few BROAD statements were about the weight of the toy or marketing considerations such as the need to produce a lot of units.

3.5 *Looking across the episodes*

Our analyses focuses primarily on the three episodes since the preamble section of the meeting was a catch-up time in which Tommy reiterated the design decisions made in the previous meeting. As described in Sections 3.2–3.4, the episodes contain different codes with different frequencies of occurrence. These differences not only helped identify the episodes but also highlighted patterns in the meeting's conversation. The highest frequency of BROAD segments co-occurred with the onset of GENERATING IDEAS segments in Episode I, and then the frequency of both codes began to decrease at the beginning of and throughout Episode II. The team's discussion of INTERFACE was almost non-existent before the targeted user group was identified. The Conversation Topic segments show several other patterns in relation to Agenda Item segments. USER/USAGE becomes the primary focus only when the discussion concerns FEATURES, INTERFACE or APPLICATIONS and is nearly absent during the technical ARCHITECTURE discussions. MEDIA and PEN/MEDIA show a similar pattern as well. Only PEN is consistently associated with ARCHITECTURE, hence its dominance in Episode III. GROUP is the single Conversation Focus code that shows remarkable consistency throughout the E2 meeting (see Section 3.8 for more discussion).

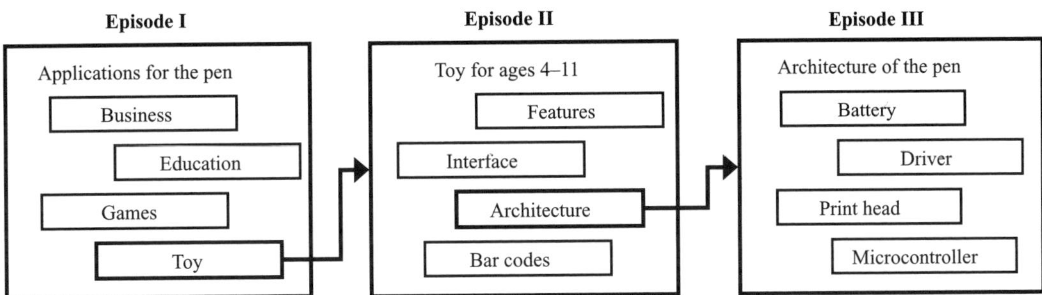

Figure 2. Diagram showing hierarchical structure of the E2 meeting.

Several segments in Design Activity also reveal expected patterns. The appearance of PROBLEM DEFINITION at the beginning of the meeting and once more as a trigger for Episode II is unsurprising as an early, although brief, aspect of a design process[41]. FEASIBILITY, which is the assessment of possible or planned solutions, took place throughout the three episodes, but occurred with the highest frequency in Episode II. EVALUATION, the comparing and contrasting of different solutions, occurred mainly in Episode II. This middle episode is primarily concerned with implementing the pen for children, suggesting the need to decide between different implementation choices. Similarly, MODELING (the determination of how to build the solution) occurred mostly in Episode III during the ARCHITECTURE discussion.

41. Atman et al. (2007a), *op. cit.*

A pattern can also be seen across the Context Focus timelines. Both BROAD and CLOSE segments occurred throughout the meeting; however, as the frequency of BROAD segments diminished, the emphasis on CLOSE consistently increased (see Section 3.7 for more discussion). Relating to this BROAD to CLOSE transition, Figure 2 illustrates a hierarchical pattern of discussion activity showing a progression from more general issues to more specific issues. Similar patterns were identified in an earlier study by Morozov, Kilgore, and Atman[42] where designers were observed to serially 'nest' design discussions and decisions within slightly higher-level discussions. Figure 2 displays the relationship of one of these nested discussions to the three episodes (see Section 4 for further discussion).

42. Morozov, A., Kilgore, D., and Atman C.J. (2007) Breadth in Design Problem Scoping: Using Insights from Experts to Investigate Student Processes in *Proceedings of ASEE Annual Conference*, session 3430.

3.6 *Transition rates*

The transition rates for the four coding schemes (see Table 2) also reveal striking differences between the episodes. In all episodes, Conversation Topic has the highest rate, largely due to the continuous attention towards the group process (GROUP) throughout the meeting. When segments coded as GROUP were removed from the transition calculation, the transition rate for Conversation Topic dropped considerably. For every coding scheme except Context

Table 2. Transition rates (per minute) of the four coding schemes.

	Transition Rates (per minute)			
Coding Scheme	Episode I	Episode II	Episode III	Overall
Agenda Item	1.4	2.4	0.0	1.3
Conversation Topic	8.8	12.1	7.2	8.5
Conversation Topic (w/o *Group*)	1.9	3.1	0.4	1.9
Design Activity	2.6	4.2	4.0	3.1
Context Focus	2.5	2.2	0.3	1.8

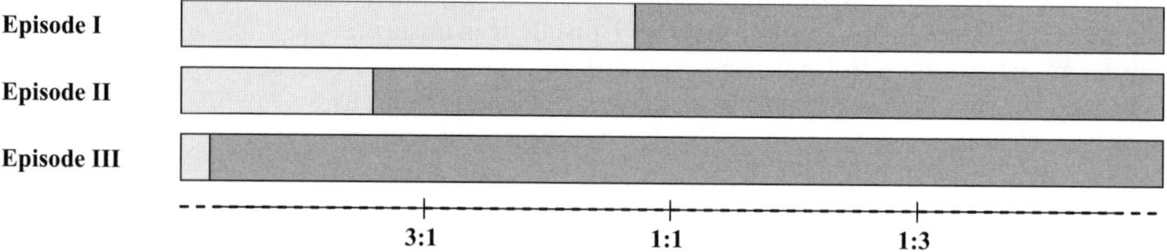

Figure 3. Ratios of BROAD (lighter) to CLOSE (darker) context focus by episode.

Focus, the transition rate reaches its maximum during Episode II, as is visually presented in Figure 1. The increased transition rate for Design Activity in Episode II appears to correspond with the shift from discussing product applications to discussing how to implement the product. In Episode III, the transition rate decreases significantly for most of the categories, particularly for Agenda Item and Context Focus. This drop reflects the narrowing of the discussion to the pen's architecture. The Design Activity rate, however, remains largely unchanged.

3.7 *Shifting from broad context focus to close context focus*

The Context Focus timelines and transition rates show that the discussion shifted from BROAD to CLOSE over time (Figure 1). In Episode I, slightly less time was spent on BROAD (33.8%) than CLOSE (39.2%). The ratio shifts more toward CLOSE in Episode II (14.2% BROAD to 58.0% CLOSE). The ratio then becomes overwhelmingly a CLOSE discussion in Episode III (2.4% BROAD to 76% CLOSE), and parallels the change in conversation from FEATURES and INTERFACE to ARCHITECTURE (see Figure 3).

This BROAD to CLOSE shift also relates to Design Activity. In Episode I, of the 40 segments coded FEASIBILITY, 47.5% were also labelled BROAD and 42.5% were labelled CLOSE. In Episode II, the FEASIBILITY ratio is 37.2% BROAD to 51.2% CLOSE, and by Episode III, the ratio is 9.1% to 79.5%, respectively. Two other Design Activity codes, GATHERING INFORMATION and GENERATING IDEAS, are in all episodes and display the same shift. Thus, the design process activities appear linked to the increased emphasis towards CLOSE.

3.8 *Stability of group process*

Despite all the differences between the episodes, attention to the group process (GROUP) continued throughout the meeting. The proportion of time, spent on GROUP per episode, was also similar (16.3% to 17.5% to 15%). A comparison of the two overall transition rates reported in Table 2 for the Conversation Topic scheme, one rate that included GROUP segments and one rate in which GROUP segments were removed, also demonstrates the importance that group process had throughout the meeting. This finding is consistent with Goldschmidt and Eshel's view[43] that: "collaboration cannot be built and maintained without a massive investment in social constructs. Apart from humour the actors must express frequent approval of what is being proposed; they must ask many questions in order to be totally involved, and they must side-track to foster a friendly and supportive atmosphere". The kind of social atmosphere that develops is a consequence of the attention being paid to group process, and according to Lloyd[44] may have a direct bearing on whether, when, and by whom, broader contextual issues are discussed.

43.
Goldschmidt, G. and Eshel, D. (Chapter 18) Behind the Scenes of the Design Theatre: Actors, Roles and the Dynamics of Communication.

44.
Lloyd, P. (chapter 5) Ethical Imagination and Design.

4 DISCUSSION

As in previous studies,[45] our approach to coding and timelines proved successful for exploring an engineering team design process. The episode analysis revealed much about how the designers in the E2 meeting approached design, including the important role that context played in the process. These results reinforce our current understanding of the design process and support the idea that educational efforts meant to emphasise and integrate contextual issues more firmly in the teaching and learning of design can benefit from research into the design practices of professional engineers. As this study demonstrated, for these professional designers contextual issues were essential considerations in the design process.

45.
For example, Atman et al. (1999), *op. cit.*

In Atman et al.[46] it was noted that expert engineers exhibited a consistent pattern (evident in timelines) showing a 'cascade' of activities through the design process. The Design Activity timeline for the E2 meeting (Figure 1) exhibits the beginning of this cascade pattern, consistent with those earlier results. The meeting began with

46.
Atman et al. (2007a), *op. cit.*

the team leader delineating the design problem and gathering information relevant to solving it. After about 15 minutes, the cascade began as the discussion transitioned into developing alternative solutions with the onset of GENERATING IDEAS and FEASIBILITY. The cascade continues to be evident in Episode III with the emphasis on MODELING and FEASIBILITY. GATHERING INFORMATION activities, which included both asking for and providing each other with information, were not only present in the early phase of the design process, but were also frequent throughout the E2 meeting and appear to accompany (by necessity) other design activities in the cascade. In the early problem scoping phase, which occurred principally in the first 15 minutes (preamble) of the meeting, the information being gathered and discussed was focused on setting the parameters for the design problem, and in later phases the information focused on explicating aspects of potential solutions and enabling decision-making. Meeting E2 was the second of two brainstorming sessions, and some problem scoping activity had taken place in the first E1 meeting. This led to a truncated problem scoping phase in the E2 meeting. Additionally, had the design process continued beyond brainstorming, we would expect the team to cascade into project realisation and begin finalising a design solution.

Related to this potential expert activity pattern is another one of this study's findings. Although the Context Focus timeline (in Figure 1) shows that BROAD aspects were considered throughout the meeting, BROAD was emphasised most in Episode I as the team began developing new solutions. Since the problem scoping activity in the preamble was affected by the conversation that took place in the E1 meeting, there ensued a lack of emphasis on BROAD contextual issues, which are generally associated with problem scoping. As noted earlier, Restreppo and Christiaans[47] found that during early problem scoping designers accessed contextual information such as users and environment, which is the type of information that we classify as BROAD. However, once the discussion cascaded into developing new ideas, broad issues were emphasised. This is consistent with the findings of Ennis and Gyeszly[48] and others, who show that broad contextual issues influence concept generation. In a previous study[49] researchers speculated on whether a consideration of more broad issues early in the design process was linked to greater expertise, and if so, whether analysing the design processes of practicing engineers would support that linkage. Our findings appear to add credence to the idea that experienced engineers explore a design problem's broader contextual aspects before considering the close context in depth.

47. Restreppo and Christiaans (2004), *op. cit.*

48. Ennis and Gyeszly (1991), *op. cit.*

49. Atman et al. (2007b), *op. cit.*

The three episodes of the E2 meeting also support the shifting emphasis from BROAD to CLOSE by structuring the discussion hierarchically. As shown earlier in Figure 2, Episode II constrains the general application discussion of Episode I by considering an interface for one specific user group. Episode III continues this constraint by focusing on the interface's architecture. Morozov, Kilgore, and Atman[50] found that some expert engineers approach problems in this way: first framing the problem in a broad sense and then focusing on specific factors in more detail. This process can be repeated with a specific factor, creating a nested discussion process that further narrows its focus with each nesting. Through conscious effort or not, the choices made for structuring the design discussion in meeting E2 created a hierarchical structure emphasising BROAD issues first and CLOSE issues later.

50. Morozov, Kilgore and Atman (2007), *op. cit.*

An interesting aspect of this nested discussion is that, although the design group decided in Episode I to focus their discussion on applications of the pen for a four-to-eleven age group, at one point in Episode II the discussion took a sidetrack into possible business and security applications. The conversation then swung back to children's applications before narrowing further into architectural considerations in Episode III. However, the possible business and security ideas were not completely shelved and later became a part of the architecture conversation with regard to the degree of flexibility that should be designed into the pen so that it might fulfil multiple uses. One interpretation of this observation is that it may be possible for parallel nested discussions to take place in a design meeting. This could be related to the ideas discussed by Mabogunje et al.[51] regarding a pattern of interaction they call 'resumption'. They argue that the act of initiating, pausing, and resuming multiple threads in parallel in design conversation is consistent with other examples in natural communication.

51. Mabogunge, A., Eris, O., Sonalkar, N., Jung, M. and Leifer, L. (Chapter 3) Spider Webbing: A Paradigm for Engineering Design Conversations during Concept Generation.

Additionally, although there was clearly a BROAD to CLOSE transition that took place across the overall meeting, there were other Context Focus changes taking place within the episodes. These changes might be characterised as role-reversals. In Episode I, BROAD issues were the driving/defining concepts of the conversation that would at times initiate more detailed, CLOSE discussion to further explore and assess an idea. In Episodes II and III, CLOSE issues dominated the conversation but were interrupted by the injection of a BROAD topic to assess and potentially refocus the current discussion. When one aspect of context (whether BROAD or CLOSE) was the focus of the design discussion, the other aspect of context was not ignored but served as a means to explore, assess, or redirect the current design. Although the emphasis on BROAD versus CLOSE context shifted over time, the engineers in the meeting appeared to continually consider both aspects of context throughout the meeting. This finding highlights the complex role contextual awareness plays in the design process and motivates the need for continued research in this area.

The professional engineers in this study demonstrated how contextual focus helped to shape the design process, and that consideration of a wide-range of contextual issues was integral to a rich and productive discussion. Contextual issues drove the design discussion into new arenas of creative speculation, and helped to guide that speculation by providing details, boundaries, and analytic perspectives. The results of this and other design studies illustrate some ways that context is considered in the design process, and help to frame some of the challenges faced by engineering educators and their students. Engineering students need to acquire a sense of the importance of context, they need to learn to identify and explore a full range of contextual issues, and they need to experience and understand the dynamic relationships of contextual issues and design activities. Ultimately, engineering education needs to reflect the best of what engineers actually do. Certainly the educational approaches discussed earlier, in which context issues are more integrated in student design work, are important in fostering that approach. However, the design processes taught to students should be informed by what this (and other studies) reveal about the practices of real engineers and the extent to which considering context does matter.

ACKNOWLEDGEMENTS

This study was supported by a National Science Foundation grant EEC-0639895, the Boeing Company, the Mitchell T. and Lella Blanche Bowie Fund, and the Centre for Engineering Learning and Teaching at the University of Washington. We wish to thank the organisers of DTRS7 for their hard work and dedication to this research project. In addition we would like to thank the reviewers of our manuscript and other workshop attendees who provided such thoughtful feedback on this study. We also wish to acknowledge the people in the larger CELT community who helped launch our analysis with their brainstorming ideas and encouragement. Finally we give special thanks to the undergraduate researchers whose diligent work made this research possible: Athena Epilepsia, Avram Epilepsia, Rebecca Kim, Ngoc Nguyen, Adrienne Oda, Jessica Tran and Alice Ward.

List of Contributors

ROBIN ADAMS is an assistant professor at Purdue University in the School of Engineering Education. Her research focuses on cross-disciplinary practice and ways of thinking, design knowledge and learning, epistemologies of engineering practice, and strategies for connecting research and professional practice. She also leads programs for building community in engineering education research.

SAEEMA AHMED is an associate professor in the Department of Management Engineering, Technical University of Denmark. She conducted her PhD at the University of Cambridge, where she was also a fellow of New Hall. She leads a group whose research interests include design thinking, management of design knowledge and ontology.

ÖMER AKIN is Professor of Architecture at Carnegie Mellon University. He has been researching design cognition since 1974. His more recent work on design also deals with requirements, management, and digital technologies. His publications are extensive and include *Psychology of Architectural Design* and *A Cartesian Approach to Design Rationality*.

EMINE SERAP ARIKOĞLU is a PhD student in collaborative engineering at the University of Grenoble. She graduated in Industrial Engineering from Galatasaray University, Turkey in 2006. She is currently working on utilisation of scenarios for developing a shared vision between designers of users' needs.

CYNTHIA ATMAN is Professor in Industrial Engineering and the Director of the Center for Engineering Learning and Teaching in the College of Engineering at the University of Washington. She also directs the Centre for the Advancement of Engineering Education. Her research focuses on design learning and engineering education.

PETRA BADKE-SCHAUB is Professor of Design Theory and Methodology at Industrial Design Engineering at TU Delft, Netherlands. Her work includes research methods, decision making and communication in teams, team mental models, experience and creativity in design. Her publications include over 100 refereed journal and conference papers and four books.

LINDEN J. BALL is a senior lecturer in Psychology at Lancaster University, UK. He conducts experimental research on fundamental deductive and inductive reasoning processes as well as naturalistic research examining thinking in domains such as design. His published work focuses primarily on how cognitive uncertainty engenders shifts in reasoning strategies.

ERIC BLANCO is an associate professor in product development at the Grenoble Institute of Technology. He graduated from ENS Cachan in 1993. He obtained his PhD in Industrial Engineering in 1998. His current research deals with analysis of coordination and cooperation through product representations and information flow in collaborative design.

LUCIËNNE BLESSING is Vice-Rector of the University of Luxembourg and has a chair in engineering design and methodology. Her research interests include the design process, design research methodology, and user interfaces. She is co-editor of *Research in Engineering Design* and is a member of the advisory board of the Design Society.

JIM BORGFORD-PARNELL is Assistant Director of the Centre for Engineering Learning and Teaching at the University of Washington. He taught furniture design, research methods, and educational theory and pedagogy courses for 25 years; and was a professional furniture designer. He is currently engaged in research and instructional development in engineering education.

BO T. CHRISTENSEN is an assistant professor at Copenhagen Business School in Denmark. A cognitive psychologist by training, his works include both ethnographic studies of engineering designers and experimental studies of design cognition. His theoretical focus is on creative cognitive processes such as analogy, simulation and incubation.

NIGEL CROSS is Emeritus Professor of Design Studies at The Open University, UK. He was one of the originators of the series of Design Thinking Research Symposia. His work in design cognition has addressed issues of teamwork and creativity, and the cognitive skills of outstanding designers. His most recent book is *Designerly Ways of Knowing*, Birkhauser, 2007.

SHANNA DALY is a postdoctoral researcher at the University of Michigan in Design Science and Nanotechnology Education. She recently completed a PhD in engineering education drawing on her own design experiences (engineering, education, and dance) to explore what it meant for professionals from a variety of disciplines to experience design.

CHRISTOPHER LE DANTEC is a PhD student in the Human-Centered Computing program at the Georgia Institute of Technology. His research is focused on examining how human values expressed through technology affect marginal communities like poor and homeless.

KATHERINE DEIBEL is a graduate research assistant in the Centre for Engineering Learning and Teaching and a doctoral candidate in Computer Science and Engineering at the University of Washington. Her research interests span computer science and education and include alternative literacies, educational technology, assistive technologies, research methodologies, and information visualisation.

ELLEN YI-LUEN DO is an associate professor of Architecture and Computing at Georgia Tech. She studies design cognition, the methods and processes in which people engage in design, and design computing – developing computational artifacts and processes to support design activities and enhance life.

ANDY DONG is a senior lecturer at the University of Sydney. His research explores design and its theoretical, cognitive and biological intersections with language. He was awarded the *Design Studies* prize in 2005 for his paper *The latent semantic approach to studying design team communication.*

KEES DORST is Professor of Design and associate Dean for Research at the Faculty of Design, Architecture and Building, University of Technology, Sydney. He is also senior design researcher at Eindhoven University of Technology. He is author of numerous papers and several books, including most recently, *Understanding Design*.

CHRIS EARL is Dean of the Faculty of Mathematics, Computing and Technology at the Open University, UK. He studied mathematics at Cambridge before doing a PhD in Design at the Open University. After working at Bristol Polytechnic and Newcastle University, he returned to the OU as Professor of Engineering Product Design in 2000.

CLAUDIA ECKERT is a senior lecturer in Design at the Open University, UK. Her education encompassed mathematics, philosophy, computer science and artificial intelligence, before she completed a PhD in Design and postdoctoral research at the Open University. After nearly ten years at the Cambridge University Engineering Design Centre, she rejoined the OU in 2008.

OZGUR ERIS is an assistant professor of Design and Mechanical Engineering at Franklin W. Olin College of Engineering. He has published on the role of inquiry in design, design knowledge generation and capture, and data mining. He is the author of *Effective Inquiry for Engineering Design*.

DORON ESHEL is pursuing a Master's degree in architecture at the Technion – Israel Institute of Technology in Haifa. Working as an architect for more than 15 years, he has his own design practice and teaches architectural design at the National Technicians' School. His main interest is design education.

KILIAN GERICKE studied mechanical engineering. Currently he works as a research associate at the Technische Universität Berlin in the Engineering Design and Methodology Group. He is interested in design methodology, project risk management and creativity. Besides his research activities, he supervises students in engineering design and systematic product development.

JOHN GERO is Research Professor at the Krasnow Institute for Advanced Study, George Mason University. Formerly, he was Professor and Co-Director of the Key Centre of Design Computing and Cognition, University of Sydney. He has authored/edited 46 books and published over 550 research papers on computational and cognitive studies of designing.

FRIEDRICH GLOCK is a sociologist and mechanical engineer. He is a researcher and lecturer in the Multidisciplinary Design Group at the Institute for Technology Assessment and Design, University of Technology Vienna. He has a professional background and publications in the fields of educational research, science and technology studies, and design research.

GABRIELA GOLDSCHMIDT is Professor of Architecture and Town Planning at the Technion – Israel Institute of Technology in Haifa. She has an extensive background in both the practice and the teaching of design. This informs her research on design cognition and in particular design reasoning, visual thinking and representation in design.

SHAWN JORDAN is a doctoral candidate in the School of Engineering Education at Purdue University. His research interests include geographically-distributed design teams, technology-supported design environments, interdisciplinarity, creativity, and innovation. He holds bachelors and master's degrees in Electrical and Computer Engineering. He also founded a contest-winning interdisciplinary Rube Goldberg team.

MALTE JUNG is a PhD candidate at the Centre for Design Research at Stanford University. In his research, he is exploring the role of emotions in design team interaction.

JEFF KAN completed his PhD in design computing and cognition at the University of Sydney. His research focuses on developing quantitative methods to study the cognitive behaviour of designers. He formerly taught design studio and computer-aided design at the Department of Architecture, Chinese University of Hong Kong.

ALLISON KANG is a doctoral student in Science Education at the University of Washington. She was a graduate researcher in engineering design at the Centre for Engineering Learning and Teaching. She is currently working for the Genomics Outreach for Minorities Program in the College of Engineering.

DEBORAH KILGORE is a research scientist in the Centre for Engineering Learning and Teaching at University of Washington. She holds a PhD in Educational Human Resource Development from Texas A&M University. Her interests and expertise include educational research methods, adult learning theory, student development, and women in education.

MAAIKE KLEINSMANN is an assistant professor at Delft University of Technology. She studies collaborative design processes in industry. Her research focuses on exploring how actors in the field collaborate and on the development of tools and techniques to improve collaborative design. Her work has been published in *Design Studies and Co-design*.

KRISTINA LAUCHE is an associate professor in the Department of Product Innovation Management in Industrial Design Engineering at Delft University of Technology. Her research and teaching address product innovation as social practice, understanding user needs for system development, activity theory in design, and distributed teams in complex work environments.

LARRY LEIFER is Professor of Mechanical Engineering at Stanford University. His design thinking research program deals with global multi-disciplinary teams. He teaches *Design Entrepreneurship*. He is founding director of the Stanford Centre for Design Research and a member of the Hasso Plattner Institute of Design at Stanford.

PETER LLOYD is Head of the Design Group at The Open University, UK. His research is focussed on the processes and discourse of designing in all fields, with a recent interest in design ethics. He has taught in the area of design process and design ethics and is co-editor of the book you are reading.

RACHAEL LUCK teaches and undertakes research in the School of Construction Management and Engineering, at the University of Reading, UK. A theme in her research is to better understand practices of design through the study of language use in design interactions.

ADE MABOGUNJE is the Associate Director of the Centre for Design Research at Stanford University. He conducts empirical studies of the engineering design process. He has published in the areas of design theory and methodology, knowledge management, design protocol analysis, design education, and simulation studies of design teams.

LLEWELLYN MANN is a lecturer and Foundations coordinator at Central Queensland University in Rockhampton, Australia. His research focuses on developing professional skills in engineers needed for 21st century practice, including sustainable design, cross-disciplinary practice and cross-cultural awareness. He recently completed a Postdoctoral position at Purdue University's School of Engineering Education.

BEN MATTHEWS is an associate professor of Design Studies, sharing his time between the Mads Clausen Institute and the SPIRE Centre for Participatory Innovation at the University of Southern Denmark. His principal work concerns research methodology for design studies, and applications of ethnomethodology and post-Wittgensteinian philosophy to the understanding of design as a human activity.

JANET MCDONNELL is Professor of Design Studies at Central Saint Martins College of Art and Design, University of the Arts, London. She is also a chartered electrical engineer. Her research focus is design practice in natural settings, motivated by an interest in supporting reflective practitioners and enabling user engagement in design.

ANDRE NEUMANN studied Psychology and Cognitive Science at the Radboud University in Nijmegen. Since 2005 he has been a PhD candidate at the Faculty of Industrial Design Engineering at Delft University of Technology.

WAI HO NG is a supply chain management analyst on the 787 program at the Boeing Company in Everett Washington. He has a Masters degree in Industrial Engineering and conducted engineering design research while a graduate research coordinator at University of Washington's Centre for Engineering Learning and Teaching.

ARLENE OAK is an assistant professor of Material Culture at the University of Alberta. Her background as a designer and historian informs her research in the social psychology of design practice. Her current work explores the relationship between talk, social identity, and the design process in the contexts of design education and professional design practice.

FRANCK POURROY is an associate professor in the Faculty of Mechanics, University of Grenoble. He obtained a PhD degree in Mechanical Engineering in 1992 and carries out his research at G-SCOP laboratory. His main research interests are focused on managing technical knowledge and on collaborative work in mechanical engineering.

FRASER REID is Head of the School of Applied Psychosocial Studies at the University of Plymouth. His published work has focused on human communication, group interaction and social influence, most recently in settings using new keypad technologies, such as mobile telephones, electronic mail, and instant messaging.

ISABELLE REYMEN is an assistant professor in Design Processes in the Department of Technology Management at Eindhoven University of Technology in the Netherlands. Her research interests include design science methodology and the organisation of artefact creation processes.

MICHAEL SCHMIDT-KRETSCHMER has worked as a designer as well as manager and head of several R&D departments. He works as a chief engineer in the field of design research and education at the Technische Universität Berlin. His main research covers design processes, the usability of design methodologies and requirements engineering.

FRIDO SMULDERS is an assistant professor of Product Innovation Management at Delft University of Technology. His research and publications concentrate on describing and understanding the socio-interactive dimension of the product and architectural design and realisation processes. He supervises research degree projects and lectures on Technology Based Entrepreneurship, Project Leadership and Creativity.

NEERAJ SONALKAR is a PhD candidate at the Centre for Design Research at Stanford University. His research interests include studying group collaborative ideation and innovation in developing communities.

MARTIN STACEY has been a senior lecturer in Computer Science at De Montfort University since 1996. He studied cognitive psychology at Oxford and Carnegie Mellon before doing a PhD in artificial intelligence at the University of Aberdeen and postdoctoral research on intelligent computer support for design at the Open University.

JENNIFER TURNS is an associate professor of Technical Communication at the University of Washington. She holds a PhD in industrial engineering from the Georgia Institute of Technology. Her research interests include cognitive processes in design and teaching, user-centred and learner-centred design, and engineering education.

RIANNE VALKENBURG is an expert on human-centred innovation and is a business consultant at *Dutch* in Amsterdam. Formerly, as Professor of Human Technology at the Hanzeuniversity Groningen she researched new methodologies for user-research in product innovation. She is co-author of the books *Integrated Product Development* and *Human Technology Interaction*.

WILLEMIEN VISSER is Senior Researcher of Cognitive Ergonomics at LTCI, UMR 5141 CNRS TELECOM ParisTech and INRIA. In her book *Cognitive Artifacts of Designing* she presents both a critical review of her cognitive design studies research conducted over 30 years and her view of design as the construction of representations.

Author Index